INTRODUCTION TO MACHINE ARITHMETIC

MARVIN L. STEIN, WILLIAM D. MUNRO, *University of Minnesota*

INTRODUCTION TO MACHINE ARITHMETIC

ADDISON-WESLEY PUBLISHING COMPANY

Reading, Massachusetts · Menlo Park, California · London · Don Mills, Ontario

This book is published under the editorship of
MICHAEL A. HARRISON

Copyright © 1971 by Addison-Wesley Publishing Company, Inc.
Philippines copyright 1971 by Addison-Wesley Publishing Company, Inc.
All rights reserved. No part of this publication may be reproduced, stored in a retrieval system, or transmitted, in any form or by any means, electronic, mechanical, photocopying, recording, or otherwise, without the prior written permission of the publisher. Printed in the United States of America. Published simultaneously in Canada. Library of Congress Catalog Card No. 78-111956.

PREFACE

This book is an attempt to provide, for those who are not content to deal with a computer as a black box, some fundamental ideas and an underlying and unifying point of view about what actually takes place in the arithmetic section of a digital computer. The book explains the basic algorithms of computer arithmetic from a mathematical, not an engineering, point of view. It enables the person interested in software to understand what the engineering does and why it does it without his having to understand the physical elements of circuitry. At the same time, the information contained in this text is basic for a designer who approaches computer arithmetic from an engineering perspective. Even the software-oriented person, however, may be faced with design problems in attempting to provide computer arithmetic by means of interpretative programs or microprogramming. We believe that the material in this book will also serve him well.

Understanding the ideas underlying the algorithms of machine arithmetic is not enough. The physicist may consider the computer as an instrument to do the calculations involved in solving a differential equation. The social scientist may approach computer arithmetic as an easy means of obtaining the results of a statistical analysis. In most cases, what these users are probably interested in is the arithmetic of real numbers. Hence they must understand in what sense the results of the arithmetic algorithms built into a computer are related to, and can be understood as, the results of real arithmetic. The book also addresses itself to this problem.

To us the subject matter is of interest in itself. Because it represents an essential component in understanding why computers do what they do and how they do it, we believe it will also be of interest and value to those concerned with design structure or programming. It should be useful to both computer science students and those whose primary concern is with the computer as a simple tool capable of providing useful information.

The mathematical level for understanding the material in the book is not high. One reason for this is that in its most elemental abilities a computer is not capable of high-level mathematics. On the other hand, just as the simple tasks a computer can perform can be combined into a more sophisticated whole, the reader whose level of mathematical sophistication extends beyond the minimum requirements will gain more from this text.

For this reason, while the material can certainly be taught at the sophomore level, the authors use it in courses given to mixed audiences of upper-division undergraduate and first-year graduate students.

The book is functional in many ways. In itself, it is appropriate for a one-semester course covering the theory involved. At the same time, although the book is about neither the hardware of computers nor programming, it is suitable in part or *in toto* for inclusion in a full-year course in either of these subjects. We use the material as a considerable portion of a full-year course in the basic theory of programming, offered for juniors, seniors, and early graduate students. These students are not exclusively computer-science majors, but represent many different fields, including even the fine arts. The material can also be used in a course in computer design. With appropriate selections the book can be valuable at an early level in a survey course on computers, in introducing concepts of accumulators, floating-point arithmetic, etc.

We have included over two hundred exercises. These are primarily designed to promote understanding and to give the student an indication of the depth of his own comprehension. Some of these exercises are marked with one or two asterisks. However, this device is not an attempt on the part of the authors to formally graduate the difficulty of the exercises. Rather, this notation is to warn the student that exercises so marked may require more consideration and depth of thought than may appear superficially.

The combined experience of the authors in the use of, teaching about, and research in computers stretches over more than forty years. To a considerable extent this book reflects our attempt to compile in one source the answers to many questions which we have asked ourselves and which we consider to be important to our understanding but for which we could not reach to the bookshelf for ready information.

We express our thanks to the many students and colleagues who have broadened our points of view. In particular, we would like to thank William Franta, Richard Hotchkiss, and Evelyn Munro for many helpful suggestions and criticisms of the manuscript. Finally, we express our appreciation to Joyce Hakala Jore for care, patience, and adherence to deadlines in typing the manuscript.

Minneapolis, Minnesota M. L. S.
January 1971 W. D. M.

CONTENTS

Chapter 1   **The Digital Representation of Numbers**

     1.1   Introduction 1
     1.2   Some properties of the integers; operations 2
     1.3   The digital representation of integers 6
     1.4   Arithmetic, base r 12
     1.5   Rational and real numbers 14
     1.6   Modular arithmetic 23

Chapter 2   **Addition and Subtraction**

     2.1   Introduction 32
     2.2   The n-position, base-r accumulator 37
     2.3   Arithmetic modulo r^n 39
     2.4   Negative numbers, complements 42
     2.5   Interpreting residue-class arithmetic as ordinary arithmetic . . 43
     2.6   Forming complements 46
     2.7   The closed accumulator 49
     2.8   Addition and subtraction instructions 57

Chapter 3   **Nonnumeric Aspects of Arithmetic**

     3.1   Introduction 67
     3.2   Boolean algebra 68
     3.3   The logic of binary addition 75
     3.4   The logic of binary subtraction 83
     3.5   Flip-flops 87
     3.6   Arithmetic in any base 95

Chapter 4   **Shifting**

     4.1   Introduction 105
     4.2   Definition of shifting operations 106
     4.3   Left shifting 108
     4.4   Circular right shifting 113
     4.5   Open or end-off right shifting 116

Chapter 5 Multiplication

- 5.1 Introduction 121
- 5.2 Multiplication by shifting the multiplicand 127
- 5.3 Shifting partial products relative to the multiplicand 130
- 5.4 Negative factors 137
- 5.5 Multiply-add 144
- 5.6 Serial multiplication 146

Chapter 6 Division

- 6.1 Introduction 149
- 6.2 The application of the basic division algorithm to machine division 152
- 6.3 Division with negative operands 165

Chapter 7 Fixed and floating-point arithmetic; scaling

- 7.1 Introduction 188
- 7.2 Scale factors 193
- 7.3 Scaling integral arithmetic operations 198
- 7.4 Scaling for nonintegral operations 211
- 7.5 Floating-point operation 217

Chapter 8 Error in Computation

- 8.1 Introduction 234
- 8.2 The fundamental round-off error 235
- 8.3 Round-off error and the arithmetic operations 248
- 8.4 Exact computations; increased precision 250

Answers 263

Index 289

CHAPTER 1

THE DIGITAL REPRESENTATION OF NUMBERS

1.1 INTRODUCTION

Probably the most primitive concept of number which the human race has developed is incorporated in the positive whole numbers or integers, which we often refer to as the natural numbers. As implied by the word "natural," the system of positive integers is so normal and "obvious" to the human being that its properties were known and utilized many centuries before mathematicians made them a precise logical result of an assumed set of axioms. A child's first introduction to the ideas of computation is by means of these "counting" numbers, and it is no accident that the methods of representing the integers considered in this chapter utilize symbols called digits.

    The positive whole numbers were enough for the earliest needs of man, but as his sophistication grew so did his need to describe things which they could not represent. The necessity for dividing a man's estate among several sons, the computation of the hypotenuse of a forty-five degree right triangle, and, most important, the requirement of a symbol for *nothing* led to a broader and broader base of numerical description. Thus the natural integers were extended to include all integers, the rational numbers, and the irrational numbers, to finally yield the set of real numbers. It is with computations on real numbers that we are concerned here and in the later chapters.

    Mathematical sophistication gives us a description of real numbers in which we can consider the number we call π as a precise ratio of the circumference of a circle to its diameter. In practice when we are computing the circumference of a circle from a given diameter we are forced to use an approximation to π such as 3.1416. Furthermore, when we deal with such a finite set of digits in a computational problem we can break down any of the arithmetic operations into three stages:

1) the digit-by-digit algorithm for finding the digits of the result;
2) the final location of the decimal point; and
3) the application of the appropriate rule of signs.

    If we separate stage (1) from the other two, we return to the arithmetic of integers. Therefore, all arithmetic computation basically is achieved by calculation on integers, followed by the determination of sign and the location of the decimal point. Thus, in multiplying 3.5 by −2.41, we first find the product of the integers 35 and 241 to obtain 8435. We then count digits to the right of the decimal point to locate its position as being just after the eight, to form 8.435. Finally, we apply the rule of signs which tells us that the product is negative, to obtain the result of the total computation, −8.435. In the earlier chapters we will be concerned with the algorithms for achieving the first of these steps, namely, the arithmetic of integers. Therefore we shall be using certain properties of the integers, which we will describe and develop in the remainder of this chapter.

1.2 SOME PROPERTIES OF THE INTEGERS; OPERATIONS

In a precise, mathematical definition of the real numbers we can take as a starting point an abstract description of the natural integers by means of a set of axioms from which we can deduce the properties of integers that we will utilize here. In turn, these properties lead to extensions to the set of all integers and the rational and irrational numbers. It is not our purpose to develop such an axiomatic approach and all of its ramifications step by step. Rather, we will assume that the reader is familiar with the final outcome of such an approach, and so we will dwell on only a few of the results which we have occasion to use most often.

Many of the properties we use are common to the complete set of integers and will be stated for it. The immediate properties which we want for the natural or positive integers, include the following:

PROPERTY 1   The number one, written 1, is a natural integer.

PROPERTY 2   Every natural integer has a unique sequel of which it is the antecedent.

PROPERTY 3   The integer 1 has no antecedent.

PROPERTY 4   If two sequels are equal, then so are their antecedents.

PROPERTY 5   If a set of natural integers A has the two properties:

  a) A contains 1, and
  b) if A contains any positive integer, it also contains its sequel,
then A is the entire set of natural integers.

Properties 1 through 4 simply describe the familiar counting characteristics of the whole numbers; that is, they point out that there is a first one, the integer 1, and that it is always possible to go from an integer n to its sequel, which we may think of as $n + 1$. They further assert the familiar fact that antecedents are unique, and thus that we can also count backward. Property 5 asserts the validity of *mathematical induction*, on which so many proofs depend. [Note that, in practice, Property 5 may be modified by replacing the integer 1 in (a) by any subsequent integer N, in which case the set contains all integers from N on.]

The binary operations of addition, subtraction, multiplication, and division are primitively defined for the natural integers. The fact that the result of a given operation is not defined for all pairs of natural integers (for example, the result of $3 - 5$) is an important motivation for extending the set of positive whole numbers. Thus in what follows we will deal with all integers, including zero and the negative integers. Indeed, although we are concerned with integers at this point, we note that much of the discussion is valid for all real numbers.

We assume that for any integers a and b the operation of *addition* is well defined and produces a unique result called the *sum*, denoted by $a + b$, and that the operation of *multiplication* is well defined and produces a unique result called the *product*, denoted by $a \times b$ or $a \cdot b$, or ab. These operations satisfy:

1. **The commutative law:**

$$a + b = b + a \quad \text{and} \quad ab = ba.$$

2. **The associative law:**

$$(a + b) + c = a + (b + c) \quad \text{and} \quad (ab)c = a(bc).$$

3. **The distributive law:**

$$a(b + c) = ab + ac.$$

The inverse of addition, *subtraction*, is always possible; that is, for any two integers a and b there exists an integer c for which $a + c = b$, and we write $c = b - a$, called the *difference* of b and a. The inverse of multiplication, or division, will be treated in the next section.

The integers are *ordered*. Complementary to the idea of *equality* we have the idea of *inequality*: we say that a is less than b, and write $a < b$, if and only if there exists a positive integer c such that $a + c = b$. Alternatively, we may say that b is greater than a, and write $b > a$. These ideas permit us to order the integers in a conveniently arranged array,

$$\cdots < -3 < -2 < -1 < 0 < 1 < 2 < 3 < \cdots,$$

in which any number is less than its neighbor to the right. This gives the familiar classification of positive integers as those greater than zero (>0) and of negative integers as those less than zero (<0). For inequalities,

$$a < b \quad \text{implies} \quad a + c < b + c \quad \text{for all} \quad c,$$

and

$$a < b \quad \text{implies} \quad ac < bc \quad \text{for} \quad c > 0,$$

but
$$ac > bc \quad \text{for} \quad c < 0.$$
Finally, we recall that the *absolute value*, denoted by $|a|$, is defined by
$$|0| = 0,$$
$$|a| = a \quad \text{for} \quad a > 0,$$
$$|a| = -a \quad \text{for} \quad a < 0.$$
The absolute value satisfies
$$|ab| = |a|\,|b|$$
and the triangle inequality
$$|a + b| \le |a| + |b|.$$

1.2.1 The Division Algorithm

Although the inclusion of zero and the negative integers permits us to assert that subtraction is always possible, the same is not true for division. For example, there is no integer c for which $17 = 3c$. To assign a correct value to c we need to introduce the rational numbers. However, since we are currently concerned with integers, our present discussion of division is phrased exclusively in terms of them. The following theorem is fundamental to this discussion and to much of what follows in this book.

Theorem 1.1 (*Division Algorithm*). *For any integer a (called the dividend) and any integer $b \ne 0$ (called the divisor) there exist unique integers Q (called the quotient) and R (called the remainder) such that*
$$a = bQ + R, \quad \text{and} \quad 0 \le R < |b|.$$

Proof. If $b > 0$, we have the ordering
$$\cdots < -3b < -2b < -b < 0 < b < 2b < 3b < \cdots$$
Either $a = Qb$ for some element of the ordering or it lies between two such elements with
$$Qb < a < (Q + 1)b.$$
In the first case, $a = Qb$ and $R = 0$. In the second case, we let $a - Qb = R$, and it follows that $a = Qb + R$ and
$$0 < R < b.$$
If $b < 0$, we apply the same reasoning to $|b| = -b$ to get
$$a = q|b| + R \quad \text{with} \quad 0 \le R < |b|.$$

We then define $Q = -q$ to get
$$a = -Q|b| + R = Q(-|b|) + R = Qb + R.$$
To show uniqueness we assume that there exist two pairs (Q_1, R_1) and (Q_2, R_2) such that
$$a = Q_1 b + R_1$$
and
$$a = Q_2 b + R_2,$$
with $0 \le R_1 < |b|$ and $0 \le R_2 < |b|$. Then,
$$a - a = (Q_1 - Q_2)b + (R_1 - R_2) = 0.$$
Thus, $R_1 - R_2$ must be a multiple of b, but
$$-|b| < R_1 - R_2 < |b|,$$
and hence $R_1 - R_2$ cannot be a multiple of b unless $R_1 - R_2 = 0$. Therefore $R_1 = R_2$ and $(Q_1 - Q_2)b = 0$. Since $b \ne 0$,
$$Q_1 - Q_2 = 0 \quad \text{and} \quad Q_1 = Q_2.$$

We note that the condition that $0 \le R < |b|$ is the one which provides uniqueness. If this restriction is removed, there are infinitely many pairs Q and R for which $a = Qb + R$.

The condition of the remainder, $0 \le R < |b|$, of Theorem 1.1 is not the only one which will guarantee a unique result. We might, for example, replace it by the condition $|R| < |b|$, and choose R to have the same sign as the dividend, a. If a is nonnegative, the condition is the same as before, giving the uniqueness of Q and R of Theorem 1.1. If $a < 0$, the remainder condition becomes
$$-|b| < R \le 0.$$
If $R \ne 0$, we can always modify the unique Q and R of the theorem to yield unique results R^* and Q^*, satisfying $-|b| < R^* < 0$. For example, if
$$a = Qb + R,$$
where $0 < R < |b|$, we define $R^* = R - |b|$ and $Q^* = Q + 1$ for $b > 0$, and $Q^* = Q - 1$ for $b < 0$. Thus for $b > 0$,
$$a = Qb + R = (Q^* - 1)b + R^* + b = Q^* b + R^*,$$
and for $b < 0$,
$$a = Qb + R = (Q^* + 1)b + R^* - b = Q^* b + R^*.$$
In what follows we assume that the division problem is always uniquely

solved by finding the quotient and remainder of Theorem 1.1 or, as explained above, the quotient and remainder which result from assigning to the remainder the same sign as that of the dividend. The matter of which choice is appropriate will be explored in the chapter on division.

1.3 THE DIGITAL REPRESENTATION OF INTEGERS

It is not hard to envisage how primitive man, in counting 15 objects, counted out ten on his fingers, perhaps made a mark in the sand to record this amount, and then counted out another 5 on the fingers of one hand. In precisely the same way, we make a mark of 1 in the "tens column" followed by a 5, to count the number 15. Thus we use a *positional* and *digital* representation of the integers. Because of the implicit association of the idea of counting with the positive integers, we give a first formulation of the positional concept based on a description of how we go from a nonnegative integers to its sequel. Since the concept of zero is fundamental to the digital representation of numbers, we adjoin zero to the positive integer to yield the class of nonnegative integers, in which zero is the antecedent of 1, but zero has no antecedent.

Definition 1.1 Consider an infinite sequence of positions ordered according to the nonnegative integers from right to left:

$$\ldots 3\ 2\ 1\ 0.$$

In each position let a mark correspond to any of the integers d, where $0 \leq d < r, r \geq 2$. These marks are called digits, and the usual ordering of the integers applies, so that zero is the minimal digit and $r - 1$ the maximal. The number zero is represented by an infinite sequence of zeros

$$\ldots 0\ 0\ 0\ 0$$

and the number one by

$$\ldots 0\ 0\ 0\ 1,$$

and if the integer n is represented, its sequel $n + 1$ is obtained by replacing all maximal digits on the right of the rightmost nonmaximal digit by zeros and the rightmost nonmaximal digit by the next digit in order.

In practice the infinite string of leading zeros is omitted. We see that the set of integers which can be represented is the entire set of nonnegative whole numbers, since a representation of zero is given, a representation of one is given, and since we can go from n to $n + 1$; the principle of mathematical induction (Property 5) of Section 1.2 guarantees the remaining integers. The restriction $r \geq 2$ eliminates the case $r = 1$, in which the only digit is zero and the only number represented is zero.

1.3 THE DIGITAL REPRESENTATION OF INTEGERS 7

Example 1.1 If $r = 10$, we have the *decimal* system, and application of the rule to the number $n = 12$ yields $n + 1 = 13$. Since there are no maximal digits on the right, we simply replace the rightmost nonmaximal digit 2 by its successor 3. For the number $n = 3999$ the rule gives $n + 1 = 4000$. Since the 9's are maximal for $r = 10$, we replace them by zeros. The first nonmaximal digit, 3, is then replaced by the next one in order, 4.

The number r in Definition 1.1 is called the *base* or *radix* of the positional representation which, in turn, is called a *number system*.

Example 1.2 We generate the first ten positive integers for base $r = 2$ (binary), $r = 3$ (trinary), $r = 5$ (quinary), and $r = 8$ (octal), using the rule of Definition 1.1. In each case, zero is represented by 0.

$r = 2$		$r = 3$		$r = 5$		$r = 8$	
1	(m)	1		1		1	
10		2	(m)	2		2	
11	(m)	10		3		3	
100		11		4	(m)	4	
101	(m)	12	(m)	10		5	
110		20		11		6	
111	(m)	21		12		7	(m)
1000		22	(m)	13		10	
1001	(m)	100		14	(m)	11	
1010		101		20		12	

(m) = maximal

To represent an arbitrary integer N in the positional notation we will use $N = d_n d_{n-1} d_{n-2} \ldots d_2 d_1 d_0$ with subscripts on digits that correspond to the ordering of the position. The contribution of each digit is given by the following theorem.

Theorem 1.2 *If, for base $r \geq 2$ and digits $0 \leq d_i < r$, the nonnegative integer N is represented by*

$$N = d_n d_{n-1} \ldots d_2 d_1 d_0,$$

then

$$N = d_n r^n + d_{n-1} r^{n-1} + \cdots + d_2 r^2 + d_1 r + d_0.$$

Proof. The assertion is trivial for $N = 0$ and 1. For $N = d_n d_{n-1} \ldots d_1 d_0$ we assume that the theorem is true. Then

$$N = d_n r^n + \cdots + d_j r^j + \cdots + d_0 = \sum_{k=0}^{n} d_k r^k.$$

Let d_j be the rightmost nonmaximal digit. The counting rule yields
$$X = d_n r^n + \cdots + (d_j + 1)r^j + 0r^{j-1} + \cdots + 0.$$
However, since d_j is the first nonmaximal digit,
$$\begin{aligned} N &= d_n r^n + \cdots + d_j r^j + (r-1)r^{j-1} + \cdots + (r-1) \\ &= d_n r^n + \cdots + d_j r^j + r^j + (r^{j-1} - r^{j-1}) + \cdots + (r-r) - 1 \\ &= d_n r^n + \cdots + (d_j + 1)r^j - 1 = X - 1. \end{aligned}$$
Thus $X = N + 1$, and the principle of mathematical induction applies, to give the desired result for all nonnegative integers.

This theorem recapitulates the familiar facts that each position corresponds to a power of the base and that the integer is represented as a polynomial in the base r, with the restriction on the coefficients that they be digits, $0 \leq d_i < r$. This restriction enables us to show the uniqueness of the representation of Theorem 1.2.

Theorem 1.3 *For any nonnegative integer N, if*
$$N = d_n r^n + \cdots + d_1 r + d_0 = \sum_{k=0}^{n} d_k r^k$$
and $r \geq 2$ and $0 \leq d_k < r$, the digits d_0 through d_n are unique.

Proof. Suppose
$$N = \sum_{k=0}^{n} d_k r^k = \sum_{k=0}^{m} c_k r^k.$$
If $m \neq n$, we can fill in with $c_k = 0$ or $d_k = 0$ until $m = n$. Thus,
$$N = \sum_{k=0}^{n} d_k r^k = \sum_{k=0}^{n} c_k r^k$$
so that
$$\sum_{k=0}^{n} (d_k - c_k) r^k = 0.$$
Assume there exists a first digit from the right such that $c_j \neq d_j$. Then, eliminating low-order zero terms from the summation, we may write
$$\sum_{k=j}^{n} (d_k - c_k) r^k = r^j \sum_{k=0}^{n-j} (d_{k+j} - c_{k+j}) r^k = 0.$$

1.3 THE DIGITAL REPRESENTATION OF INTEGERS

Since $r^j \neq 0$, this last equation implies that

$$\sum_{k=0}^{n-j} (d_{k+j} - c_{k+j})r^k = 0.$$

Therefore

$$d_j - c_j = -\sum_{k=1}^{n-j} (d_{k+j} - c_{k+j})r^k$$

$$= -r \sum_{k=0}^{n-j-1} (d_{k+j+1} - c_{k+j+1})r^k.$$

This implies that $d_j - c_j$ is divisible by r, but if d_j and c_j are digits, then

$$-(r-1) \leq d_j - c_j \leq (r-1).$$

The only integer in this range divisible by r is zero. Hence, $d_j - c_j = 0$. This contradicts the assumption that there exists a first digit from the right such that $c_j \neq d_j$. Hence $c_j = d_j$ for all $j = 0, 1, \ldots, n$.

The extension of the digital representation to negative integers is straightforward. If $N < 0$, we may write $N = -|N|$. We follow this convention, obtaining first the unique digital representation for $|N| > 0$ and then prefixing the representation with a minus sign. Thus we have, for negative fifteen, the unique representation -15 in the decimal system.

1.3.1 Determining the Digits for Base r

While the counting technique described above for a number system, base r, provides us with a rule for obtaining the unique digits of the representation of any integer, it is obviously an impractical scheme for finding the binary digits to represent a large integer, say the (decimal) integer 9,234,567. A more satisfactory way to obtain these digits is outlined below, but before considering this matter, we comment on notations for the digits themselves. The conventional symbols we use are those for the decimal system and are probably Arabic in origin. It is apparent that as long as $r < 10$ and the number of digits for base r is equal to r, we can continue to use these familiar notations. For values of $r > 10$, however, we shall have to devise symbols for all digits whose values exceed nine. In base twelve (duodecimal), the number ten is a digit, and we cannot use the notation 10, since this would mean twelve. For bases larger than ten it has become the convention to use alphabetic symbols to represent digits greater than nine, although no standard notation has evolved. Thus, if we agree that the digits, base twelve, are 0, 1, 2, 3, 4, 5, 6, 7, 8, 9, u, and v, we can write $2u$ for the digital representation of $34_{10} = 2$ (twelve) + (ten). We have here introduced the convention of appending a decimal subscript to a sequence of digits to define the base.

Whatever the symbols we use, the digits can be calculated from the following theorem.

Theorem 1.4 *For any integer $N > 0$ and base $r \geq 2$, if the digital representation of N is*

$$N = d_n d_{n-1} \ldots d_2 d_1 d_0$$

the digits d_i, $i = 0, 1, \ldots, n$, can be computed according to the following algorithm. We apply the basic division algorithm to divide repetitively by the base r to form in order the sequence

$$N = q_0 r + d_0,$$
$$q_0 = q_1 r + d_1,$$
$$\vdots$$
$$q_k = q_{k+1} r + d_{k+1},$$
$$\vdots$$
$$q_{n-1} = q_n r + d_n = 0 \cdot r + d_n = d_n.$$

The remainders which result are the digits we seek.

Proof. From the division algorithm. Theorem 1.1, q_k and d_k are uniquely determined and $0 \leq d_k < r$. Further, since $N > 0$, each $q_k \geq 0$, and since $r \geq 2$, the q_k are monotonically decreasing in order:

$$N > q_0 > q_1 > \cdots > q_k > \cdots \geq 0.$$

Such a sequence must generate some first zero element, say $q_n = 0$. Successive substitutions then give

$$N = d_0 + rq_0$$
$$= d_0 + r(d_1 + rq_1)$$
$$= d_0 + rd_1 + r^2(d_2 + rq_2)$$
$$\vdots$$
$$= d_0 + rd_1 + r^2 d_2 + \cdots + r^n(d_n + rq_n)$$
$$= d_0 + rd_1 + r^2 d_2 + \cdots + r^n d_n.$$

The uniqueness of such a representation, given by Theorem 1.3, guarantees that the remainders d_i, are the digits we seek. We note that continuing the procedure with $q_n = 0 = 0 \cdot r + 0$ would generate the first of the leading infinity of zeros in the representation.

1.3.2 Conversion of Bases

If a number is represented in decimal and $r = 10$, the result of Theorem 1.4 appears trivial. For example, if $N = 12345$, division by 10 gives, by inspection, a quotient $q_0 = 1234$ and a remainder $d_0 = 5$. The true utility of the algorithm is in the conversion of the representation from one base to another. For conversion from base ten to any other base we represent all numbers in decimal (including the digits for the new system) and do the arithmetic in decimal.

In the following example we illustrate the conversion of the decimal representation of an integer to new base notation.

Example 1.3 The decimal number 417 can be represented in binary, octal, and duodecimal as follows:

Binary	Octal	Duodecimal
$417 = 2(208) + 1$	$417 = 8(52) + 1$	$417 = 12(34) + 9$
$208 = 2(104) + 0$	$52 = 8(6) + 4$	$34 = 12(2) + 10$
$104 = 2(52) + 0$	$6 = 8(0) + 6$	$2 = 12(0) + 2$
$52 = 2(26) + 0$		
$26 = 2(13) + 0$		
$13 = 2(6) + 1$		
$6 = 2(3) + 0$		
$3 = 2(1) + 1$		
$1 = 2(0) + 1$		

Thus the binary representation of 417_{10} is 110100001_2 and the octal representation is 641_8. For the duodecimal representation we can convert the decimal notation 10 for the duodecimal digit ten to, say, u, giving $2u9_{12}$.

In Example 1.3 all preliminary notations are in decimal and the arithmetic is in decimal. If we were to reverse the conversion, however, we would see that conversion of the octal number 641_8 to 417_{10} requires the performance of successive divisions by ten in octal, namely by 12_8. We examine this approach in the next section. However, for those who feel more at home with decimal arithmetic, we show how conversions from base $r \neq 10$ to decimal can be accomplished by means of the polynomial representation of Theorem 1.2.

Example 1.4 The numbers 641_8, 1011101_2, and $28v_{12}$ can be represented in decimal as follows, with all notations to the right of the equal signs in decimal.

$$641_8 = 6 \cdot 8^2 + 4 \cdot 8^1 + 1 \cdot 8^0 = 417 = (6 \cdot 8 + 4)8 + 1,$$
$$1011101_2 = 1 \cdot 2^6 + 0 \cdot 2^5 + 1 \cdot 2^4 + 1 \cdot 2^3 + 1 \cdot 2^2 + 0 \cdot 2^1 + 1 \cdot 2^0$$
$$= 93,$$
$$28v_{12} = 2 \cdot 12^2 + 8 \cdot 12^1 + (11) \cdot 12^0 = 395.$$

This scheme is also useful for checking conversions from decimal to other bases. The nested method of computing 641_8 requires fewer operations.

1.4 ARITHMETIC, BASE r

In later chapters we will consider in detail the structure of algorithms for doing digital arithmetic in any legitimate base r. Here we content ourselves with a brief consideration of how the formalities of decimal arithmetic can be extended to other bases. The techniques which we use for digit-by-digit computation are based on the polynomial representation of integers and the basic laws of combination given in Sections 1.2 and 1.3. Since the determination of the digits of the result of an operation is independent of the algebraic signs, we assume without loss of generality two nonnegative integers M and N, represented in base r by

$$M = c_m c_{m-1} \ldots c_0 = c_m r^m + c_{m-1} r^{m-1} + \cdots + c_1 r + c_0,$$
$$N = d_m d_{m-1} \ldots d_0 = d_m r^m + d_{m-1} r^{m-1} + \cdots + d_1 r + d_0.$$

We then have

$$M + N = (c_m + d_m) r^m + \cdots + (c_0 + d_0),$$
$$M - N = (c_m - d_m) r^m + \cdots + (c_0 - d_0),$$
$$MN = (c_m r^m + \cdots + c_0) d_0 + [(c_m r^m + \cdots + c_0) d_1] r$$
$$\qquad + [(c_m r^m + \cdots + c_0) d_2] r^2 \cdots + [(c_m r^m + \cdots + c_0) d_m] r^m.$$

To achieve these results in practice, we align the digits for addition and subtraction so that c_k will correspond to d_k by position. If each $c_k + d_k$ is a digit or each $c_k - d_k$ is a digit, the results for addition and subtraction are complete. Note that polynomial algebra is independent of the variable, and so the form of the operations does not depend on the base r. It is simply the formal manipulation of coefficients which is always the same. Thus, decimal forms for manipulating digits carry over to other bases. The value of the base r is of concern in coefficient manipulation only when we reduce coefficients to digits to propagate borrows and carries properly. For addition, if $c_k + d_k \geq r$, *carry* of one to the next higher order is generated, as in decimal. Similarly for subtraction, if $c_k - d_k < 0$, a *borrow* is generated to the next higher order. This will be considered in detail in later chapters. Here we have merely outlined how the formalities of digital decimal arithmetic can be applied to arithmetic in other base number systems. The decimal techniques apply in other bases as illustrated in the following example where carries and borrows are circled.

Example 1.5 In binary we wish to form the sum, difference, and product of $M = 1011$ and $N = 101$.

```
                ① ① ① ①      carries
      M + N:    1 0 1 1
                  1 0 1      align digits
              ─────────────
              1 0 0 0 0      sum

                  ①          borrows
```

1.4   ARITHMETIC, BASE r

$$
\begin{array}{llllll}
M - N: & 1 & 0 & 1 & 1 & \\
 & & 1 & 0 & 1 & \text{align digits} \\
\hline
 & 0 & 1 & 1 & 0 & \text{difference}
\end{array}
$$

$$
\begin{array}{lllllll}
MN: & & & 1 & 0 & 1 & 1 \\
 & & & & 1 & 0 & 1 & \text{align digits} \\
\hline
 & & & 1 & 0 & 1 & 1 & \text{digit-by-digit product} \\
 & 1 & 0 & 1 & 1 & & & \text{digit-by-digit product (shifted)} \\
\hline
 & 1 & 1 & 0 & 1 & 1 & 1 & \text{product}
\end{array}
$$

In octal these would appear as

$$
\begin{array}{llllllll}
 & & \textcircled{1} & & & \textcircled{-1} & & \\
M + N: & 13 & & M - N: & 13 & & MN: & 13 \\
 & \underline{5} & & & \underline{5} & & & \underline{5} \\
 & 20 & & & 06 & & & 57 \\
 & & & & & & & \textcircled{1} \\
 & & & & & & & \overline{67}
\end{array}
$$

and in decimal as

$$
\begin{array}{llllllll}
 & & & & & \textcircled{-1} & & \\
M + N: & 11 & & M - N: & 11 & & MN: & 11 \\
 & \underline{5} & & & \underline{5} & & & \underline{5} \\
 & 16 & & & 06 & & & 55
\end{array}
$$

Division represents a more complex procedure, but fundamentally we use the division algorithm digit by digit to form partial quotients and partial remainders. We illustrate the application of this technique in other bases in the following examples.

Example 1.6 We wish to find Q and R in octal with $M = QN + R$ and $M = 226$ and $N = 15$ in octal.

$$
\begin{array}{r}
1 \text{partial quotient} \\
15\overline{)226} \\
\underline{15} \\
5 \text{partial remainder}
\end{array}
$$

$$
\begin{array}{r}
13 \text{quotient} \\
15\overline{)226} \\
\underline{15} \\
56 \\
\underline{47} \\
7 \text{remainder}
\end{array}
$$

We can now use the conversion algorithm of Section 1.3 for any base.

Example 1.7 We wish to express the binary number 110010011 in decimal. We form

$$110010011 = (1010)(101000) + 11,$$
$$101000 = (1010)(100) + 0,$$
$$100 = (1010)(0) + 100,$$

so that the digits in decimal are $4 = 100_2$, $0 = 0_2$, $3 = 11_2$ to give 403_{10}. The arithmetic is done as follows:

```
           101000                          100
   1010 )110010011              1010 )101000
         1010                         1010
         ----                         ----
          1010                         000 = d₁ = 0
          1010
          ----
           1010
           1010
           ----
           0011 = d₀ = 3
```

```
             0
   1010 )100
         0
         ---
         100 = d₂ = 4
```

($d_1 = 0$, $d_0 = 3$, $d_2 = 4$)

1.5 RATIONAL AND REAL NUMBERS

The extension of the positional and digital representation of numbers from the nonnegative integers to the set of all integers was accomplished very easily by means of the inclusion of an algebraic sign. In this section we briefly review how we can also deal with rational numbers and all real numbers by a simple extension of the system of positional and digital representation. Indeed, while many methods have been used to extend the basis given by the integers to include the set of all rational and irrational numbers, it is quite possible to do this entirely in terms of digital representations, thus including all real numbers. We make no attempt to explore this avenue in depth, assuming many of the properties of reals which can be proved through digital representation. The interested reader, however, may want to pursue this topic in more detail. It is interesting to note that the mathematical motivation for extending the number system from the natural integers to all integers, to the rationals, and then to the irrationals, to include all real numbers, was an attempt to achieve what mathematicians now call "completeness." Among other things, completeness implies that all arithmetic operations can be performed on any real number operands to produce a result which is also a real number. Clearly, division of two integers does not always produce a single integral result even if we include the negative integers. An obvious extension of the number system so as to include the result of division is to *ratios* of integers or, what amounts to the same thing, an *ordered pair* of

1.5 RATIONAL AND REAL NUMBERS

integers. These numbers are called the *rationals*. A rational number is thus an ordered pair of integers a and $b \neq 0$ to which we can apply the division algorithm of Section 1.2.1 to get

$$a = Qb + R.$$

In the following section we consider how to extend the digital representation base $r \geq 2$ to these ordered pairs for which we use the conventional notations of

$$a/b, \quad \frac{a}{b}, \quad \text{or} \quad a \div b.$$

1.5.1 The Digital Representation of Rational Numbers

We again restrict our initial discussion to nonnegative integers and base our representation on the following theorem.

Theorem 1.5 *If a and b are any two positive integers and $r \geq 2$ a whole number, there exist digits d_j, $0 \leq d_j < r$, such that a/b can be represented as a series*

$$\frac{a}{b} = \sum_{j=n}^{-\infty} d_j r^j.$$

Proof. We form the following infinite sequence of applications of the division algorithm (Theorem 1.1):

$$a = bQ + R_0,$$
$$rR_0 = bd_{-1} + R_1,$$
$$rR_1 = bd_{-2} + R_2,$$
$$\vdots$$
$$rR_{k-1} = bd_{-k} + R_k,$$
$$\vdots$$

where each R_k satisfies $0 \leq R_k < b$. Since $a > 0$, $b > 0$, we have

$$0 \leq d_{-k} = r\frac{R_{k-1}}{b} - \frac{R_k}{b} \leq r\frac{R_{k-1}}{b} < r.$$

Therefore d_{-k} are all digits in base r. By successive substitutions and a straightforward induction we get

$$\frac{a}{b} = Q + \frac{R_0}{b} = Q + \frac{d_{-1}}{r} + \frac{R_1}{rb} = \cdots = Q + \frac{d_{-1}}{r} + \frac{d_{-2}}{r^2}$$
$$+ \cdots + \frac{d_{-k}}{r^k} + \frac{R_k}{r^k b}.$$

Hence, the partial sum

$$S_k = Q + \frac{d_{-1}}{r} + \cdots + \frac{d_{-k}}{r^k}$$

from the infinite series

$$Q + \sum_{j=-1}^{-\infty} d_j r^j$$

is

$$S_k = \frac{a}{b} - \frac{R_k}{r^k b}.$$

Since $R_k/r^k b < 1/r^k$ and $1/r^k \to 0$, it follows that $S_k \to a/b$. Finally, by Theorem 1.2, the nonnegative integer Q is representable as

$$Q = d_n r^n + \cdots + d_0.$$

Since we have

$$\frac{a}{b} = \sum_{j=n}^{-\infty} d_j r^j, \quad 0 \le d_j < r,$$

we can again use the positional and digital system. We introduce the *base point* or *radix point* to distinguish between the positions ordered by the negative and nonnegative powers of r. Thus, if we rewrite the series as

$$\frac{a}{b} = \sum_{j=0}^{n} d_j r^j + \sum_{j=1}^{\infty} d_{-j} r^{-j},$$

we can write correspondingly

$$\frac{a}{b} = d_n d_{n-1} \ldots d_0 \cdot d_{-1} d_{-2} d_{-3} \ldots,$$

with the base point separating the two sets of coefficients in the series.

Corollary. *The sum of the terms containing negative powers of r is less than or equal to one.*

Proof. Since $0 \le d_{-j} \le r - 1$, we have

$$0 \le \sum_{j=1}^{\infty} d_{-j} r^{-j} \le \sum_{j=1}^{\infty} (r-1) r^{-j} = 1$$

from the well-known properties of the geometric series.

If we eliminate the dual representation of one as

$$1 = \sum_{j=1}^{\infty} (r-1)r^{-j}$$

by always using the notation 1 for one, then the series of negative powers $\sum_{j=1}^{\infty} d_{-j} r^{-j} < 1$ is known as the *fractional part* of the rational number a/b. The nonnegative integer

$$Q = \sum_{j=0}^{n} d_j r^j$$

is called the *integral part* or *whole part* of a/b.

Because of the restriction $0 \leq R_k < b$ in Theorem 1.5, there are only a finite number of possible values for the infinite set of remainders. Thus there must be a first one, say R_s, such that j steps later $R_s = R_{s+j}$. Because of the recursive nature of the digit-generating algorithm, the digits from this point on will repeat in blocks of j. Such a representation is called *periodic* or *repeating*, and every rational number has such a representation. Conversely, every periodic sequence of digits interpreted to be a positional series converges to a rational number, for if N is such a number then $(r^{s+j} - r^s)N$ is an integer. Thus a rational number is characterized by having a periodic representation. In particular, a rational number may repeat by having all zero digits from some point on. Such a representation is called *terminating* (for example, $\frac{1}{4} = 0.250\ldots$ in decimal). Terminating representations of rationals are not unique, since they can also be represented in a form that gives maximal digits from some point on (for example, $\frac{1}{4} = 0.24999\ldots$). Thus, if $d_{-s} \geq 1$ and $N = \cdot d_{-1} d_{-2} \ldots d_{-s} 000 \ldots$, then

$$N = d_{-1} r^{-1} + \cdots + d_{-s} r^{-s}$$
$$= d_{-1} r^{-1} + \cdots + (d_{-s} - 1) r^{-s} + r^{-s}$$
$$= d_{-1} r^{-1} + \cdots + (d_{-s} - 1) r^{-s} + \sum_{j=s+1}^{\infty} (r-1) r^{-j}.$$

A reversal of the procedure shows that any representation culminating with an infinite sequence of maximal digits can be represented as terminating. If we agree to use the terminating form of the representation exclusively, then, as with integers, the digital representation base r is unique.

Since *every* series $\sum_{j=n}^{-\infty} d_j r^j$, where $0 \leq d_j < r$, converges by the same comparison test as used in the corollary to Theorem 1.5, and since those series which are periodic are one-to-one with the rationals, the remainder which are nonperiodic must converge to the irrational numbers. It is possible to define the set of real numbers from this point of view. It is beyond the scope of this book to consider either this approach or to show, whatever the method of defining the reals, that the inclusion of all such series gives a representation for all real numbers. This fact we will assume. We note that regardless of the rational or irrational character of the numbers involved, in the practice of computation we will always use terminating rational approximations to them.

If we assume that any nonnegative real number can be represented by an infinite sequence of digits of appropriate form, and if we eliminate the case of a final infinite sequence of maximal digits, then the representation is unique.

Theorem 1.6 *If* $x = \sum_{j=n}^{-\infty} d_j r^j$ *with* $0 \le d_j < r$, *and if for any* k, *there is at least one* $j > k$ *for which* $d_j \ne (r - 1)$, *then the* d_j *are unique.*

Proof. Let

$$x = \sum_{j=n}^{-\infty} d_j r^j = \sum_{j=n}^{-\infty} c_j r^j.$$

Assume that there is a first digit $d_k \ne c_k$. Let $d_k > c_k$ without any loss of generality. Then

$$x - x = \sum_{j=k}^{-\infty} (d_j - c_j) r^j$$

$$= (d_k - c_k) r^k + \sum_{j=k-1}^{-\infty} (d_j - c_j) r^j.$$

But $d_k - c_k \ge 0$, so $(d_k - c_k) r^k \ge r^k$. Furthermore, if d_j and c_j are digits,

$$\left| \sum_{j=k-1}^{-\infty} (d_j - c_j) r^j \right| \le r^k.$$

In the last expression the equality can hold only if $d_j - c_j = (r - 1)$ for $j = k - 1, k - 2, \ldots$, which implies $d_j = r - 1$ and $c_j = 0$ for these values of j, contrary to hypothesis. Thus,

$$\sum_{j=k-1}^{-\infty} (d_j - c_j) r^j > -r^k,$$

and it follows that

$$x - x = (d_k - c_k) r^k + \sum_{j=k-1}^{-\infty} (d_j - c_j) r^j > 0,$$

a contradiction. Therefore there is no $d_k \neq c_k$; that is, all $d_k = c_k$.

Mathematically, we use the same positional and base point notation for the irrationals as the rationals. That is, if $x = \sum_{j=n}^{-\infty} d_j r^j$, we write

$$x = d_n d_{n-1} \ldots d_1 d_0 \cdot d_{-1} d_{-2} \ldots \quad \text{if} \quad n \geq 0$$

and

$$x = 0.00 \ldots 0 d_n d_{n-1} \ldots \quad \text{if} \quad n < 0.$$

Practically, we will use a *truncated* or *rounded* version of x given by $\bar{x} = \sum_{j=n}^{n-s} \bar{d}_j r^j$ where $\bar{d}_j = d_j$ for truncation, and d_{n-s} may be $d_{n-s} + 1$ in a rounded version (which may produce a carry). Thus, for $n \geq 0$ if $x = d_n d_{n-1} \ldots d_0 \cdot d_{-1} d_{-2} \ldots$ we might use $\bar{x} = d_n d_{n-1} \ldots d_1 d_0 \cdot d_{-1} \ldots d_{-k}$ as a terminating approximation. A more complete discussion of this procedure and the errors involved will be given in a later chapter.

We use the same scheme for the digital representation of negative real numbers that we do for negative integers. That is, if $x < 0$ and

$$|x| = d_n d_{n-1} \ldots d_1 d_0 \cdot d_{-1} d_{-2} \ldots,$$

we write

$$x = -d_n d_{n-1} \ldots d_1 d_0 \cdot d_{-1} d_{-2} \ldots$$

If $x = a/b$ is rational, its sign is of course determined by the usual rules: if a and b are of like sign $x > 0$ and if they have unlike signs $x < 0$.

The algorithms which we use for digital computation on integers are based on the polynomial format of the positional representation. If we use a terminating rational approximation to a real number in the form

$$x \sim \bar{x} = \sum_{j=0}^{n} d_j r^j + \sum_{j=-1}^{-s} d_j r^j,$$

we can write it as

$$\bar{x} = r^{-s} \left[\sum_{j=0}^{n} d_j r^{j+s} + \sum_{j=-1}^{-s} d_j r^{j+s} \right]$$

$$= r^{-s} \left[\sum_{k=s}^{n+s} d_{k-s} r^k + \sum_{k=s-1}^{0} d_{k-s} r^k \right]$$

$$= r^{-s} \sum_{k=0}^{n+s} d_{k-s} r^k = r^{-s} N,$$

where $N = \sum_{k=0}^{n+s} d_{k-s} r^k$ is an integer if $s \geq 1$. If no negative powers appear, $s = 0$ and $\bar{x} = r^0 N$ with N an integer. In either case, if we use a separate scheme to handle the arithmetic involving the factor r^{-s} in $\bar{x} = r^{-s} N$, we can use the previously considered polynomial-based algorithms on N. That is, the arithmetic of real numbers is in practice reduced to the arithmetic of integers.

We note that writing $\bar{x} = r^{-s} N$ is a special case of multiplying by a power of the base. In general, if

$$x = \sum_{j=n}^{-\infty} d_j r^j, \quad \text{then} \quad r^k x = \sum_{j=n}^{-\infty} d_j r^{j+k},$$

and this *shifts* the base point k places to the right if $k > 0$ and to the left if $k < 0$. Thus, if $\bar{x} = r^{-s} N$ so that $N = r^s \bar{x}$ is an integer, the factor r^s is the one required to shift the base point to the right-hand end of the sequence of digits retained. If $\bar{x}$ is an integer, $s = 0$ and no shifting is required. In a later chapter we will consider some of the ramifications of shifting the relative position of the digits and the base point, and we will show the way in which this procedure gives us the separate scheme for the arithmetic on the factor r^{-s}. In the earlier chapters it will be the digital aspects of arithmetic on the integer N with which we will be concerned, and hence, our discussions will again center around the integers.

1.5.2 Calculation of the Digits for Real Numbers

We have seen that the algorithm in the proof of Theorem 1.5 provides a first step in which

$$a = bQ + R_0,$$

with Q the integral part of a/b and R_0 the fractional part. The computation of the digits of Q, any base r, is the same as given in Section 1.3.1 for integers. The digits to the right of the base point are then generated as the sequence of quotients in the remaining steps of the algorithm.

Example 1.8 Convert the fraction $\frac{14}{3}$ to binary. We form, in decimal notation,

$$14 = 3(4) + 2,$$
$$2(2) = 3(1) + 1,$$
$$2(1) = 3(0) + 2,$$
$$2(2) = 3(1) + 1,$$
$$\ldots \text{etc.}$$

Setting the integral part $4 = 100_2$ we have $\frac{14}{3} = 100.101010\ldots_2$.

1.5 RATIONAL AND REAL NUMBERS 21

Example 1.9 Convert $\frac{29}{7}$ to octal. We form (in decimal)

$$29 = 7(4) + 1,$$
$$8(1) = 7(1) + 1,$$
$$8(1) = 7(1) + 1,$$
$$\vdots$$

and as above, $\frac{29}{7} = 4.111\ldots_8$.

The technique illustrated in the last two examples can always be used if the arithmetic is done exclusively in the base in which a and b are represented.

We can see from the next example that the steps of the algorithm have the same form as the steps of ordinary division done in decimal.

Example 1.10 To find the decimal digits of $\frac{14}{3}$ we use

```
      4.66...
3 ) 14.00000
    12
    ──
     20
     18
     ──
      20...
```

Similarly, in binary $(\frac{14}{3})_{10} = (\frac{1110}{11})_2$, and we use

```
         1 0 0.1 0 1
11 ) 1 1 1 0.0 0 0 0
     1 1
     ───
     0 1 0 0
         1 1
         ───
         1 0 1
             1 1
             ───
             1 0 0
                 1 1 ...
```

In other words, we get in binary the same sequence as given in Example 1.8. The relation between multiplying by r and bringing down the next zero is obvious.

To obtain the decimal digits for a rational number represented in some other base and still do the arithmetic in decimal, we can utilize the power-of-r interpretation of the representation.

Example 1.11 To convert to decimal the rational repeating octal number $26.333\ldots$, we consider it as

$$2(8^1) + 6(8^0) + 3(8^{-1}) + 3(8^{-2}) + 3(8^{-3}) + \cdots$$

$$= 2(8^1) + 6(8^0) + \frac{3}{8}\left[1 + \frac{1}{8} + \frac{1}{8^2} + \cdots\right]$$

$$= 16 + 6 + \frac{3}{8} \cdot \frac{1}{1 - \frac{1}{8}} = 22 + \tfrac{3}{7} = 22.428571428571\ldots$$

The calculation of the digits in the infinite sequence for an irrational number would necessarily be based on the definition of the number. Thus, to compute the binary digits of π we would generally use the same mathematical formulation that we use for obtaining the decimal digits. Since in computation we will deal only with terminating approximations, which are rational, the same computational methods will apply as given above. Thus, we might use $\tfrac{22}{7}$ or $\tfrac{355}{113}$ as an approximation to π which could be handled in the same way as the fractions in Examples 1.7 and 1.9. Usually, however, the terminating rational approximations which we wish to convert to a new base will be of the form of mixed numbers, that is, terminating digital representations with digits on both sides of the base point. In this case the rational numbers have a very specialized character in which the division can be done easily. Thus, if in base r

$$d_n d_{n-1} \ldots d_1 d_0 \cdot d_{-1} \ldots d_{-k} = \frac{d_n d_{n-1} \ldots d_0 \ldots d_{-k}}{r^k},$$

the division is always by a power of r. The initial quotient is the integral part (converted as before) given by $d_n \ldots d_0$ and the first remainder is $d_{-1} \ldots d_{-k}$. Since each successive division is by r^k, it can be accomplished by shifting the base point if the arithmetic is done in base r. We achieve the equivalent by leaving the base point in place and multiplying the fractional part of the number by r^*, the new base. The integral part of the product is the desired quotient in the division and thus the required digit, base r^*.

Example 1.12 To convert the decimal number 23.247 to binary and octal, we note that the integral part is 23_{10}, which is converted in the usual manner to $10111_2 = 27_8$. The digits to the right of the point are given by

0.247		0.976		0.247	
$\underline{2}$		$\underline{2}$		$\underline{8}$	
0.494	$d_{-1} = 0$	1.952	$d_{-4} = 1$	1.976	$d_{-1} = 1$
0.494		0.952		0.976	
$\underline{2}$		$\underline{2}$		$\underline{8}$	
0.988	$d_{-2} = 0$	1.904	$d_{-5} = 1$	7.808	$d_{-2} = 7$
0.988		0.904		0.808	
$\underline{2}$		$\underline{2}$		$\underline{8}$	
1.976	$d_{-3} = 1$	1.808	$d_{-6} = 1$	6.464	$d_{-3} = 6$

Thus, $23.247_{10} = 10111.001111\ldots_2 = 27.176\ldots_8$.

If the original base is not decimal, the same procedure applies, but the arithmetic is done in the appropriate base.

Example 1.13 To convert 1101.101 in binary to decimal, we note that the integral part is $1101_2 = 13_{10}$. We then form

$$
\begin{array}{r} 0.101 \\ \underline{1010} \\ 1010 \\ 1010 \\ \underline{1010} \\ 110.010 \quad d_{-1}=6, \end{array}
\qquad
\begin{array}{r} 0.101 \\ \underline{1010} \\ 0100 \\ 010 \\ \underline{010} \\ 010.100 \quad d_{-2}=2, \end{array}
\qquad
\begin{array}{r} 0.1 \\ \underline{1010} \\ 101.0 \quad d_{-3}=5, \end{array}
$$

with all zeros from that point on. Hence, $1101.101_2 = 13.625_{10}$.

In example 1.13 representation in both binary and decimal is terminating since the fractional part is $\tfrac{5}{8}$ and 8 is exactly divisible into $8 = 2^3$ and $1000 = 10^3$. The termination of both digital sequences is not always the case as is seen in Example 1.12, and since in practice we use only a finite number of digits in the new base, this phenomenon may introduce a secondary truncation or round-off error.

1.6 MODULAR ARITHMETIC

After real numbers have been approximated by rational numbers, represented by terminating sequences of digits, and after the base points have been relocated by the introduction of scale factors, arithmetic processes reduce to operations on integers. For this reason the properties of integers are important in the study of arithmetic operations in computing machines. Of particular importance is the property of congruence or belonging to a residue class. It is this property of an integer which we will look into in this section.

Definition 1.2 (Congruence) If m is an integer, we say that any two integers N and M are congruent, modulo m, if and only if there exists an integer k such that $N - M = km$. We write

$$N \equiv M \pmod{m}$$

and call m the modulus.

The notation for congruence modulo m is due to Gauss, and, as we shall see, the idea of congruence will prove to be a powerful tool for considering the basic properties of the operations of arithmetic. The notation for congruence is similar to that for equality, and congruence has many of the properties of equality. For example, for all integers N, M, and P the congruence relation is:

1. DETERMINATIVE: Either $N \equiv M \pmod{m}$ or $N \not\equiv M \pmod{m}$
2. REFLEXIVE: $N \equiv N \pmod{m}$
3. SYMMETRIC: If $N \equiv M \pmod{m}$, then $M \equiv N \pmod{m}$
4. TRANSITIVE: If $N \equiv M \pmod{m}$ and $M \equiv P \pmod{m}$ then $N \equiv P \pmod{m}$

Properties 1 through 4 easily follow from Definition 1.2. For (1) the difference $N - M$ is either divisible by m or not. For (2) $N - N = (0)m$, while for (3) if $N - M = km$, then $M - N = (-k)m$. Finally, for (4) if $N - M = k_1 m$ and $M - P = k_2 m$, then $N - P = N - M + M - P = (k_1 + k_2)m$.

Restricting the definition of congruence to a positive modulus m causes no loss in generality. If $N \equiv M \pmod{m}$, then $N \equiv M \pmod{-m}$, for $N - M = km = (-k)(-m)$. With the modulus $m = 0$, congruence would become equality, which, while interesting, affords nothing new. With the modulus $m = 1$, all integers are congruent to each other, which is of no current interest. Accordingly, in what follows, we will restrict ourselves to moduli $m > 1$.

We have seen that the notation for congruence is similar to that for equality, and, as a relationship, congruence has the same properties as the relationship of equality. The following theorem shows that congruence also has the same properties with respect to the operations of addition, subtraction, and multiplication as does equality.

Theorem 1.7 *If $N \equiv \bar{N} \pmod{m}$ and $M \equiv \bar{M} \pmod{m}$, then*

$$N + M \equiv \bar{N} + \bar{M} \pmod{m},$$
$$N - M \equiv \bar{N} - \bar{M} \pmod{m},$$

and

$$NM \equiv \bar{N}\bar{M} \pmod{m}.$$

Proof. By assumption, $N - \bar{N} = k_1 m$ and $M - \bar{M} = k_2 m$ where k_1 and k_2 are integers. Hence,

$$(N \pm M) - (\bar{N} \pm \bar{M}) = (N - \bar{N}) \pm (M - \bar{M})$$
$$= k_1 m \pm k_2 m = (k_1 \pm k_2)m$$

and

$$NM - \bar{N}\bar{M} = NM - N\bar{M} + N\bar{M} - \bar{N}\bar{M}$$
$$= N(M - \bar{M}) + \bar{M}(N - \bar{N})$$
$$= N k_2 m + \bar{M} k_1 m = (N k_2 + \bar{M} k_1)m.$$

The carrying over of the properties of the arithmetic operations from equality to congruence does not extend quite so simply to division. For example, $6 \equiv 4 \pmod{2}$ and $2 \equiv 2 \pmod{2}$, but if we divide the congruent integers 6 and 4 by the integers 2 and 2, congruent in the same modulus, we do not get congruent results, since $3 \not\equiv 2 \pmod{2}$. This is not surprising if we consider the fact that the sum, difference, and product of any two integers is an integer, but the ratios N/M and $\bar{N}/\bar{M}$ may not be integers. Even if they are, with $N = Q_1 M$ and $\bar{N} = Q_2 \bar{M}$, we have

$$N - \bar{N} = k_1 m = Q_1 M - Q_1 \bar{M} + Q_1 \bar{M} - Q_2 \bar{M} = Q_1 k_2 m + \bar{M}(Q_1 - Q_2).$$

Thus, $\bar{M}(Q_1 - Q_2) = (k_1 - Q_1 k_2)m$ and we see that $Q_1 - Q_2$ is necessarily divisible by m only if $\bar{M}$ is not (said to be relatively prime to m). However, in a later chapter we shall see that division in modular arithmetic can be treated by other techniques.

If we use a fixed value of the modulus $m > 0$ as a divisor, then for any integral dividend a, the division algorithm, Theorem 1.1, tells us that there is a unique remainder R, and it must have one of the m possible values $R = 0, 1, 2, \ldots, m - 1$. This suggests the following definition.

Definition 1.3 (Residue Class) The set of all integers having the same remainder on division by the modulus m in the sense of the division algorithm is called a residue class (mod m).

We see that there are exactly m residue classes (mod m) and that each integer belongs to one and only one residue class (mod m). Thus, a modulus $m > 0$ subdivides the set of all integers into m distinct and disjoint subsets which we have called the residue classes. If we agree that we will not deal with individual integers but only with the residue class of which an integer is a member (that is, make no distinction between integers that are in the same residue class), we will have reduced the problem of working with an infinite set to one of working with a finite set. In an arithmetic machine which is necessarily finite, this idea can be of great value if properly implemented, and this is what we will be concerned with in the following discussion.

We use the notation $R_i, i = 0, 1, 2, \ldots, m - 1$, to stand for the m residue classes (mod m). Here the subscript identifies the value of the remainder to which the particular residue class corresponds. Clearly, the word residue is used in the same sense as the word remainder.

Two integers N and M may be related either by virtue of being congruent modulo m or by virtue of belonging to the same residue class modulo m. In the following theorem we show that, if the two integers are related in either of these two ways, they must be related in both, that is, the two relationships are the same.

Theorem 1.8 *Let N and M be any two integers and $m > 0$ be an integer; then $N \equiv M \pmod{m}$ if and only if N and M are members of the same residue class (mod m).*

Proof. Assume that $N \equiv M \pmod{m}$. Let $0 \le R_1 < m$ and $0 \le R_2 < m$ be the unique remainders, defined by Theorem 1.1, obtained by dividing N and M respectively by m. Then

$$N = Q_1 m + R_1, \qquad M = Q_2 m + R_2,$$

and

$$N - M = (Q_1 - Q_2)m + R_1 - R_2 = km.$$

Hence, $R_1 - R_2 = (k + Q_2 - Q_1)m$ is divisible by m, but since R_1 and R_2 are remainders, they must satisfy

$$-(m - 1) \le R_1 - R_2 \le m - 1.$$

However, the only integer divisible by m in this range is zero. Therefore, $R_1 = R_2$; that is, M and N are members of the same residue class (mod m). Now assume that N and M are members of the same residue class (mod m). Let R be the common remainder. Then on division by m,

$$N = Q_1 m + R, \qquad M = Q_2 m + R.$$

Therefore,

$$N - M = (Q_1 - Q_2)m;$$

that is, $N \equiv M \pmod{m}$.

For any integer there are only a finite number of possible remainders $R = 0, 1, 2, \ldots, m - 1$ on division by m. If we divide an integer R having one of these values by m, the result is

$$R = (0)m + R;$$

that is, R is a member of the residue class defined by itself. Since every integer is in a unique residue class, and belonging to the same residue class implies congruence, *every* integer must be congruent to exactly one of the m possible remainders. The set of all integers congruent to a particular one of the restricted set of remainders forms the residue class. The complete class is determined by any member and can be generated by adding all possible multiples of the modulus m, that is, km for all possible integral values of k, to any member. Any collection of m integers, no two of which are congruent modulo m, determines the m residue classes and is referred to as a *complete residue system*. In particular, the restricted set of remainders $0 \le R < m$, no two of which are congruent, forms the *least nonnegative complete residue system*. We will refer to this system as the *residues* modulo m.

The relationship of congruence or (essentially the same thing) membership in the same residue class will prove to be of value in working with arithmetic using digital representations of numbers. For example, the successive digits provided by the computational algorithm given in the proof of Theorem 1.4 can be thought of as the representatives of the residue classes generated by the restricted remainder set $0 \leq R < r$, or residues, modulo r, for base r. We shall see that the relation between digits base r and congruences modulo r is also connected with the carry in addition and the borrow in subtraction. The concept will also afford a simple means of treating the arithmetic of negative integers.

Another case in which residue class concepts are immediately suggested is that of the ordinary desk-model of mechanical adding machine. Suppose we consider a simple desk machine capable of representing up to five decimal digits and of adding or subtracting. If we start at zero and continue to add one, we successively represent on the machine the integers $N = 0, 1, 2, \ldots, 99999$. These integers, $0 \leq N < 10^5$, comprise precisely the least nonnegative complete residue system modulo 10^5. If we continue to count, that is, add one again and again, the machine will return to zero and begin to regenerate, in order, $\bar{N} = 0, 1, 2, \ldots$ where again $0 \leq \bar{N} < 10^5$. The numbers which we have generated by counting, however, will be $N = 100{,}000, 100{,}001, 100{,}002, \ldots$, and in all cases $N \equiv \bar{N} \pmod{10^5}$. Similarly, if we start at zero and count backwards by subtracting ones, the machine will again show $\bar{N} = 99999, 99998, 99997, \ldots$ while we are counting $N = -1, -2, -3, \ldots$. Again $N \equiv \bar{N} \pmod{10^5}$. Thus, the machine automatically categorizes all integers into the residue classes modulo 10^5 represented by the least nonnegative complete residue system $0 \leq \bar{N} < 10^5$ where $\bar{N}$ is representable on the machine. In this sense, while we cannot represent an arbitrary integer directly, we can always represent its *equivalent* (that is, an alternate number of the residue class to which it belongs) from the set of residues which can appear on the machine. The extension of this idea to an n-position, base-r machine is obvious if we consider the modulus r^n. This suggests that we can actually extend the representational properties of a machine if we think in terms of replacing arbitrary integers by more convenient equivalents in the congruence relation.

We saw in Theorem 1.7 that for the operations of addition, subtraction, and multiplication we achieve congruent results, if we operate on congruent operands. Thus, if we think of $\bar{N}$ and $\bar{M}$ as members of the residue system representable on a machine modulo r^n, we can do the operations of addition, subtraction, and multiplication on any two integers N and M by carrying out the operation on their congruent machine representatives $\bar{N}$ and $\bar{M}$ and achieve congruent results. In effect, we will be adding, subtracting, and multiplying residue classes. We will explore the implications of this concept in later chapters. In the balance of this chapter we will give a precise definition and brief discussion of residue-class arithmetic.

We define operations of addition, subtraction, and multiplication in which both the operands and the results are chosen from among the residue classes modulo m. We use the conventional symbols $+$, $-$, and $\cdot$ (or juxtaposition) for these operations. To form $R_i + R_j = R_k$, we select an arbitrary representative from each of the residue classes R_i and R_j and add them in the usual manner. The sum lies in a unique residue class and we take this class as the result R_k. By a representative of a residue class we mean any element of the class. Similarly, we form $R_i - R_j = R_k$ and $R_i R_j = R_k$ by selecting an arbitrary representative from each of the residue classes R_i and R_j. Since the representatives are integers, we have no difficulty in forming their difference or their product. The difference or product of the representatives is also an integer and is thus a representative of a unique residue class, which we take to be the result R_k.

The residue-class operations that we have defined reduce to operations on integers and so present no difficulty. However, we must verify that these operations are well defined, that is, that they produce unique results which are independent of the particular representatives chosen from R_i and R_j. To test the definitions we choose two representatives from each of the operand classes, x and $\bar{x}$ from R_i and y and $\bar{y}$ from R_j. We form $x + y = z$ and $\bar{x} + \bar{y} = \bar{z}$. The results z and $\bar{z}$ each represent a unique residue class. We must show that the residue class represented by z and that represented by $\bar{z}$ are identical. Since elements of the same residue class are congruent, we have

$$x \equiv \bar{x} \pmod{m}.$$
$$y \equiv \bar{y}$$

It follows from Theorem 1.7 that

$$x + y \equiv \bar{x} + \bar{y} \pmod{m}$$

that is

$$z \equiv \bar{z} \pmod{m}.$$

If z and $\bar{z}$ are congruent, they must belong to the same residue class. This shows that the resultant residue class, derived from addition of two residue classes, is unique and depends in no way on the particular representative chosen from the operand residue classes. We will leave it to the reader to establish the same result for the subtraction and multiplication operations on residue classes.

The foregoing indicates the path we take in solving a most fundamental problem of machine arithmetic. The continuum of real numbers to which arithmetic applies must be dealt with in terms of a finite set of numbers which can be represented on the machine. We first approximate any real number by a terminating rational number. Then, after relocating the base point by use of scale factors, we reduce the problem to that of dealing with the integers, an infinite set. By introducing residue classes and considering all integers in a class to be equivalent, we reduce the problem of dealing with the infinite set of integers to that of dealing with a finite set of representatives of residue classes.

So, in some sense, in machines we replace the arithmetic of real numbers by the arithmetic of residue classes. How we do this and why, and when we can consider the results of residue-class arithmetic as equivalent to the results of ordinary arithmetic, will be considered in detail in the later chapters of this book.

EXERCISES

SECTION 1.2

1.1 Reformulate Properties 1–5 of the positive integers in terms of nonnegative integers. That is, consider that in the set of nonnegative integers, zero is the "first" one.

In Exercises 1.2 through 1.6, let A be the set of all integers n satisfying the given relation. Prove, using the format of Property 5, that A is the set of all positive integers.

1.2 $\sum_{k=1}^{n} k = \dfrac{n(n+1)}{2}$

1.3 $\sum_{k=1}^{n} k^2 = \dfrac{n(n+1)(2n+1)}{6}$

1.4 $\sum_{k=1}^{n} (2k-1)^2 = \dfrac{n(4n^2-1)}{3}$

1.5 $\sum_{k=1}^{n} k^3 = \dfrac{n^2(n+1)^2}{4}$

1.6 *The Binomial Theorem.* For natural integers a and b, show that

$$(a+b)^n = a^n + \sum_{k=1}^{n-1} \binom{n}{k} a^{n-k} b^k + b^n,$$

where

$$\binom{n}{k} = \dfrac{n!}{k!(n-k)!}$$

is a binomial coefficient and, as usual, $n! = 1 \cdot 2 \cdot 3 \cdots n$.

1.7 Define $0! = 1$ and for any positive integer a (or b), define a^0 (or b^0) $= 1$. Establish the formula for Exercise 6 in the form

$$(a+b)^n = \sum_{k=0}^{n} \binom{n}{k} a^{n-k} b^k,$$

using the formulation of induction established in Exercise 1.1.

1.8* Suppose we assume only the positive integers and Properties 1–5 and the properties of addition. Show that for any pair a and b one of the conditions

$$a + c = b, \quad a = b, \quad a = b + c$$

30   THE DIGITAL REPRESENTATION OF NUMBERS

(where c is also a positive integer) must hold. Thus, we have
$$a < b, \quad a = b, \quad \text{or} \quad a > b$$
for any pair.

1.9* Assume that for any nonempty set of positive integers S, there is a first integer; that is, assume there is an a in S such that $a < b$ for all $b \neq a$ in S. With this assumption show that Property 5 can be formulated (and proved) by assuming the contradiction that 5a and 5b hold but there is a nonempty set not contained in A. Apply this technique to redo Exercise 1.2.

SECTION 1.2.1

1.10 For the dividends and divisors listed below apply the Division Algorithm. In each case do it two ways, with the condition of a nonnegative remainder, and with a remainder of the same sign as the dividend.

	Dividend	Divisor
a)	7143	-17
b)	-6080	42

1.11 Suppose that nonnegative dividends are represented in decimal and that divisors are always positive powers of ten. Devise a scheme for telling by inspection the decimal representation of the unique quotient and remainder.

SECTION 1.3

1.12 Use the counting definition to write the first 20 integers in bases 4, 7, 11, and 13.

1.13 For each of the numbers in Exercise 1.12 show that your representation is correct by expansion in powers of the base.

1.14 Determine what condition must hold in any base r in order that a digital representation may be divisible by the base.

SECTION 1.3.2

1.15 For the following decimal numbers, find the digits in the base r.
- a) 12345, base 8
- b) 54321, base 16
- c) 2971, base 7
- d) 345, base 2

1.16 Make the following conversions to the base indicated.
- a) 10110101_2 to decimal
- b) 10110101_2 to octal
- c) 2346_8 to decimal
- d) 456_7 to decimal

SECTION 1.4

1.17 Do the conversions of Exercise 1.16 by division in the initial base.

SECTION 1.5.1

1.18 Show that every terminating basimal representation represents a rational number, that is, that

$$\sum_{j=n}^{m} d_j r^j$$

with $m < n$, is rational. Consider all combinations of algebraic sign for n and m.

1.19 Find the base 2 and base 8 representations of each of the following decimal numbers:
    a) 6/5   b) 12.34   c) 27.125

1.20 Find the decimal representations for the following numbers given in base 2 or 8.
    a) $(\frac{6}{5})_8$ b) $(\frac{1101}{10})_2$
    c) 3.214_8 d) 1101.11101_2

SECTION 1.6

1.21 Suppose the greatest common divisor of a and m is d, denoted $(a, m) = d$. If $a \equiv b \pmod{m}$ show that $(b, m) = d$.

1.22 Show that any positive integer represented decimally is congruent to the sum of its digits modulo 9. This is the basis for the arithmetic check known as "casting out nines."

1.23 Construct an addition and multiplication table for residue classes modulo 5. Use the representatives 0, 1, 2, 3, and 4 as symbols for the classes.

1.24 Find the remainder on dividing 3^{30} by 11.

1.25* Prove the restricted cancellation law for congruences. If $(a, m) = d$ (see Exercise 1.21) and $m = Md$, then if $ab \equiv ac \pmod{m}$,

$$b \equiv c \pmod{M}.$$

Thus, the cancellation law modulo m is valid in unmodified form for common factors a such that $(a, m) = 1$.

CHAPTER 2

ADDITION AND SUBTRACTION

2.1 INTRODUCTION

Human beings ordinarily perform the operations of addition and subtraction by representing the operands digitally in some base r and aligning the representations so that the coefficients of corresponding powers of r lie one above the other. They then operate on the correspondingly located digits to form the sequence of sum or difference digits. The basic operation of addition is that of forming the sum of two correspondingly located digits. This may also require adding a possible carry from a lower-order position. The residue of the sum modulo r is the sum digit, and the quotient from the application of the division algorithm determines a possible carry to the next higher order. This quotient is one (a carry) or zero (no carry). The operation of subtraction is that of forming the difference of two correspondingly located digits and applying a possible borrow from a lower order position. The residue modulo r is the difference digit, and the quotient propagates a borrow, if any, to the next higher position. These processes will be described analytically below.

Suppose that the operands to be dealt with are X and Y and that these are to be represented in the base r by the n-digit sequences $\{x_i\}$ and $\{y_i\}$, $i = 0, 1, \ldots, n - 1$. Then

$$X = \sum_{i=0}^{n-1} x_i r^i,$$

$$Y = \sum_{i=0}^{n-1} y_i r^i,$$

(2.1)

and the sum S can be expressed as

$$S = X + Y = \sum_{i=0}^{n-1} (x_i + y_i) r^i = \sum_{i=1}^{n-1} (x_i + y_i) r^i + (x_0 + y_0).$$

The term $x_0 + y_0$, which has been split off from the sum, represents the first

2.1 INTRODUCTION

pair of correspondingly located digits to be handled. By applying the division algorithm, we can represent this term as

$$x_0 + y_0 = c_0 r + s_0, \qquad 0 \le s_0 < r,$$

where c_0 and s_0 are the unique quotient and remainder obtained by division of $x_0 + y_0$ by r. Because $0 \le x_0 + y_0 < 2r$, the quotient c_0 must satisfy $c_0 = 0$ or $c_0 = 1$. On substitution of these results into the expression for S and collecting coefficients of powers of r, we obtain

$$S = \sum_{i=2}^{n-1} (x_i + y_i) r^i + (x_1 + y_1 + c_0) r + s_0.$$

If we now divide $x_1 + y_1 + c_0$ by r, we determine a unique quotient c_1 and remainder s_1 such that

$$x_1 + y_1 + c_0 = c_1 r + s_1, \qquad 0 \le s_1 < r.$$

Since the dividend is nonnegative and must be less than $2r$, the quotient c_1 satisfies $c_1 = 0$ or $c_1 = 1$. We then take c_1 as the carry, the remainder s_1 as the sum digit and proceed to the next digital position. This gives, as above

$$S = \sum_{i=3}^{n-1} (x_i + y_i) r^i + (x_2 + y_2 + c_1) r^2 + s_1 r + s_0.$$

The same procedure can be applied to the coefficient of r^2 that was applied above to the coefficient of r and, in general, the same procedure can be applied to the coefficient $(x_{i+1} + y_{i+1} + c_i)$ of r^{i+1} that was applied to $(x_i + y_i + c_{i-1})$, the coefficient of r^i. If we deal with coefficients of successively higher powers of r in this manner, we finally obtain

$$S = X + Y = c_{n-1} r^n + \sum_{i=0}^{n-1} s_i r^i. \qquad (2.2)$$

We note that we must add a term involving r^n to the sum in order to account for the possibility of a nonzero c_{n-1}.

Example 2.1 For addition of the two, three-digit decimal numbers 643 and 372, ordinary long-hand technique yields

```
  ① ①
    6 4 3
    3 7 2
  ─────────
  1 0 1 5
```

where carries have been circled. According to the analytic procedure indicated, this breaks down into:

$$x_0 = 3, \quad y_0 = 2,$$
$$x_0 + y_0 = 5 = (0)10 + 5,$$

with $c_0 = 0$ (no carry), $s_0 = 5$. The next step gives

$$x_1 = 4, \quad y_1 = 7$$
$$x_1 + y_1 + c_0 = 11 = (1)10 + 1,$$

so that $c_1 = 1$ (a carry), $s_1 = 1$. Finally, since

$$x_2 = 6, \quad y_2 = 3,$$

we have

$$x_2 + y_2 + c_1 = 10 = (1)10 + 0,$$

with $c_2 = 1$ (a carry) and $s_2 = 0$. The value $c_2 = 1$ generates a carry to the fourth position, corresponding to 10^3, and we must add $c_2 10^3 = 10^3$ for the final result

$$c_2 10^3 + s_2 10^2 + s_1 10^1 + s_0 = (1)10^3 + (0)10^2 + (1)10 + 5$$
$$= 1015.$$

For each step of the addition procedure, we have

$$(x_i + y_i + c_{i-1}) - s_i = c_i r,$$

with c_{-1} always equal to zero. Therefore, s_i is congruent to $x_i + y_i + c_{i-1}$ modulo r; that is, these two integers are in the same residue class modulo r. In fact, s_i will be the least nonnegative integer in this class and so may be called the *residue* modulo r.

The difference $D = X - Y$ may be represented as

$$D = \sum_{i=0}^{n-1} (x_i - y_i) r^i = \sum_{i=1}^{n-1} (x_i - y_i) r^i + (x_0 - y_0).$$

Consider the term $x_0 - y_0$ which has been split off from the difference. We generate the difference digit d_0 and the possible borrow b_0 from this term through division by the base r. The division algorithm yields

$$x_0 - y_0 = b_0 r + d_0, \quad 0 \le d_0 < r.$$

Since x_0 and y_0 are digits to the base r, their difference must satisfy

$$-(r - 1) \le (x_0 - y_0) \le (r - 1)$$

In the case $0 \le (x_0 - y_0) \le (r - 1)$, it is clear that the quotient $b_0 = 0$. In the case $-(r - 1) \le (x_0 - y_0) < 0$, however, in order to place the remainder in the range $0 \le d_0 < r$ the unique quotient must be $b_0 = -1$. Substituting for $x_0 - y_0$ in the expression for D and collecting coefficients of like powers of r, we obtain

$$D = \sum_{i=2}^{n-1} (x_i - y_i)r^i + (x_1 - y_1 + b_0)r + d_0.$$

We again apply the division algorithm to express

$$x_1 - y_1 + b_0 = b_1 r + d_1, \qquad 0 \le d_1 < r.$$

By an analysis parallel to that just given, it follows that if $x_1 - y_1 + b_0 \ge 0$, $b_1 = 0$; otherwise $b_1 = -1$. Thus, we can write

$$D = \sum_{i=3}^{n-1} (x_i - y_i)r^i + (x_2 - y_2 + b_1)r^2 + d_1 r + d_0.$$

As in addition, the same procedure can be applied to the coefficient of r^2 that was applied to the coefficient of r and, in general, the same procedure can be applied to the coefficient $(x_{i+1} - y_{i+1} + b_i)$ of r^{i+1} that was applied to $(x_i - y_i + b_{i-1})$, the coefficient of r^i. If we proceed to coefficients of successively higher powers of r in this manner, we finally obtain

$$D = b_{n-1} r^n + \sum_{i=0}^{n-1} d_i r^i. \tag{2.3}$$

Although X and Y contained terms through r^{n-1} only, we must include a term involving r^n in the difference to account for the possibility of a nonzero quotient b_{n-1}. Since $\sum_{i=0}^{n-1} d_i r^i < r^n$,

$$b_{n-1} = -1$$

implies $D < 0$; that is, a nonzero borrow to the nth position occurs only if $D < 0$. For each step of the subtraction procedure, we will have (assuming $b_{-1} = 0$)

$$(x_i - y_i + b_{i-1}) - d_i = b_i r.$$

Therefore, $x_i - y_i + b_{i-1}$ and d_i are congruent modulo r, that is, in the same residue class modulo r. In fact, d_i will be the least nonnegative integer in this class and so may be called the *residue* modulo r.

Example 2.2 We wish to subtract the three-digit decimal number 297 from the three-digit decimal number 648. In long-hand,

```
     -①
      6 4 8
    - 2 9 7
      3 5 1
```

with the only borrow circled. To illustrate the current discussion; we set

$$x_0 = 8, \quad y_0 = 7,$$

with

$$(x_0 - y_0) = 1 = (0)10 + 1,$$

to obtain $b_0 = 0$ (no borrow) and $d_0 = 1$. The next step yields,

$$x_1 = 4, \quad y_1 = 9,$$

with

$$(x_1 - y_1 + b_0) = -5 = (-1)10 + 5,$$

to give $b_1 = -1$ (a borrow) and $d_1 = 5$. Finally,

$$x_2 = 6, \quad y_2 = 2$$

so that

$$(x_2 - y_2 + b_1) = 3 = (0)10 + 3,$$

with $b_2 = 0$ (no borrow) and $d_2 = 3$. The difference D is

$$D = (0)10^3 + (3)10^2 + (5)10 + 1 = 351.$$

Since $b_2 = 0$ there is no borrow propagated to the fourth position. This is because the difference is positive.

To summarize, we form the sum (2.2) of two operands X and Y as given in (2.1) by repeated applications of the division algorithm to generate the carries c_i as quotients and the sum digits s_i as remainders on division by the base r in the following way:

If

$$c_{-1} = 0,$$
$$x_i + y_i + c_{i-1} = c_i r + s_i (i = 0, 1, 2, \ldots, n - 1). \tag{2.4}$$

Similarly, we form the difference (2.3) of the two operands by repeated application of the division algorithm to generate the borrows b_i as quotients and the difference digits d_i as remainders on division by the base r in the following way:

If

$$b_{-1} = 0$$

then

$$x_i - y_i + b_{i-1} = b_i r + d_i (i = 0, 1, 2, \ldots, n - 1). \tag{2.5}$$

For the carries and sum digits in addition, and the borrows and difference digits in subtraction, uniqueness follows from the division algorithm. It also follows from the remainder inequality that, for both operations, the various s_i and d_i obtained are indeed digits in the number system base r.

It is not too surprising that when we build addition and subtraction into a digital computing machine, we try to do so in a manner similar to that

2.2 THE n-POSITION, BASE-r ACCUMULATOR

The total digital storage capacity of a particular computing machine is necessarily finite and fixed. In practice this digital storage capacity is dealt with in either of two ways. It may be divided into units of fixed digital capacity called *registers*, and all operands will then be represented by exactly the number of digits stored in a register or by some integral multiple thereof. On the other hand, operands may be represented by varying numbers of digits, and the exact quantity of digits required by a particular operand will be assigned to it from the total digital storage capacity. In either case, we will assume that a special register, called the accumulator, has been designated for the purpose of forming and storing sums and differences and that this register is of fixed capacity. Thus, all operands for this register and all results must be of precisely this capacity. When using pencil and paper techniques, we do not ordinarily limit to one fixed value the number of digits which we utilize to represent operands and results. Thus, the assumption of fixing the number causes some variations from the way in which arithmetic is done by hand.

For definiteness in the discussion which follows and in accordance with what has been said above, we will assume that *all results are to be represented by precisely n digits in the number system base r*. We will also assume that, no matter how many digits from the total digital storage have been assigned to represent it, *at the time an operand is operated upon it is represented by precisely n digits in the base r*. Thus, any integer N which is to be used as an operand in the operations of addition and subtraction or which arises as a result of such an operation must take the form

$$N = \sum_{i=0}^{n-1} \delta_i r^i, \quad 0 \le \delta_i < r.$$

This immediately raises the question of how we are to handle operands and results N which normally have representations

$$N = \sum_{i=0}^{m} \delta_i r^i, \quad 0 \le \delta_i < r, \quad m \ge n.$$

At this point we will take a simple and direct approach to this problem. We will *discard all digits δ_i for $i \ge n$.*

Before turning to an investigation of the consequences of throwing away high-order operand or result digits, we will first introduce a fundamental computing element for dealing with n-digit operands and producing n-digit results. We will call the element an *n-position, base-r, open accumulator*. This device has two properties:

1. It stores or holds a sequence of n base-r digits,
2. It accepts a second sequence of n base-r digits and forms from them and stores a third sequence of n base-r digits, s_i or d_i, $i = 0, 1, 2, \ldots, n - 1$, in accordance with the rules given by (2.4) or (2.5).

For an example we will form the sum and the difference in this accumulator for two operands X and Y. Originally the digits of X are assumed to be stored in the accumulator and each digit y_i ($i = 0, 1, 2, \ldots, n - 1$) is transmitted into the ith position of the accumulator as illustrated schematically below.

$n-1$	$n-2$	$n-3$		i		2	1	0
x_{n-1}	x_{n-2}	x_{n-3}	$\cdots$	x_i	$\cdots$	x_2	x_1	x_0
$\uparrow$	$\uparrow$	$\uparrow$		$\uparrow$		$\uparrow$	$\uparrow$	$\uparrow$
y_{n-1}	y_{n-2}	y_{n-3}		y_i		y_2	y_1	y_0

Depending on the operation, a carry c_{i-1} or a borrow b_{i-1} is propagated into the ith position ($i = 0, 1, 2, \ldots, n - 1$) where it is understood that c_{-1} and b_{-1} are zero. Then

$$x_i + y_i + c_{i-1} \quad \text{or} \quad x_i - y_i + b_{i-1}$$

are respectively replaced by s_i or d_i, and c_i or b_i are propagated to position $i + 1$ as described in Section 2.1. The final state of the accumulator for addition may be pictured as:

$n-1$	$n-2$	$n-3$		i		2	1	0
s_{n-1}	s_{n-2}	s_{n-3}	$\cdots$	s_i	$\cdots$	s_2	s_1	s_0

and for subtraction as:

$n-1$	$n-2$	$n-3$		i		2	1	0
d_{n-1}	d_{n-2}	d_{n-3}	$\cdots$	d_i	$\cdots$	d_2	d_1	d_0

Since coefficients of powers r^m for $m \geq n$ are to be discarded, c_{n-1} and b_{n-1}

have not been retained. We may think of them as having fallen out of the left-hand end of the accumulator. It is for this reason that the accumulator is called *open*.

Example 2.3 For the operations of Examples 2.1 and 2.2 and a three-digit accumulator we would have, for the sum

```
Initial   | 6 | 4 | 3 |   X
            ↑   ↑   ↑
            3   7   2     Y

Final     | 0 | 1 | 5 |   S
```

The carry outside the accumulator is lost. For subtraction

```
Initial   | 6 | 4 | 8 |   X
            ↑   ↑   ↑
            2   9   7     Y

Final     | 3 | 5 | 1 |   D
```

2.3 ARITHMETIC MODULO r^n

Consider an operand X which has $m \geq n$ digits in its representation base r. We may write

$$X = \sum_{i=0}^{m} x_i r^i = \sum_{i=n}^{m} x_i r^i + \sum_{i=0}^{n-1} x_i r^i.$$

In accordance with the decision to discard digits of order greater than $n - 1$, the actual machine operand X^* that we will deal with is

$$X^* = \sum_{i=0}^{n-1} x_i r^i.$$

We will examine the difference between X and X^*. We have

$$X - X^* = \sum_{i=n}^{m} x_i r^i = r^n \sum_{i=0}^{m-n} x_{n+i} r^i = kr^n,$$

where

$$k = \sum_{i=0}^{m-n} x_{n+i} r^i$$

is an integer. Thus, it follows that $X \equiv X^* \pmod{r^n}$. We see that when we discard the high-order digits of the actual operand X, retaining only the lower-order n digits which fit into the accumulator, we are, in effect, replacing X by a congruent machine representative X^*. The modulus is formed by raising the base of the number system to the number of digits in the accumulator, that is, a modulus dictated by the "geometry" of the machine. Since each of the n-digit, base-r integers lies in a different residue class modulo r^n, X^* must be the least nonnegative integer in the same residue class as is X. Thus, the effect of discarding the high-order digits is to eliminate the distinction between operands which are members of the same residue class modulo r^n. Such operands are lumped together and replaced by a machine operand which is the least nonnegative integer in the residue class.

Let X and Y be any two nonnegative integers, let X^* and Y^* be their n-digit machine representations, and let R stand for the true result obtained by adding X to Y in the usual way. Then

$$R = X + Y = k_1 r^n + X^* + k_2 r^n + Y^*, \tag{2.6}$$

where k_1 and k_2 are appropriate integers. In the accumulator we can form only the n-digit sum of X^* and Y^*. Call this result R^*. Then

$$R^* = \sum_{i=0}^{n-1} s_i r^i. \tag{2.7}$$

However, the true sum of the machine representations is

$$X^* + Y^* = c_{n-1} r^n + \sum_{i=0}^{n-1} s_i r^i. \tag{2.8}$$

On subtraction of (2.7) from (2.8), it follows that

$$R^* \equiv X^* + Y^* \pmod{r^n}.$$

It also follows from (2.6) that

$$R \equiv X^* + Y^* \pmod{r^n}.$$

Accordingly, by the transitivity property of the congruence relationship

$$R \equiv R^* \pmod{r^n}.$$

Example 2.4 In a three-position accumulator, the addition of Example 2.1 was carried out in machine form in Example 2.3, yielding 015 as the result of adding 643 and 372. From the above discussion $X = 643$, $Y = 372$, so

$$X + Y = 643 + 372 = (0)10^3 + 643 + (0)10^3 + 372$$
$$= (0)10^3 + X^* + (0)10^3 + Y^*,$$

and $X = X^* = 643$ and $Y = Y^* = 372$. The sum

$$X^* + Y^* = 1015 = (1)10^3 + 015 = (1)10^3 + R^*,$$

while $R = 1015$. Thus

$$R = 1015 \equiv 015 = R^* \pmod{10^3}.$$

Let R be the true result obtained by subtracting Y from X in the usual way and let R^* be the machine result obtained by subtracting the machine representatives in the accumulator. We have

$$R = X - Y = k_1 r^n + X^* - k_2 r^n - Y^*,$$

where k_1 and k_2 are integers and it follows that

$$R \equiv X^* - Y^* \pmod{r^n}.$$

However, it is easy to see that

$$(X^* - Y^*) - R^* = b_{n-1} r^n.$$

Accordingly,

$$X^* - Y^* \equiv R^* \pmod{r^n}$$

and, by transitivity,

$$R \equiv R^* \pmod{r^n}.$$

Example 2.5 Suppose we have a three-position decimal accumulator and we wish to subtract $Y = 6542$ from $X = 3256$. We have

$$X - Y = (3)10^3 + 256 - (6)10^3 - 542$$
$$= (3)10^3 + X^* - (6)10^3 - Y^*,$$

with $X^* = 256 \equiv X \pmod{10^3}$ and $Y^* = 542 \equiv Y \pmod{10^3}$. Formation of $X^* - Y^*$ gives

$$\begin{array}{r} 2\ 5\ 6 \\ -5\ 4\ 2 \\ \hline \underline{(-1)}7\ 1\ 4 \end{array}$$

in which the borrow to the fourth place is ignored, so $R^* = 714$. The true result is $R = -3286$ and

$$R = -3286 \equiv 714 = R^* \pmod{10^3},$$

since $R^* - R = 4000 = (4)10^3$.

The decision to discard digits of orders greater than or equal to n, which was prompted by the constraints to deal only with operands and results having precisely n digits, has led to substitution of the *addition and subtraction of residue classes modulo r^n* for our ordinary addition and subtraction. In this arithmetic the properties of the accumulator are such that the least nonnegative integer in a residue class is always selected as the representative. However, the human being, in dealing with residue class arithmetic, is not constrained in this way. He can, in his operations external to the accumulator, freely substitute one representative of a residue class for another without changing the results of the residue class arithmetic finally performed by the accumulator.

2.4 NEGATIVE NUMBERS, COMPLEMENTS

Let N be any whole number. We define the *complements of N modulo r^n* to be the nonnegative integers of the form

$$kr^n - |N| \geq 0, \qquad k \text{ integral}.$$

We will single out the smallest of the complements modulo r^n for special attention. It will be referred to as *the complement*.

We note at once that a *negative integer and all of its complements are congruent modulo r^n*. This means that in residue class addition and subtraction modulo r^n we can avoid all operations with negative representatives simply by replacing each negative integer with any of its complements. This has no effect on the operation of the accumulator whose result cannot be altered by replacing an operand with a different representative from its residue class. Therefore, we can, if we wish, modify the accumulator so that it performs only the operation $+|Y|$, where

$$|Y| = \sum_{i=0}^{n-1} y_i r^i,$$

and y_i are the digits transmitted to the accumulator. The absolute-value sign here and those used below are unnecessary but are introduced for emphasis. An accumulator restricted in this manner will be called *additive*. When called upon to perform the operation $-|Y|$ in an additive accumulator, we replace the digits y_i by the digits of any of the complements modulo r^n of $-|Y|$ and perform instead the equivalent residue class operation

$$+|(kr^n - |Y|)|.$$

On the other hand, the accumulator can be restricted to perform only the operation $-|Y|$. In this case the accumulator will be called *subtractive*. In the subtractive accumulator it is the operation $+|Y|$ which offers difficulty. We avoid this by replacing the operation $+|Y|$ with the operation sequence $-(-|Y|)$ and then $-|Y|$ with one of its complements. Thus, finally, we perform the equivalent operation

$$-|(kr^n - |Y|)|.$$

Example 2.6 In a three-position decimal accumulator, complements (mod 10^3) of the positive number 123 would be numbers of the form

$$k(10^3) - 123 \geq 0,$$

with k an integer, satisfying $k \geq 0.123$, that is, $k \geq 1$. Thus for $k = 1, 2, 3, \ldots$ the complements are 877, 1877, 2877 . . . . If the number were -123, complements would be

$$k(10^3) - |-123| = k(10^3) - 123,$$

or the same as above. We note, however, that we might represent the positive number 123 directly, but would probably choose to use one of the positive equivalents 877, 1877 . . . to represent the negative -123.

We see that if we are willing, when necessary, to take the trouble to replace a negative integer by one of its complements, we can choose either the additive or the subtractive accumulator as fundamental, and eliminate the other. Although there is no particular theoretical reason for doing so, we will, in our further discussions, assume that the subtractive accumulator has been eliminated and that addition is the basic operation. We adopt the additive accumulator because human beings tend to find addition easier to think about than subtraction. As we will see below, however, the subtractive accumulator has advantages over the additive one, if the accumulator is constructed to perform residue-class arithmetic relative to a slightly different modulus.

2.5 INTERPRETING RESIDUE-CLASS ARITHMETIC AS ORDINARY ARITHMETIC

There is probably little demand for a device which carries out only addition and subtraction operations on residue classes as such. Thus we are faced with the problem of interpreting operations on residue classes in terms of ordinary arithmetic. Since in operations on residue classes the accumulator naturally selects the least nonnegative representative, it appears that by restricting magnitudes of results to be residues, we can generate answers consistent with ordinary arithmetic. From this point of view we will investigate the consequences of the assumption

$$|R| < r^n, \tag{2.9}$$

where, as before, R represents the true result, that is, the result of carrying out the operations on the original operands by means of ordinary arithmetic.

We know that R and the accumulator result R^*, obtained by operating on machine representatives of the original operands, satisfy

$$R \equiv R^* \pmod{r^n}.$$

This means that there exists an integer k such that

$$R = R^* + kr^n.$$

Since the accumulator stores only n digits, we have

$$0 \leq R^* < r^n. \tag{2.10}$$

It follows from (2.9) and (2.10) that $-2 < k < 1$. Since k is an integer, k must either be -1 or 0. Therefore

$$R = R^*, \tag{2.11}$$

or

$$R = R^* - r^n. \tag{2.12}$$

From (2.10), Equation (2.11) holds, if and only if $R \geq 0$, while (2.12) holds if and only if $R < 0$. In the latter case we can replace R by $-|R|$ and rewrite (2.12) in the form

$$|R| = r^n - R^*, \tag{2.13}$$

or

$$R^* = r^n - |R| \geq 0. \tag{2.14}$$

Thus, if R satisfies (2.9) and is nonnegative, the number R^*, read in the accumulator, gives the value of R. If R satisfies (2.9) and is negative, the number R^* gives the complement of R, and the absolute value of the true result can be obtained as the complement of R^* by (2.13).

Example 2.7 For a five-position decimal accumulator, the restriction imposed by (2.9) is that

$$|R| < 10^5,$$

which is satisfied, for example, by either $R = 23456$ or $R = -23456$. For $R > 0$, (2.11) holds and $R^* = 23456$. For $R = -23456$, (2.12) holds and so by (2.14) we have

$$R^* = 10^5 - 23456 = 76544,$$

the complement of R. The complement of R^* then gives $|R|$ as

$$10^5 - 76544 = 23456$$

from Equation (2.13).

Condition (2.9) is a severe one: violation of this condition leads to a computational error of the type usually referred to as *overflow*. If (2.9) is satisfied, however, we can interpret the accumulator result as giving either the actual result or the complement of the actual result. *We can tell which, provided that we know the sign of R in advance.* If we do not have *a priori* knowledge of the sign of R, the reading of R^* yields an indeterminate answer. For example, with $r = 10$, $n = 1$, we will perform the operations "six minus three" and "two minus nine." For both, $|R| < 10$. Carrying out the computations, we get

$$6 - 3 \equiv 3 \pmod{10},$$
$$2 - 9 \equiv 2 + (10 - 9) \equiv 3 \pmod{10}.$$

In both cases, $R^* = 3$. Without advance knowledge that our result is negative, we could not determine from R^* that the correct result in the second case is $-(10 - 3) = -7$.

When we do not know the sign of R in advance, we can impose the stronger condition

$$|R| < r^{n-1}. \tag{2.15}$$

If R satisfies (2.15), it must also satisfy (2.9), and so all of the previous conclusions remain valid. Thus, if R is nonnegative, $R^* = R$ and so by (2.15)

$$0 \leq R^* < r^{n-1}. \tag{2.16}$$

It is easy to see that an n-digit integer in the range (2.16) must have its highest-order digit equal to zero. Conversely, an n-digit integer R^* starting with zero must satisfy (2.16). If R^* satisfies (2.16) and R has the form $R = R^* - r^n$, then the smallest absolute value of R would occur when $R^* = r^{n-1} - 1$. In this case

$$R = -((r - 1)r^{n-1} + 1),$$

and it follows that $|R| > r^{n-1}$, which contradicts (2.15), and so the assumption that R has the form (2.12) is false. Therefore, if (2.15) is satisfied, $R = R^*$ if and only if the leading digit of R^* is zero.

In case R is negative, $R^* = r^n - |R|$ and it follows from (2.15) that

$$r^{n-1}(r - 1) + 1 \leq R^* \leq r^n - 1. \tag{2.17}$$

It is easy to verify that an n-digit integer R^* which satisfies (2.17) must have a highest-order digit equal to $r - 1$. Conversely, assume that the leading digit of R^* is $r - 1$. Then, if (2.15) also holds, $R^* \neq R$, for under these circumstances a necessary and sufficient condition for equality is that the leading digit of R^* be zero, and so the alternative equation (2.12) must hold. Therefore, if (2.15) holds, $R^* = r^n - |R|$ if and only if the leading digit of R^* is $r - 1$.

Example 2.8 In Example 2.7 on a five-position machine the values of $R^* = 23456$ and $R^* = 76544$ corresponded to $R = 23456$ and $R = -23456$. For a six-place accumulator either value of R would satisfy the more restrictive condition (2.15) with $n = 6$. This would give representations $R^* = 023456$ for the positive number, and $R^* = 976544$ for the negative. The leading zero in the first case tells us $R = R^*$. For the negative-valued R, the leading digit is $9 = 10 - 1 = r - 1$ so the true value is

$$R = -(10^6 - R^*) = -23456.$$

Thus, with the restriction (2.15) on the result R, we can readily distinguish between negative and nonnegative R by observing the highest-order digit of R^*. If this digit is zero, R is nonnegative, and $R = R^*$. If this digit is $r - 1$, R is negative and its absolute value can be obtained by subtracting R^* from r^n. In every case R is uniquely determined by R^*. When condition (2.15) is imposed, so that the highest-order digit can be used in this way, it is frequently called the *sign digit*.

2.6 FORMING COMPLEMENTS

Let A be a negative integer. Then $A = -|A|$. Assume that on discarding digits of orders greater than or equal to n (if any) from the representation of $|A|$, the integer $|B|$ results. Then from our previous results

$$|A| \equiv |B| \pmod{r^n} \quad \text{and} \quad A \equiv B \pmod{r^n},$$

where $B = -|B|$. Since B is a negative integer, it is congruent to all of its complements modulo r^n, that is, for all integers k, sufficiently large,

$$B \equiv kr^n - |B| \pmod{r^n}.$$

Therefore, by the transitivity of the congruence relation

$$A \equiv kr^n - |B| \geq 0 \pmod{r^n}.$$

We see from the foregoing discussion that we may replace a negative integer with a nonnegative n-digit representative of its residue class in either of two ways:

1. Form a complement of the original integer and then discard digits of orders greater than or equal to n from the complement.
2. Discard digits of orders greater than or equal to n from the absolute value of the original integer and then form a complement from the remaining n-digits.

Of these two methods, the second is preferable because it requires only the formation of the complements of operands which are represented by exactly

n digits. Let X be such an operand. Then
$$|X| < r^n.$$
The complements of X have the form
$$kr^n - |X| \geq 0.$$
We see that any integers k such that
$$k \geq \frac{|X|}{r^n}$$
will result in a correct complement, but
$$\frac{|X|}{r^n} < 1.$$
Therefore, any $k \geq 1$ is admissible, and we may take $k = 1$ for simplicity. Since R and R^* are also less than r^n in magnitude, we will be able to simplify considerations with regard to the complements of all numbers we deal with by always taking $k = 1$. This yields the least nonegative of the complements which we have previously defined to be *the complement modulo r^n*, or more briefly, *the complement*. With $k = 1$, the complement of the complement gives the absolute value of the integer originally complemented. For example, given R^* the complement of R, we complement it to obtain
$$r^n - R^* = r^n - (r^n - |R|) = |R|.$$

In order to avoid either the subtraction or the addition operation, we form the complement of an integer, an operation which itself involves subtraction. It appears that no great advantage has been gained. An examination of the complementation operation on a digit-by-digit basis, however, reveals that it is very simple. Suppose the digits of $|X|$ are $x_i, i = 0, 1, \ldots, n-1$, and we perform the operation $r^n - |X|$ digitally in the conventional way. We have

$$\begin{array}{rcccccc} r^n = & 1 & 0 & 0 & \ldots & 0, \\ -|X| = & -0 & x_{n-1} & x_{n-2} & \ldots & x_0. \end{array}$$

Let $c_j, j = 0, 1, \ldots, n-1$, represent the digits obtained by performing the above subtraction, and let x_i be the lowest-order digit of X that is different from zero. Then it follows that

$$\begin{aligned} c_j &= 0, & j &= 0, 1, \ldots, i-1 \\ c_j &= r - x_j, & j &= i \\ c_j &= r - 1 - x_j, & j &= i+1, i+2, \ldots, n-1. \end{aligned}$$

(If all digits are zero we take $i = n$.) The digit c_j is thus directly determinable from the corresponding digit x_j without reference to any other digits of X, once the lowest-order nonzero digit has been determined. To emphasize the fact that c_i is determined in a different manner than any of the other complement digits, namely by subtracting x_i from the radix r, we call this complement *radix-complement* (in decimal the ten's-complement, in binary the two's-complement, in octal the eight's-complement, etc.).

Example 2.9 In a six-position decimal machine, we form the ten's complement of 012340 by leaving the right-hand zero alone, subtracting the 4 from ten and all remaining digits from 9 to obtain 987660. Similarly, if the machine were octal, $r - 1 = 7$ and we subtract 4 from 8 and 3, 2, 1, 0 from 7 to form the eight's complement 765440.

The operation $r - 1 - x_j$ is particularly simple in the binary system in which many machines operate. In this system $r = 2$ and $x_j = 0$ or $x_j = 1$. Thus, if

$$x_j = 0, \quad r - 1 - x_j = 1,$$

and if

$$x_j = 1, \quad r - 1 - x_j = 0,$$

and the operation is merely that of changing ones to zeros and zeros to ones.

Since the form $r - 1 - x_j$ never requires a borrow, its use can be maximized and a simpler rule given by the device of representing the complement as $(r^n - 1 - |X|) + 1$. If $x_j^*, j = 0, 1, \ldots, n - 1$ represent the result of $r^n - 1 - |X|$, we see that

$$x_j^* = r - 1 - x_j, \quad j = 0, 1, \ldots, n - 1.$$

Therefore, we may state that the radix complement of X can be obtained by subtracting all digits from $r - 1$ and then adding one. This means we no longer need to separate the digits into three categories, and all digits including zeros are treated alike.

Example 2.10 The two's complement of the eight-position binary number 01101100 is obtained by leaving the final zeros alone, subtracting the four's-place 1 from 2 to obtain 1 and interchanging all remaining ones and zeros to get 10010100. We note that this can also be obtained by initially interchanging all ones and zeros and then adding one

$$\begin{array}{r} 1\,0\,0\,1\,0\,0\,1\,1 \\ +1 \\ \hline 1\,0\,0\,1\,0\,1\,0\,0 \end{array}$$

with all digits treated the same.

From both aesthetic and engineering points of view (and particularly, if $r = 2$), it is advantageous to have a complementation rule in which the x_j^* given above would be the digits of the complement, that is, a rule that would eliminate the necessity of the final step of adding one to the integer $r^n - 1 - |X|$. We can utilize such a rule if the accumulator performs the equivalent of addition and subtraction of the residue classes modulo $r^n - 1$ instead of modulo r^n. In the following section we will show how to modify the accumulator so that it performs this kind of arithmetic.

We have imposed the restriction $R < r^{n-1}$ on the true result. As a consequence, we know:

1. that the accumulator result R^* equals R if and only if the leading digit of R^* is zero, and

2. that R^* equals the complement of R if and only if the leading digit of R^* is $r - 1$.

Thus, as far as the machine or accumulator representation of results is concerned, we can easily tell whether or not a result is represented by its value or by the value of its complement from the first digit of its n-digit representation. To this point we have placed no restrictions whatsoever on the individual operands from which R is formed. Without additional information on the operands we are not able to tell merely from an inspection of its digits whether we are dealing with the operand or its complement modulo r^n. If needed, we can, for convenience, impose the same condition on operands as on results, namely that their magnitude be bounded by r^{n-1}. Then the n-digit representation of the magnitude of an operand must start with zero, while the n-digit representation of its complement must always start with the digit $r - 1$.

2.7 THE CLOSED ACCUMULATOR

In the preceding section we saw that the complement modulo r^n of an n-digit integer N can be formed by adding one to the number $r^n - 1 - |N|$. We define this latter number to be the *complement modulo $r^n - 1$ of N* or the radix less one complement (in particular the nine's complement, one's complement, or seven's complement). It has a simpler rule of formation than the complement modulo r^n, particularly if $r = 2$. If the accumulator were to perform the addition and subtraction of residue classes modulo $r^n - 1$, we could make use of such complements and take advantage of their simpler digital rule of formation. We see intuitively that in order to accomplish this change in the operation of the accumulator, the results it produces should somehow be displaced by one. This can be achieved by the introduction of *end-around carry* or *end-around borrow*. End-around carry or borrow is the propagation of a carry or borrow generated from digital position $n - 1$ into digital position zero. If we arrange this, the lowest-order position will no longer differ from the other positions by virtue of receiving no carry or borrow, and the highest-order position will no longer differ from the others by virtue of sending no carry or borrow to another position. One can imagine that this modification to the accumulator has been accomplished by bending the original accumulator around so that the highest-order position is adjacent on the right to the lowest order. For this reason, the modified accumulator will be referred to as *closed* or *circular*, in contrast to the previously defined open accumulator. The closed accumulator proves to have properties which are quite similar to those of the open one. Thus, we will confine ourselves to a summary of these properties, dwelling only on details where points of difference arise. The summary will also provide a review of the features of the open accumulator, and so the reader will have an opportunity to consolidate his understanding of both types of accumulating devices.

In the open accumulator, we discarded carries or borrows from the highest-order position in order to meet the restriction that the result be an n-digit integer. In accordance with this we also discarded from operands all digits from positions of order n and greater. In the closed accumulator we no longer discard highest-order carries or borrows, but rather apply them to the lowest-order position; and, in accordance with this change, we must make a corresponding change in the technique for reducing an arbitrary operand to n-digit form. To see how this can be done, consider a number N represented in base r^n with digits $a_0, a_1, \ldots, a_{k-1}$ $0, \leq a_i < r^n$. Thus,

$$N = a_0 + a_1 r^n + a_2 r^{2n} + \cdots + a_{k-1} r^{(k-1)n}.$$

Now note that

$$r^n \equiv 1 \ (\text{mod} \ r^n - 1).$$

Since residue-class multiplication is independent of the representatives used, repeated multiplication on the left by r^n and on the right by one gives

$$r^n \cdot r^n = r^{2n} \equiv 1 \ (\text{mod} \ r^n - 1)$$
$$\vdots$$
$$r^{jn} \equiv 1 \ (\text{mod} \ r^n - 1)$$

for any nonnegative integer j. Similarly for each term in N,

$$a_j r^{jn} \equiv a_j \ (\text{mod} \ r^n - 1),$$

so that

$$N = \sum_{j=0}^{k-1} a_j r^{jn} \equiv \sum_{j=0}^{k-1} a_j \ (\text{mod} \ r^n - 1).$$

That is, N is congruent to the sum of its base r^n digits modulo $r^n - 1$.

Any integer N, represented in base r, can also be considered to be represented in base r^n. We use the following device (filling in leading zeros if necessary). We group the digits from the right in blocks of n each. Thus

$$N = \sum_{j=0}^{kn-1} d_j r^j = \sum_{j=0}^{n-1} d_j r^j + \sum_{j=n}^{2n-1} d_j r^j + \cdots + \sum_{j=(k-1)n}^{kn-1} d_j r^j.$$

For a typical block we have

$$\sum_{j=(i-1)n}^{in-1} d_j r^j = \sum_{j=0}^{n-1} d_{j+(i-1)n} r^{j+(i-1)n}$$

$$= \left(\sum_{j=0}^{n-1} d_{j+(i-1)n} r^j \right) r^{(i-1)n}$$

$$= a_{i-1} r^{(i-1)n}.$$

The number

$$0 \le a_{i-1} = \sum_{j=0}^{n-1} d_{j+(i-1)} r^j < r^n, \qquad i = 1, 2, \ldots, k,$$

is a legitimate digit, base r^n. Any number base r, grouped in blocks of n digits in base r, can be considered as represented in base r^n with these blocks as digits. Therefore, the original number is congruent to the sum of these block numbers modulo $r^n - 1$.

Let the result of forming the first sum of these n-digit, base-r numbers be N_1. It is obvious that

$$N_1 = \sum_{j=0}^{k-1} a_j < N = \sum_{j=0}^{k-1} a_j r^{jn}.$$

If N_1 has more than n digits base r, we can repeat the operation to obtain a sum $N_2 < N_1$, and we can continue until we generate $N^* < r^n$ with $N^* \equiv N \pmod{r^n - 1}$. We obtain N^* digitally in the following way:

1. Starting from the right, we group the digits of the operand into blocks of n, filling out with leading zeros if necessary.

2. Considering each group of n-digits as representing the coefficients of the powers r^0 through r^{n-1} of an integer, we line the blocks up and form their sum.

3. If the sum exceeds n digits, we repeat the procedure.

Example 2.11 Suppose we have a three-position decimal machine so that $n = 3$ and $r = 10$. Consider the decimal number 634271922. We can group this into blocks of three as follows [634] [271] [922]. The numbers in the square brackets are the digits base 10^3, which we add to obtain 1827. This is still a four-digit number, so we repeat the process with $[827] + [001] = 828$ which is the required three-digit number with

$$634\ 271\ 922 \equiv 828 \pmod{10^3 - 1}.$$

We note that this procedure is an extension of the rule of adding digits to test for divisibility by nine. Thus, if we use the procedure with $n = 1$ and continue from 828 we have $8 + 2 + 8 = 18$ and $8 + 1 = 9$. Thus, the original number is congruent to nine (mod 9) and therefore divisible by nine.

Reducing an accumulator result to n digits by means of end-around borrows or carries can be considered as a special case of the process just described. If the sum described in step 2 above is formed in a closed n-digit base-r accumulator, the end-around effect will reduce the sum to n digits, automatically handling any necessary repetitions of the procedure, as described in step 3.

An end-around carry or borrow will never produce a second end-around carry or borrow. To see this, assume that in the original sum the lowest-order position to produce a carry was the ith. Then, since $c_{i-1} = 0$,

$$x_i^* + y_i^* = c_i r + s_i, \qquad c_i = 1.$$

Therefore,

$$s_i = x_i^* + y_i^* - r \leq r - 2.$$

Thus, a carry resulting from end-carry propagated to the ith position will produce in that position the result $r - 1$, and no further carry can propagate. Similarly, if the ith position is the first position of the original difference to produce a borrow, $b_{i-1} = 0$, and we have

$$x_i^* - y_i^* = b_i r + d_i, \qquad b_i = -1.$$

Therefore,

$$d_i = x_i^* - y_i^* + r \geq 1.$$

This last inequality follows from the fact that

$$-(r - 1) \leq x_i^* - y_i^* < 0.$$

Accordingly, any further borrow due to an end-around borrow which goes as far as the ith position cannot propagate a second borrow from that position, since it must be applied to a digit greater than zero.

We now turn to an examination of the relationship between the true result and the result given by the closed accumulator. To show that the final machine result is congruent to the true result modulo $r^n - 1$, let the operands be X and Y and their machine representations be X^* and Y^*. Then we can write $R = X + Y$. Since $X \equiv X^* \pmod{r^n - 1}$, and $Y \equiv Y^* \pmod{r^n - 1}$, we also have $R \equiv X^* + Y^* \pmod{r^n - 1}$. The sum $X^* + Y^*$ and the difference $X^* - Y^*$ are congruent to the accumulator result. Let

$$X^* + Y^* = c_{n-1}r^n + \sum_{i=0}^{n-1} s_i r^i,$$

$$X^* - Y^* = b_{n-1}r^n + \sum_{i=0}^{n-1} d_i r^i.$$

Since the closed accumulator handles the highest-order carries and borrows in an end-around fashion, the machine result R^* becomes, for addition,

$$R^* = \sum_{i=0}^{n-1} s_i r^i + c_{n-1},$$

and, for subtraction,

$$R^* = \sum_{i=0}^{n-1} d_i r^i + b_{n-1}.$$

We can form the differences to get

$(X^* + Y^*) - R^* = c_{n-1}(r^n - 1)$ and $(X^* - Y^*) - R^* = b_{n-1}(r^n - 1)$.

It follows that, for addition,

$$X^* + Y^* \equiv R^* \pmod{r^n - 1},$$

and, for subtraction,

$$X^* - Y^* \equiv R^* \pmod{r^n - 1}.$$

If we apply the transitivity property of congruence relationships, we obtain, finally, for either operation,

$$R \equiv R^* \pmod{r^n - 1}.$$

We have shown that end-around borrow or carry produces no further end-around borrow or carry. Therefore, in both cases the value for R^* shown above is actually the n-digit accumulator result.

Since a closed accumulator yields sums and differences congruent to the true sums and differences modulo $r^n - 1$, we may consider that it performs the addition and subtraction of the residue classes modulo $r^n - 1$; and, as always, we can freely substitute for any result or operand any (congruent) integer from the same residue class. These results are parallel to the results obtained for the open accumulator, with one important difference. In the open accumulator, all results are represented by the least nonnegative representative of the residue class. For the modulus $r^n - 1$, this is not necessarily true for operands and results which are congruent to zero. Consider a result congruent to $k(r^n - 1)$, and thus to zero, where $k > 0$. Such a result can be obtained by successive additions of $r^n - 1$ to the accumulator set initially to zero. On the first addition the result will stand at $r^n - 1$. The following computation illustrates the action of the accumulator for all succeeding additive transmissions of the digits of $r^n - 1$.

$$
\begin{array}{l}
\overline{}^{\,n\text{ digits}} \\
r^n - 1 = r - 1 \quad r - 1 \quad \ldots \quad r - 1 \qquad \text{(accumulator reading after first addition)} \\
r^n - 1 = r - 1 \quad r - 1 \quad \ldots \quad r - 1 \qquad \text{(transmit to accumulator)} \\
\,\,r - 1 \quad r - 1 \quad \ldots \quad r - 2 \\
\,+1 \qquad \text{(end-around carry)} \\
r^n - 1 = r - 1 \quad r - 1 \quad \ldots \quad r - 1 \qquad \text{(final reading).}
\end{array}
$$

Thus, for positive multiples of the modulus we do not get the least nonnegative representative from the residue class. In the case of results congruent to $k(r^n - 1)$, where $k \leq 0$, however, the machine representation of the result will indeed be the least nonnegative member of the residue class, namely zero. We can see that this is so by following the action of the accumulator in subtracting $r^n - 1$ from zero.

$$
\begin{array}{l}
\overline{}^{\,n\text{ digits}} \\
\,0 \quad\;\; 0 \quad\;\; 0 \quad \ldots \quad\; 0 \quad\;\; 0 \\
-(r^n - 1) = -(r - 1 \quad r - 1 \quad r - 1 \quad \ldots \quad r - 1 \quad r - 1) \\
\,0 \quad\;\; 0 \quad\;\; 0 \quad \ldots \quad\; 0 \quad\;\; 1 \\
\, -1 \qquad \text{(end-around borrow)} \\
\,0 \quad\;\; 0 \quad\;\; 0 \quad \ldots \quad\; 0 \quad\;\; 0 \qquad \text{(final result.)}
\end{array}
$$

In summary: When we add, the closed accumulator never returns to the machine result $R^* = 0$, in which all digits are zero. When we subtract, the closed accumulator never attains the machine result $R^* = r^n - 1$, in which all digits are $r - 1$. Thus, all positive multiples of $r^n - 1$ are represented in the accumulator by $r^n - 1$, while all negative multiples of $r^n - 1$ are represented in the accumulator by zero. Hence, the residue class of integers congruent to zero modulo $r^n - 1$ will not have a unique representative in the closed accumulator. The remaining residue classes modulo $r^n - 1$, R_j, $j = 1, 2, \ldots, r^n - 2$, will, as before, be represented in the accumulator by a unique integer which is the least nonnegative in the residue class.

Example 2.12 The fact that the machine representation of zero may be a sequence of digits all of which are $r - 1$ causes no difficulty in arithmetic. Thus, in a three-position decimal machine, if we perform the addition $6 + 0 = 6$, we would have:

$$\begin{array}{r} 0\ 0\ 6 \\ 9\ 9\ 9 \\ \hline \boxed{1}\ 0\ 0\ 5 \\ \hookrightarrow \quad\quad 1 \\ \hline 0\ 0\ 6 \end{array} \quad \text{end-around carry.}$$

Similarly, $6 \times 0 = 0$ is given by:

$$\begin{array}{r} 9\ 9\ 9 \\ 6 \\ \hline \boxed{5}\ 9\ 9\ 4 \\ \hookrightarrow \quad\quad 5 \\ \hline 9\ 9\ 9 \end{array} \quad \text{end-around carry,}$$

which is the machine version of zero.

Just as with the modulus r^n, we define the *complements modulo* $r^n - 1$ of any integer N to be the set of integers of the form

$$k(r^n - 1) - |N| \geq 0, \quad k \text{ an integer.}$$

We have already pointed out that the smallest integer in this set is called *the complement of N modulo* $r^n - 1$. Since an n-digit integer will be less than or equal to $r^n - 1$ in magnitude, we can always take $k = 1$ provided we agree to form complements of operands only after reducing them to n-digit form. Results are automatically reduced to n-digit form by the accumulator. As we have seen, this complement is particularly easy to construct digit by digit.

The complement modulo $r^n - 1$ of a negative integer is congruent to it modulo $r^n - 1$. Thus, by use of complements we can dispense with either the addition or subtraction operation of a closed accumulator. Subtraction can be eliminated by replacing operations of the form $-|N|$ with operations $+ \left| (r^n - 1) - |N| \right|$ and addition by replacing operations of the form $+ |N| = -(-|N|)$ with $- \left| (r^n - 1) - |N| \right|$. As before, this is permissible since it amounts to replacing a particular representative of a residue class with another representative from the same class. If addition is chosen as the basic accumulator operation, zero will be represented by the congruent integer $r^n - 1$ consisting of n digits $r - 1$, since, as we have seen, addition will never result in $R^* = 0$. This can also be verified by noting that the operations

$$|a| - |a| = 0$$

will be replaced by

$$|a| + (r^n - 1 - |a|) = (r^n - 1).$$

Thus, practically, a closed additive accumulator implies two different representations for zero since the human operators will not ordinarily care to replace so-called "true" zeros with the integer $r^n - 1$. On the other hand, if the device is basically subtractive, an operation of the type

$$|a| - |a| = 0$$

will generate only "true" zeros. Thus, only "true" zeros will arise in a closed subtractive accumulator, provided direct copying of $r^n - 1$ into the accumulator is not permitted.

For the closed accumulator we can interpret residue-class arithmetic as ordinary arithmetic by imposition of essentially the same conditions used for the open accumulator. In summary: If the result R satisfies $|R| < r^n - 1$, then

1. If $R > 0$, $R^* = R$.

2. If $R < 0$, $R^* = r^n - 1 - |R|$; that is, $|R| = r^n - 1 - R^*$.

3. If $R = 0$, $R^* = 0$ or $R^* = r^n - 1$, depending on the basic operation of the accumulator.

If the result R is not zero and satisfies $|R| < r^{n-1}$, then

1. $R^* = R > 0$, if and only if the highest-order digit of R^* is zero.

2. $R^* = r^n - 1 - |R|$ and $R < 0$, if and only if the highest-order digit of R^* is $r - 1$.

If $R = 0$, statement 3 above holds. If the representation of zero is $r^n - 1$, the leading digit is $r - 1$, and the representation is sometimes referred to as negative zero.

So far, restrictions have been imposed on results R but not on operands. For convenience, we may also impose the condition that the magnitude of all operands be less than r^{n-1}. This restriction will enable us to distinguish between an operand and its complement, as we can with results by examining the value of the leading digit. Hence we can distinguish positive and negative operands as well as results. For a closed accumulator the restriction that an operand be less than r^{n-1} in magnitude affords a further convenience, because it voids the necessity for reducing an operand to n digits.

2.8 ADDITION AND SUBTRACTION INSTRUCTIONS

By an addition or subtraction *operation* we mean the act of transmitting the digits $x_i, i = 0, 1, \ldots, n - 1$, of

$$|X| = \sum_{i=0}^{n-1} x_i r^i,$$

either additively or subtractively, to an accumulator. By an addition or subtraction *instruction* we will mean a complete message specifying the particular operands and the disposition of the result, as well as the arithmetic operation. A large variety of instruction messages can be built around an addition or subtraction operation. We assume the existence of an agency called the *control* which can receive, interpret, and execute instructions. It is not our purpose here to investigate the means by which the control can receive and decode an instruction message. We will, however, consider some of the general problems encountered by the control and some of the facilities which it requires.

In an addition instruction the control must provide for the acquisition and storage of the digits of the addend, augend, and sum. In a subtraction instruction the digits of the subtrahend, minuend, and difference must be obtained and stored. Since the accumulator register eventually stores one of the operands and the result, it can be used for two-thirds of the storage function as well as for the addition or subtraction operation. Generally the control will have a number of registers available to it for the storage of operands. If these storage registers were assumed to have the properties of an accumulator, both the storage and operation functions of an instruction could be accomplished by the transmission of digits from one such register to another. Such a scheme would require that the control have the ability to connect any storage register to any other storage register, and that all storage registers have at least accumulative facilities. This is uneconomical duplication of resources. It is simpler to provide the control with a group of special arithmetic registers separate from the general storage registers. Initially, we will assume two such special registers: the A-register (or A) which is an accumulator, and the X-register (or X) from which digits can be transmitted to the A-register in either an additive or subtractive manner. We assume also that in at least one of A or X the complement of the digits stored can be formed in accordance with the modulus employed by A. Provision must be made for copying digits from any storage register into the X-register in order to allow the control to utilize any stored operand in the machine. If the reverse transmittal to copy digits from X into any storage register is permitted, the need for connecting each register to any other is eliminated. All digits from one register can be copied into another by sending them via X. If it is also possible to copy the digits in A into X, we can transmit results from A into any storage register.

We will assume that the facilities available to the control have been arranged in the manner just described. This arrangement can be schematically shown as follows.

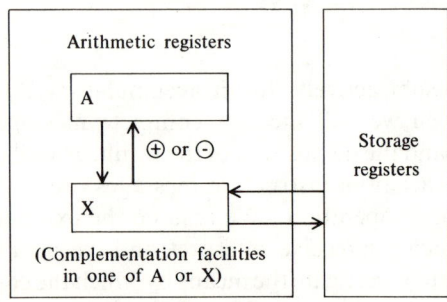

Here register A represents an additive or a subtractive accumulator which may operate either modulo r^n or modulo $r^n - 1$. The register X (or possibly A) can change the digits of the operand with respect to the particular modulus of A. The arrow from X to A represents transmission of digits in a manner equivalent to either the addition or subtraction operation. The other arrows represent the copying of digits from one register into another. This schematic is typical of the facilities available in many computing machines for executing addition and subtraction instructions. In some machines more arithmetic registers than indicated (but still few relative to the number of storage registers) will be provided. These additional operational registers can be considered to be either A or X for any particular addition or subtraction instruction.

We now consider the manner in which the control could use this typical structure to carry out an addition or subtraction instruction, and some of the problems it might encounter. We assume that the instruction is of the general form "obtain the digits of the integer in a specified storage register and add (subtract) that integer to (from) the integer presently in the accumulator register, leaving a sequence of digits representing the sum (difference) in the accumulator register." For definiteness, we assume that the accumulator is additive and that it performs the equivalent of arithmetic on residue classes modulo M, where M is either r^n or $r^n - 1$. We assume also that the necessary restrictions on magnitudes of results will be met. Thus, if the digits of $|N|$ are transmitted to the accumulator, in the result it will be the equivalent of adding an amount equal to $|N|$, and if the digits of $M - |N|$ are transmitted to the accumulator, in the result it will be equivalent to subtracting an amount equal to $|N|$.

The manner in which the control carries out the instruction depends on

2.8 ADDITION AND SUBTRACTION INSTRUCTIONS

the interpretation which is assigned to the sequences of digits *in the storage registers*. A conventional interpretation is that of *absolute value and sign*. Since we deal with complement numbers in the arithmetic registers, however, we may wish to avail ourselves of the convenience of the complement interpretation in the storage registers as well. To conform with customary human use of the absolute value and sign representation of integers, we could let a leading zero represent a plus sign and a leading digit $(r - 1)$ correspond to a minus sign. Thus, for example, in dealing with five-digit integers in the binary system, $|\text{ten}| = 01010$, and we might use

$$+|\text{ten}| = (0)\ 1010,$$
$$-|\text{ten}| = (1)\ 1010.$$

The digit used as a sign indicator has been placed in parentheses to indicate that in this case and in general the sign indicators will be in some sense kept separate from the digits of the absolute value. If complement representations are used, the digits of the absolute value of nonnegative numbers will be stored directly, while for negative numbers the digits of the complement will be stored.

If the absolute value is less than r^{n-1}, nonnegative numbers will start with zero and negative ones with $r - 1$. In the above example, for ten we store 01010 and for minus ten we store 10101 or 10110, depending on whether the modulus is $2^5 - 1$ or 2^5. For complement representation the sign indicator is in the register with the other digits, identified by the fact that it is in the highest digital location. We see that there are at least three different sequences of digits to represent minus ten in a storage register, and it is not surprising that the action of the control in carrying out an arithmetic instruction will depend on which sequence is actually employed.

2.8.1 Complement Operands

We consider the addition instruction for the case in which the storage registers contain complements for negative numbers. Let the sequence of digits obtained from the storage register specified by the instruction represent an integer N. Then, if $N \geq 0$, the operation of addition consists of adding $|N|$ to the operand represented by the digits already in A. Since, in this case, the digits obtained from the storage register will already represent $|N|$, the control copies these digits from storage into X, and transmits from X to A. On the other hand, if N is negative, the operation of addition consists of subtracting $|N|$ from the operand already in A or equivalently adding an amount $M - |N| \geq 0$ to this operand. Since complement representation is employed, the digits of $M - |N|$ will *already* be in the storage register. Thus, they need only be copied into X and then transmitted to A as before. In either case we may summarize the action of the control in the following way.

1. Copy the digits from the specified storage register into the X-register.
2. Transmit additively from the X-register to the A-register.

Example 2.13 Suppose that in a three-position, decimal additive machine using nine's complements we wish to form $6 + 5$ in one case and $6 + (-5)$ in another. In each case we assume that 6 is already in the accumulator and that the second operand is in register S. The operation of the control is given schematically by

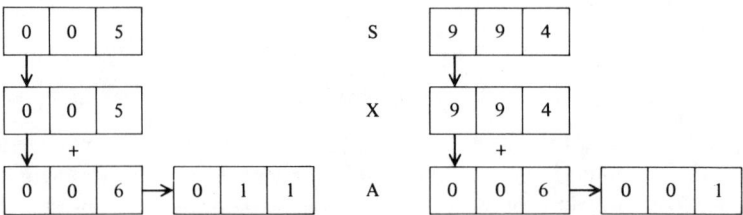

The necessary end-around carry in $6 + (-5) = 006 + 994$ is automatic in the nine's-complement accumulator.

We see that when complement representation is used in storage registers, the control need not be concerned with the sign of the number being added. Whatever the sign, it merely copies the digits from the storage register into X and then transmits them additively to A.

An entirely similar situation prevails for the subtraction instruction, and we may summarize the action of the control in this case in the following way.

1. Copy the digits from the specified storage register into the X-register.
2. Complement the digits in the X-register.
3. Transmit additively from the X-register to the A-register.

Example 2.14 In the same machine as that of Example 2.13, we wish to subtract to form $6 - 5$ and $6 - (-5)$. The schematic is then

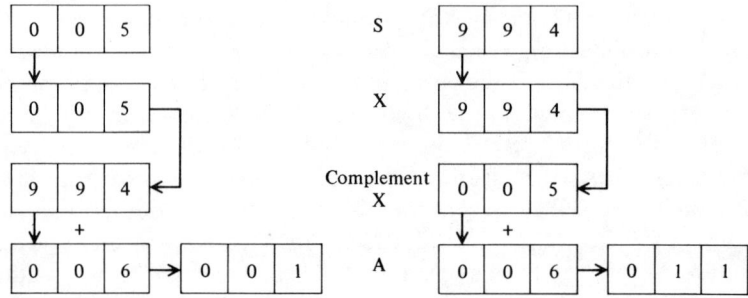

We see that for complement representations in storage, the control performs the same operation on digits regardless of sign, and, hence, *no sign*

2.8 ADDITION AND SUBTRACTION INSTRUCTIONS

discrimination is necessary in the arithmetic unit.

2.8.2 Absolute Value and Sign Operands

For addition with complements we added $|N|$ if the operand N was nonnegative and subtracted $|N|$ by adding $M - |N|$ if N was negative. In subtraction we subtracted $|N|$ by adding $M - |N|$ if N was nonnegative, and added $|N|$ if the operand N was negative. Entirely similar results could have been obtained for a subtractive accumulator. In the case of absolute-value and sign notation only the digits of $|N|$ and an associated sign indicator will be available. Thus, the control must check the sign indicator in order to determine whether or not to form a complement. For both addition and subtraction then, the control in the case of absolute-value and sign notation must carry out one sequence of steps for a positive integer and a different sequence for a negative one. In this case, *sign discrimination is necessary in the arithmetic unit.* This requires a facility beyond those which we previously assumed for the arithmetic unit. Henceforth, we will assume that for absolute-value and sign operands, storage facilities for the sign indicators and a means of testing these indicators is available.

We could avoid the necessity of complementing sequences of digits taken from storage registers by allowing the accumulator to be both additive and subtractive. This would not eliminate the requirement for dealing with complements, since a negative result would still appear in complement form in the accumulator. If absolute-value and sign representations were required in storage registers, the complement results from A would have to be recomplemented before the control could transmit them to storage. Hence, to avoid dealing with complements we must provide that only the absolute values of sums and differences be computed, with signs being handled separately. This necessitates the introduction of a *comparison* facility (such as we normally employ in pencil and paper arithmetic) which enables the control to compare the magnitudes of operands so that the order of operations can be reversed to forestall the appearance of a complement result.

To conclude this chapter we will give and discuss possible control sequences for the case in which n-digit, base-r, absolute-value operands are kept in storage with a separate sign indicator.

Control Sequence for Addition

We first copy the digits from the specified register in storage into X and record the associated sign indicator. We then have

CASE I. The sign indicators associated with the operands in A and X are alike.

1. Transmit the operand in X additively to A.
2. Leave the sign indicator associated with A unchanged.

CASE II. The sign indicators associated with the operands in A and X are not alike.

1. Complement the operand in A.
2. Transmit the operand in X additively to A.

CASE IIa. There was no carry from the highest-order position of the accumulator as a result of 2.

3. Complement the operand in A.
4. Leave the sign indicator associated with A unchanged.

CASE IIb. There was a carry from the highest-order position of the accumulator as a result of 2.

3. Change the sign indicator associated with A.

Control Sequence for Subtraction

The first step is the same as for addition. We then change the sign indicator associated with the content of X and continue as given above for addition.

Case I is obvious. In Case II, the true result R corresponding to steps (1) and (2) is

$$R = -|A| + |X|$$

where we use $|A|$ and $|X|$ to stand for the values of the operands originally in A and X respectively. In the accumulator, however, we have actually formed the sum

$$M - |A| + |X|.$$

In Case IIa, since there was no highest-order carry, this sum is equal to the accumulator result R^*. The absence of the carry also implies that

$$M - |A| + |X| < r^n;$$

that is, $R < 0$ for $M = r^n$ and $R \leq 0$ for $M = r^n - 1$. In this last case we assign equality to a negative zero. Thus, we know the sign of the true result, so to speak, in advance, and it is negative. Under the condition $|R| < M$, which we always assume to hold, the accumulator result R^* must therefore give the complement of the true result R, that is,

$$R^* = M - |R|.$$

It follows that step (3) of Case IIa results in

$$|R| = M - R^*,$$

and

$$M - (M - |A| + |X|) = |A| - |X|$$

as the value in the accumulator. In accordance with step (4), we see that for this case our final result will be

$$+|R| = |A| - |X|,$$

if the original sign indicator associated with A is positive and

$$-|R| = -\||A| - |X|\|$$

if this sign indicator is negative. When the plus sign is retained the operand in X must be negative and when the minus sign is retained, the operand in X must be positive to give the correct result.

In Case IIb, since we have carry from the highest-order position,

$$M - |A| + |X| \geq r^n;$$

that is, $R \geq 0$, and so $R = R^*$. We may then say that the result in the accumulator represents $|R|$ where

$$|R| = -|A| + |X|.$$

In Case II, if the original operand in A was negative and that in X positive, we wish the result $+|R|$ given above. We see that the sign associated with the accumulator is not correct and must be changed. If the original operand in A was positive and that in X negative, we want to compute $|A| - |X|$. We may rewrite this in terms of the result in the accumulator, following step (2) of Case II, as

$$|A| - |X| = -(-|A| + |X|) = -|R|.$$

We see that the absolute value in the accumulator is correct but that the sign associated with the accumulator is not, so that step (3) of Case IIb will give the correct absolute-value and sign result.

We leave it to the reader to work out the consequences of the rules just given for combinations of true and negative zeros given by the sign indicator and by the representation of all zeros and $r^n - 1$ where this is applicable.

The addition and subtraction process just discussed requires that facilities be provided in the arithmetic section to permit control to detect a carry from the highest-order accumulator position. This is an extension of facilities which are not necessary for the case of complement representation in storage registers. It is clear that an instruction involving only addition or subtraction is easier for the control to execute if complement representation of numbers is used in the storage registers. Unfortunately, this makes instructions involving multiplication and division more difficult for the control to execute. Since both complement and absolute-value and sign representation are used in computing machines, we will continue to consider both.

EXERCISES

SECTION 2.1

2.1 Analyse the following decimal additions and subtractions in terms of repeated applications of the division algorithm:

a) $\quad 725$
$\quad +\ 493$
$\quad \overline{1218}$

b) $\quad 627$
$\quad -492$
$\quad \overline{135}$

2.2 Show that carries and sum digits in addition and borrows and difference digits in subtraction as defined in Section 2.1 are unique.

2.3 Use the division algorithm to devise a scheme for determining the sum and carry digits for addition of three operands $X + Y + Z$ where

$$X = \sum_{i=0}^{n-1} x_i r^i, \quad Y = \sum_{i=0}^{n-1} y_i r^i, \quad Z = \sum_{i=0}^{n-1} z_i r^i.$$

What are the limitations on carries?

2.4* Generalize the results of Exercise 2.3 to formulate rules for carry and sum digits in addition of N operands.

SECTION 2.2

2.5 In each of the following additions in the base indicated, we assume a four-digit accumulator and ignore any carries generated to the fifth place. Deduce the relation between the final sum digits and the true sum.

	Decimal		Octal	Binary
X	3315	9315	3215	1011
Y	4734	2734	4734	1101
S	8049	2049	0151	1000

2.6 In each of the following subtractions, we assume a four-digit open accumulator. Deduce the relation between the digits shown and the true difference.

	Decimal		Octal	Binary
X	3315	9315	3215	1011
Y	4734	2734	4734	1101
D	8581	6581	6261	1110

SECTION 2.3

2.7 For a four-digit accumulator, reduce each of the following operands to machine form, find the machine sum digits, and show that the proper congruence relationship holds.

a) $X = 29314$, $Y = 724163$ in decimal
b) $X = 110111$, $Y = 10101$ in binary
c) $X = 27345$, $Y = 7214$ in octal

EXERCISES 65

2.8 With the operands of Exercise 2.7 do an equivalent analysis for the difference $X - Y$.

2.9* For a one-position decimal accumulator with all operands reduced to one-digit form, show that a tabulation of all possible sums represents a table of residue addition modulo ten.

2.10* Under what circumstances can the final result in an n-position accumulator be interpreted as the true sum of the original operands X and Y?

SECTION 2.4

2.11 What are the complements of the decimal 234 for a three-digit accumulator?

2.12 What are the complements of -234 for a three-digit decimal accumulator?

SECTION 2.5

2.13 For an eight-place open accumulator, determine for each base, the range of values of operands which can be utilized with a sign digit retained.
   a) Decimal   b) Binary
   c) Octal   d) Duodecimal

2.14 In a four-place binary accumulator the operands ± 3 and ± 6 satisfy $|\pm 3|$ and $|\pm 6|$ less than 2^3, and so may be represented in complement form with sign digits. Of the operations $(\pm 3) \pm (\pm 6)$, which ones are correctly represented and which cause overflow?

2.15 What condition on X and Y would have to be imposed to guarantee that every addition and subtraction operation $X \pm Y$ would have correct sign indication with no overflow?

2.16 The condition $|X| < r^{n-1}$ guarantees a sign digit of 0 or $(r - 1)$ for positive or negative numbers. If binary numbers are grouped in threes to represent octal digits, however, the binary 0 or 1 for plus or minus extends the range of leading octal digits for sign determination. Which leading octal digits correspond to positive and which to negative operands? What are the corresponding restrictions on the absolute values of operands?

2.17 Extend the results of Exercise 2.16 to decimal numbers. That is, if we do not reserve a leading zero or nine as the only sign indicator, how can we distinguish sign and what conditions must be imposed on operands?

SECTION 2.6

2.18 For a three-digit open additive decimal accumulator find the appropriate complement representation of $X = \pm 15$ and $Y = \pm 12$ and show that, with proper interpretation, addition of X and Y with every combination of signs produces a correct result.

2.19 With the operands of Exercise 2.18 and an open, six-digit, additive binary accumulator show that $X + Y$ with all combinations of signs is again correct.

2.20 The operands of Exercises 2.18 and 2.19 can be correctly represented with a five-digit binary accumulator. Would such an accumulator suffice for $X + Y$ in all cases?

SECTION 2.7

2.21 For a three-digit, closed, decimal accumulator, reduce each of the following to three-digit form:

    a) 2,397   b) 428,692   c) 329,426,712

2.22 Compare the rule for reducing operands for an n-digit closed accumulator with the method of "casting out nines" of Exercise 1.22. Generalize this idea for any base $\rho \geq 2$ and apply it specifically to $\rho = r^n$.

2.23 Redo Exercise 2.18 for a closed accumulator.

2.24 Redo Exercise 2.19 for a closed accumulator.

2.25 For the closed accumulators of Exercises 2.23 and 2.24 show that the negative form of zero produces correct results for $12 + 0 = 12$ and $12 \cdot 0 = 0$.

2.26 Redo Exercises 2.23 and 2.24, assuming closed subtractive accumulators.

SECTION 2.8.2

2.27 Apply the algorithms for addition and subtraction instructions for the cases where the sign indicator is the leading digit and both forms of zero occur.

CHAPTER 3

NONNUMERIC ASPECTS OF ARITHMETIC

3.1 INTRODUCTION

In Chapter 2 we have considered the mathematical structure of digital addition and subtraction for any base $r \geq 2$. In order to implement this structure in a digital computer we have assumed a number of basic devices such as registers to hold or store sequences of digits, and registers with specialized properties such as accumulators. We have made no attempt to describe physical entities which would have the required characteristics. A superficial description in terms of an ordinary desk calculator is straightforward. The top register is an accumulator, and we can think of the keyboard itself as an X-register containing whatever number corresponds to the keys depressed. When the add key is pushed the number in the X-register, that is, the keyboard, is added to the number in the accumulator, whose content then becomes the sum, which can be read off. The mechanical means of obtaining the sum by rotation of geared wheels may be useful for a desk calculator, but it is much too slow for the modern high-speed computer in which interrelationships of electrical impulses are utilized. Moreover, it is true that looking at a register in a modern digital computer would give no clue to the digits it contains. Other sensing devices must be used, and thus, while we may visualize the function, it is not so easy to visualize the physical reality. It is not our purpose in this chapter to discuss either the actual physical components of a computer or their relationships to each other, that is, the "networks" into which they are arranged. Rather, we will investigate something more basic, the logical components of a computer and the logical structure of the networks involved. We will pursue the answer to such questions as: "Regardless of the means used to represent digits in an accumulator and an X-register, what are the fundamental relations which must hold for addition, base r, to take place?" From this point of view the components and networks for achieving arithmetic which we discuss are logical, not physical, although they are physically realizable in many ways.

The goal of the circuitry for adding two numbers is to achieve a numerical result representing the sum. The analysis or synthesis of the circuits to produce the numerical result is itself nonnumerical, and it is this phase of arithmetic which is the topic of this chapter. To facilitate the study of these aspects of arithmetic we will introduce an algebra which is also nonnumerical. It is essentially the algebra of ordinary or Aristotelian logic. We make no attempt to give even the beginning of a rigorous (from either the philosopher's or mathematician's point of view) accounting of the relationship between a propositional logic and the algebra we introduce.[1] Instead we will take the algebra as basic to our discussion and point out its intuitive relationship to logic and the fact that it has many potential applications other than those considered here. It is this fundamental relationship, however, which gives rise to the terminology, the algebra of logic and logical circuitry.

3.2 BOOLEAN ALGEBRA

In the Boolean algebra (after George Boole[2]) which we introduce below, we consider a set consisting of two elements. Of primary concern will be the situation in which these elements can be categorized as *present* or *absent*. A little reflection, however, will point out the many instances in which we concern ourselves with precisely two possibilities: *true* or *false* (the logical association); a switch *closed* or *open* (from which we get the general reference to switching networks); a light *on* or *off*; a binary digit *one* or *zero*. With the concomitant suggestion that this open-ended view be kept in sight, we introduce a formal definition.

Definition 3.1 We shall call a *Boolean algebra* a set consisting of exactly two elements which we denote by 0 and 1, together with the following two binary and one unary operations:

1. Multiplication, the *and* operation (binary):

$$0 \otimes 0 = 0$$
$$0 \otimes 1 = 0$$
$$1 \otimes 0 = 0$$
$$1 \otimes 1 = 1$$

2. Addition, the *or* operation (binary):

$$0 \oplus 0 = 0$$
$$0 \oplus 1 = 1$$
$$1 \oplus 0 = 1$$
$$1 \oplus 1 = 1$$

---

[1] See, e. g., J. M. Anderson and H. W. Johnstone, *Natural Deduction*, Wadsworth Publishing Co., Belmont, California, 1962.

[2] George Boole, *The Mathematical Analysis of Logic* (New York: Philosophical Library, 1948). Originally published 1847.

3. Inversion, the *negation* operation (unary):

$$\bar{1} = 0$$
$$\bar{0} = 1$$

The elements or nodes of our logical networks will be devices which perform the equivalent of these operations, in the sense that they will accept inputs (one or two) and produce an output in accordance with one of the three operations defined. We shall consider an input or output to be "present" or "absent" according to a value of one or zero. We shall indicate these elements by a box containing a symbol indicating its function, together with inward-drawn arrows for input and outward-drawn arrows for output. Thus, for the three operations we use:

$$a \longrightarrow \boxed{A} \longrightarrow c = a \otimes b$$
$$b \longrightarrow$$

$$a \longrightarrow \boxed{O} \longrightarrow c = a \oplus b$$
$$b \longrightarrow$$

$$a \longrightarrow \boxed{N} \longrightarrow c = \bar{a}$$

We have employed the symbols A (for *and*), O (for *or*), and N (for *not*) to indicate the function of the element. Some authors utilize the multiplication or addition symbol. Some also use I (for *inverter*) instead of N. Any consistent symbolism which makes the purpose of the element clear may be introduced. From the definition of the Boolean operations it is not immediately apparent why the words *and, or, not*, are used, but a consideration of the output for each device makes this obvious. For the element

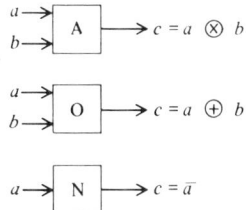

we see that an output will be present ($c = 1$) if and only if both a and b are present ($a = 1, b = 1$). Similarly, in the element

$$a \longrightarrow \boxed{O} \longrightarrow c = a \oplus b$$
$$b \longrightarrow$$

there will be an output present ($c = 1$) if and only if a is present ($a = 1$) or b is present ($b = 1$), or both ($a = b = 1$). Note that this is not the *exclusive or* which produces an output $c = 1$ if a or b is present, but not if both are present simultaneously. Finally, we see that for the element

$$a \longrightarrow \boxed{N} \longrightarrow c = \bar{a}$$

an output is present ($c = \bar{a} = 1$) if and only if an input is *not* present

($a = 0$). The bar over the symbol is read "not" followed by the name of the symbol. Thus $\bar{a}$ is read "not a."

We have used the symbols $\otimes$ and $\oplus$ to emphasize that the multiplication and addition operations are not those of ordinary arithmetic, although the function table for multiplication is identical with that for binary digit-by-digit multiplication. In practice this notation is somewhat cumbersome, and when no ambiguity is involved we shall prefer the simpler notations

$$a \otimes b = a \times b = a \cdot b = ab \quad \text{and} \quad a \oplus b = a + b.$$

Context will show whether the elements connected by the operation sign are Boolean.

The algebraic symbolism and operations corresponding to the elements of networks provide a convenient means of analysing and synthesizing complete networks without the necessity of tracing through a maze of diagrams. Since the applicable algebraic laws will prove very similar to those of real numbers, we shall have a familiar tool for manipulation of the logical relations involved. Prior to establishing these laws, however, we give an example of how logical network elements might be physically realized.

Example 3.1 A simple way to visualize Boolean and-or elements is in terms of electric switches connected in series or parallel. Thus, in series

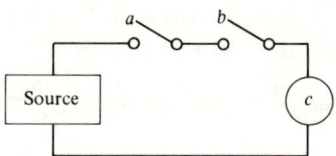

if $a = 1, b = 1$ only when closed and c is a light bulb, $c = 1$ when on, we see that the light is on if and only if both switch a *and* switch b are closed. That is, this is equivalent to

$$\begin{array}{c} a \rightarrow \\ b \rightarrow \end{array} \boxed{A} \rightarrow c = ab$$

In the same way switches in parallel produce an *or* element:

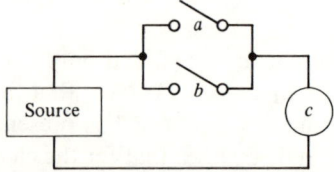

so that the light is on if either a or b (or both) is closed, giving the element

$$\begin{array}{c} a \rightarrow \\ b \rightarrow \end{array} \boxed{O} \rightarrow c = a + b$$

For inversion or negation we might use

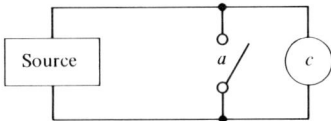

In this case as long as the switch is open ($a = 0$), the light is on ($c = 1 = \bar{a}$). If the switch is closed ($a = 1$), it shorts out the light ($c = 0 = \bar{a}$). If we assume no difficulty with overheated wires, we have

$$a \longrightarrow \boxed{N} \longrightarrow c = \bar{a}$$

It is unlikely that physical realizations such as these will be employed in modern computing devices.

3.2.1 Operational Laws of Boolean Algebra

The elements zero and one of a Boolean algebra play a somewhat similar role to their counterparts in real numbers; and the basic rules of order of operations apply, with some extensions. For any element a, we have

$$\begin{aligned} a \cdot 0 = 0 = 0 \cdot a & & a + 0 = a = 0 + a \\ a \cdot a = a & \text{and} & a + a = a \\ a \cdot 1 = a = 1 \cdot a & & a + 1 = 1 = 1 + a \end{aligned} \quad (3.1)$$

The truth of these assertions is apparent from Definition 3.1 if we assign to a each of its two possible values, 0 and 1.

$$a = 1: \quad \begin{aligned} 1 \cdot 0 &= 0 = 0 \cdot 1 \\ 1 \cdot 1 &= 1 \\ 1 \cdot 1 &= 1 = 1 \cdot 1 \end{aligned} \quad \text{and} \quad \begin{aligned} 1 + 0 &= 1 = 0 + 1 \\ 1 + 1 &= 1 \\ 1 + 1 &= 1 = 1 + 1 \end{aligned}$$

$$a = 0: \quad \begin{aligned} 0 \cdot 0 &= 0 = 0 \cdot 0 \\ 0 \cdot 0 &= 0 \\ 0 \cdot 1 &= 0 = 1 \cdot 0 \end{aligned} \quad \text{and} \quad \begin{aligned} 0 + 0 &= 0 = 0 + 0 \\ 0 + 0 &= 0 \\ 0 + 1 &= 1 = 1 + 0 \end{aligned}$$

The two binary operations of Definition 3.1 are commutative and associative, that is

$$\begin{aligned} ab = ba & \quad \text{and} \quad a + b = b + a & \text{(commutative law)} \\ a(bc) &= (ab)c \\ &= abc \quad \text{and} \quad a + (b + c) & (3.2) \\ &= (a + b) + c \\ &= a + b + c \quad \text{(associative law)} \end{aligned}$$

We establish (3.2) by assigning to a each of its two possible values 0 and 1 and using (3.1).

$$a = 1: \quad \begin{aligned} 1 \cdot b &= b = b \cdot 1 \\ 1 \cdot (bc) &= bc = (1 \cdot b)c \\ 1 + b &= 1 = b + 1 \\ 1 + (b + c) &= 1 = 1 + c = (1 + b) + c \end{aligned}$$

or

$$a = 0: \quad \begin{aligned} 0 \cdot b &= 0 = b \cdot 0 \\ 0 \cdot (bc) &= 0 = 0 \cdot c = (0 \cdot b)c \\ 0 + b &= b = b + 0 \\ 0 + (b + c) &= b + c = (0 + b) + c \end{aligned}$$

These are the equivalent of the corresponding rules for the algebra of real numbers, and they permit us to omit parentheses in successive operations without ambiguity. In forming successive Boolean additions and multiplications we can do them in any order. We can also perform combinations of the two operations in any order, since two forms of a distributive law hold.

$$\begin{aligned} a(b + c) &= ab + ac \\ a + bc &= (a + b)(a + c) \end{aligned} \quad \text{(distributive laws)} \quad (3.3)$$

For $a = 1$ we have

$$\begin{aligned} 1 \cdot (b + c) &= b + c = 1 \cdot b + 1 \cdot c, \\ 1 + bc &= 1 = (1 + b)(1 + c) = 1 \cdot 1, \end{aligned}$$

and for $a = 0$ we have

$$\begin{aligned} 0 \cdot (b + c) &= 0 = 0 \cdot b + 0 \cdot c = 0 + 0, \\ 0 + bc &= bc = (0 + b)(0 + c). \end{aligned}$$

The first distributive law, which says we can add and then multiply by first multiplying and then adding, is the familiar one. It permits us either to multiply grouped expressions term by term or to factor, for example,

$$a + ab = a \cdot 1 + ab = a(1 + b) = a \cdot 1 = a,$$

with possibly some unfamiliar but correct results. The second form of the distributive law, $a + bc = (a + b)(a + c)$ enables us to multiply, then add by first adding, then multiplying. This form does not hold for real numbers, but like its counterpart which also makes the order of two successive operations immaterial, it is of importance.

The formal application of rules (3.1), (3.2), and (3.3) clears the way for many simplifications in logical networks.

Example 3.2 Consider the following logical network:

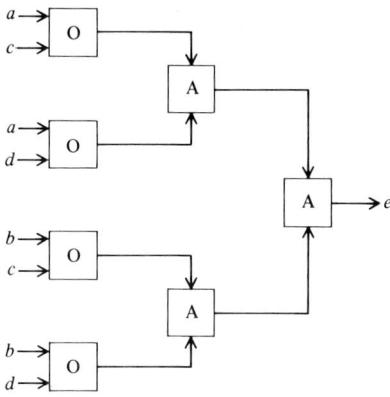

Expressed algebraically, the final output e for inputs a, b, c, and d is

$$e = [(a + c)(a + d)][(b + c)(b + d)].$$

If we apply the associative and distributive laws we can write, with the aid of Equations 3.1,

$$\begin{aligned}
e &= (a + c)(a + d)(b + c)(b + d) \\
&= (aa + ad + ca + cd)(bb + bd + cb + cd) \\
&= (a + ad + ca + cd)(b + bd + cb + cd) \\
&= (a \cdot 1 + ad + ac + cd)(b \cdot 1 + bd + bc + cd) \\
&= [a(1 + d + c) + cd][b(1 + d + c) + cd] \\
&= (a \cdot 1 + cd)(b \cdot 1 + cd) = (a + cd)(b + cd) \\
&= ab + acd + bcd + cdcd = ab + acd + bcd + cd \cdot 1 \\
&= ab + cd(a + b + 1) = ab + cd \cdot 1 = ab + cd.
\end{aligned}$$

Alternately, using the second form of distributive law we could write immediately

$$[(a + c)(a + d)][(b + c)(b + d)] = (a + cd)(b + cd) = ab + cd$$

as above. In either case, with $e = ab + cd$ we could replace the original network with the equivalent one

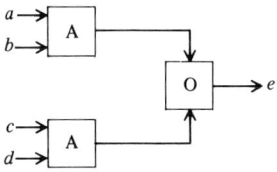

The two networks in Example 3.2 are equivalent in that for given inputs a, b, c and d, if we consider either network as a "black box" accepting these inputs, each will produce the same output e. The relative simplicity and fewer components of the second network indicate the utility of the algebraic procedure in simplifying a logical circuit already designed to produce a given result. There is no known method, given an existing circuit, for using algebraic manipulation to "optimize" it. Methods are available for "improving" the network corresponding to some special need such as reducing the number of components of a given type. We shall not consider these methods here.

The association of Boolean algebra with the logical process becomes clearer if we consider some operational properties of the inversion or negation operation. We have

$$a + \bar{a} = 1,$$
$$a\bar{a} = 0, \quad (3.4)$$
$$\bar{\bar{a}} = a,$$

since, for $a = 1$,

$$1 + \bar{1} = 1 + 0 = 1,$$
$$1 \cdot \bar{1} = 1 \cdot 0 = 0,$$
$$\bar{\bar{1}} = \bar{0} = 1,$$

and for $a = 0$,

$$0 + \bar{0} = 0 + 1 = 1,$$
$$0 \cdot \bar{0} = 0 \cdot 1 = 0,$$
$$\bar{\bar{0}} = \bar{1} = 0.$$

We illustrate the logical significance of the rules (3.4) in an example.

Example 3.3 Suppose that a is a logical proposition which can have a truth value of 1 for "true" and 0 for "false." Then

$$a + \bar{a} = 1$$

says that a true ($a = 1$, $\bar{a} = 0$, not false) or a false ($a = 0$, $\bar{a} = 1$, not true) exhaust the possibilities. This is a concise statement of the "law of the undistributed middle" of Aristotelean logic. The assertion

$$a\bar{a} = 0$$

says that a cannot be both true ($a = 1$) and false ($\bar{a} = 1$). The fact that not (not a) = a, as given by $\bar{\bar{a}} = a$ is familiar to English-speaking peoples to whom the double negative implies an affirmative.

We have given rules governing order for the binary multiplication and addition operations. If we include the unary negation operation, the rules for order are governed by the following two equations, usually referred to as deMorgan's laws. These are

$$\overline{a + b} = \bar{a}\,\bar{b},$$
$$\overline{ab} = \bar{a} + \bar{b}, \qquad (3.5)$$

since for $a = 1$,

$$\overline{a + b} = \overline{1 + b} = \bar{1} = 0 = \bar{1}\bar{b} = 0\bar{b},$$
$$\overline{ab} = \overline{1 \cdot b} = \bar{b} = \bar{1} + \bar{b} = 0 + \bar{b},$$

and for $a = 0$,

$$\overline{a + b} = \overline{0 + b} = \bar{b} = \bar{0}\,\bar{b} = 1 \cdot \bar{b},$$
$$\overline{ab} = \overline{0 \cdot b} = \bar{0} = 1 = \bar{0} + \bar{b} = 1 + \bar{b}.$$

The extension of deMorgan's laws and the associative and distributive laws to any finite number of elements is easy by induction and we leave this to the reader in the exercises. In what follows we will assume the induction has been established and use the extended versions freely.

3.3 THE LOGIC OF BINARY ADDITION

In Chapter 2 we showed that the addition operation could be achieved by a sequence of applications of the division algorithm in the form

$$x_i + y_i + c_{i-1} = c_i r + s_i, \qquad 0 \le s_i < r,$$

when $c_i = 0$ or $c_i = 1$. Here x_i and y_i represent correspondingly positioned digits of the augend and addend and c_{i-1} a possible carry from the next lower-order position. We retain the remainder s_i as the sum digit and propagate a carry when the quotient $c_i = 1$. For $r = 2$, the x_i, y_i, c_i, and s_i are all binary digits or bits. The analog between binary digits and Boolean variables is immediately obvious, and we exploit it in developing the logical relationships needed to produce the equivalent of binary addition. In any given application of the division algorithm we shall consider the bits x_i, y_i, and c_{i-1} as Boolean input variables and the sum digit s_i as a Boolean output variable. Since the goal is to achieve arithmetic addition we also supply as a second Boolean output variable the carry c_i (present or absent) to the next higher order.

A device which will accept two binary digits as input and produce proper sum and carry digits is called a *half-adder*. The reason for this name is apparent. Since the application of the division algorithm requires the acceptance of three input variables, a device which will accept these three bits as input and produce the ultimate sum digit and carry to the next higher order is termed a *full adder*. The logic of the half-adder is the simpler one of the two and we start with it, noting that it can provide the building blocks for a full adder.

3.3.1 The Binary Half-Adder

A binary half-adder accepts two binary digits as input and produces the corresponding sum and carry digits as output. Although the latter may not be the same as the ultimate digits generated in full addition, we will use no special notation at this point to distinguish the two. The possibilities are given by the following table with subscripts omitted:

x	0 0 1 1
y	0 1 0 1
s	0 1 1 0
c	0 0 0 1

The digits x and y are to be added to generate the sum digit s and the carry to the next higher order c. We start by finding some Boolean expression giving the correct values for s and c. From that point it or *any* other equivalent Boolean expression will give the relations for a half-adder. We may obtain Boolean expressions for s and c in any of several different ways. We note, for example, that the sum digit corresponds to the *exclusive or*. That is, the sum digit is one if x or y is one, but *not both x and y*. The carry digit, on the other hand, is one if and only if x is one *and* y is one. Expressed in the Boolean way these statements become

$$s = (x \text{ or } y) \text{ and not } (x \text{ and } y),$$
$$c = x \text{ and } y$$

or

$$s = (x + y)(\overline{xy}),$$
$$c = xy.$$

A network for a half-adder obtained directly from these equations is the following.

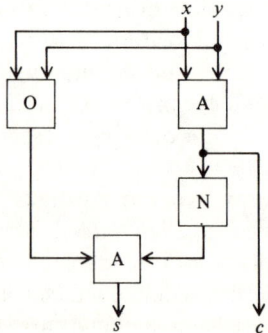

3.3 THE LOGIC OF BINARY ADDITION

The technique for obtaining a Boolean expression from a function table (sometimes called a truth table) can be as varied as the persons deriving them. We illustrate another approach in the following example.

Example 3.4 Observation of the function table for a half-adder and comparison with the basic definitions of Boolean algebra show immediately that the carry table

x	0 0 1 1
y	0 1 0 1
c	0 0 0 1

is the same as that for Boolean multiplication. Therefore, $c = xy$. The first three entries in the sum digit table

x	0 0 1 1
y	0 1 0 1
s	0 1 1 0

indicate s as $(x + y)$ from Boolean addition, but the last entry indicates $(\bar{x} + \bar{y})$. Since we require both,

$$s = (x + y)(\bar{x} + \bar{y}),$$

which is an alternative expression for s.

The expression for s in Example 3.4 can be obtained from the original expression by using the relation $\overline{xy} = \bar{x} + \bar{y}$. By means of algebraic manipulation, any number of equivalent expressions and, thus, equivalent networks can be found.

Example 3.5 Using various rules of Boolean algebra, we have for the sum and carry digits possibilities such as:

$$s = (x + y)(\overline{xy}) = (x + y)(\bar{x} + \bar{y}) = x\bar{y} + \bar{x}y$$
$$= \overline{(\bar{x} + y)} + \overline{(x + \bar{y})} = \overline{(\bar{x} + y)(x + \bar{y})}$$
$$= \overline{\bar{x}\bar{y} + xy},$$
$$c = xy = \overline{\bar{x}\,\bar{y}} = \overline{\bar{x} + \bar{y}}.$$

Thus, a network for a half-adder might be as shown below.

The network of Example 3.5 is apparently more complex than the first one developed. In terms of actual physical components, however, it is not necessarily true that what appears to be a simpler logical network may be easier or more economical to produce. The above examples do show that algebraic manipulation provides a number of alternatives from which to select a useful one.

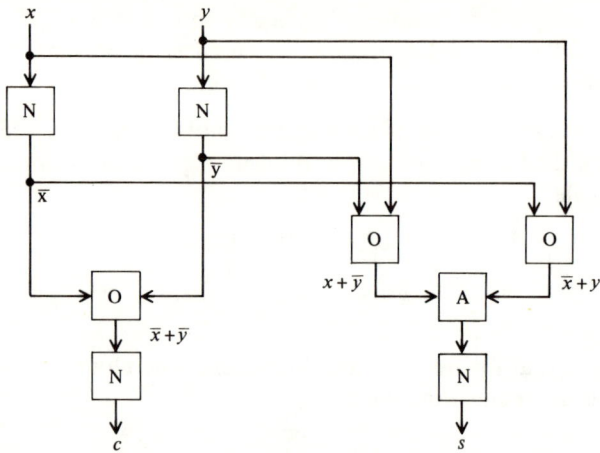

A possible network for a half-adder.

Whatever internal network is employed to generate the sum and carry digit for a half-adder, we can consider it as a single entity, accepting two input digits and providing two output digits. We shall symbolize it by the diagram which follows.

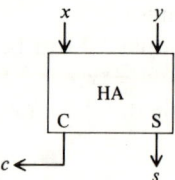

We note that because of commutativity the half-adder is symmetric with respect to its inputs. That is, x and y can be interchanged.

3.3.2 The Binary Full Adder

The Boolean relations for a binary full adder may be obtained directly from the function table or they may be obtained by utilizing half-adders. We shall consider both approaches, starting with combinations of half-adders. In either case, once obtained, the relations may be varied considerably from the original expressions to give equivalent networks.

The problem for a full adder is to combine three digits, the two addend digits and the carry digit (zero or one) from the next lower-order position. In pencil and paper arithmetic we ordinarily combine these two at a time. For example, we would usually add x_i and y_i and then add the carry c_{i-1} to the sum digit obtained from x_i and y_i. The carry from either of the two sums of digit pairs is taken as the carry. We can achieve this with two half-adders as follows:

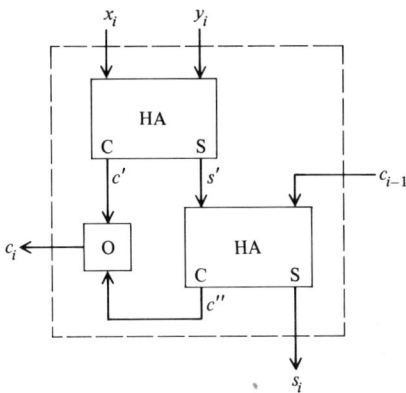

We characterize the circuit in the dotted rectangle above as a *full adder* and symbolize it by

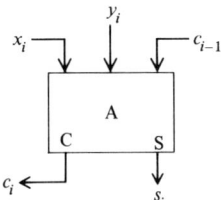

By way of illustration we have indicated the half-adders connected to provide the full addition in the order $(x_i + y_i) + c_{i-1}$. From the commutative and associative laws the order is immaterial.

Thus, we see that a full adder is completely symmetric in the three inputs. We note that it is impossible for both of the logical variables C' and C'' shown in the adder to assume the value one.

It is not necessary to use half-adders to derive logical networks for a full adder. We could proceed directly with the following function table.

c	0 0 0 0 1 1 1 1
x	0 0 1 1 0 0 1 1
y	0 1 0 1 0 1 0 1
s	0 1 1 0 1 0 0 1
c'	0 0 0 1 0 1 1 1

Here c, x, and y are to be provided as logical input variables which yield the sum s and carry c' as logical output variables. From the table, by observation, we can write Boolean relations to determine s and c':

$$s = \bar{c}(x\bar{y} + \bar{x}y) + c(\bar{x}\bar{y} + xy)$$

and
$$c' = xy\bar{c} + (x + y)c.$$
Many equivalent expressions can be formed.

Example 3.6 A simple arrangement for s is obtained by enumerating all cases in which an output variable is present.
$$s = \bar{x}y\bar{c} + x\bar{y}\bar{c} + \bar{x}\bar{y}c + xyc.$$
Similarly, we can write
$$c' = xy\bar{c} + \bar{x}yc + x\bar{y}c + xyc.$$
We leave it to the reader to show in the exercises that this is equivalent to the original version by multiplying out the parentheses.

Example 3.7 If we compose a full adder from half-adders,

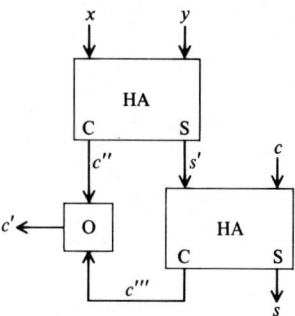

and use the circuitry for half-adders individually, we have
$$s' = x\bar{y} + \bar{x}y,$$
$$s = c\bar{s}' + \bar{c}s' = c\overline{(x\bar{y} + \bar{x}y)} + \bar{c}(x\bar{y} + \bar{x}y)$$
$$= c(\overline{x\bar{y}})(\overline{\bar{x}y}) + \bar{c}(x\bar{y} + \bar{x}y)$$
$$= c(\bar{x} + y)(x + \bar{y}) + \bar{c}x\bar{y} + \bar{c}\bar{x}y$$
$$= cxy + c\bar{x}\bar{y} + \bar{c}x\bar{y} + \bar{c}\bar{x}y,$$
and
$$c' = c'' + c''' = xy + s'c = xy + (x\bar{y} + \bar{x}y)c$$
$$= xy + x\bar{y}c + \bar{x}yc = xy(c + \bar{c}) + x\bar{y}c + \bar{x}yc$$
$$= xyc + xy\bar{c} + x\bar{y}c + \bar{x}yc.$$
These are the equivalent of those obtained in Example 3.6.

Example 3.8 Consider the network

3.3 THE LOGIC OF BINARY ADDITION

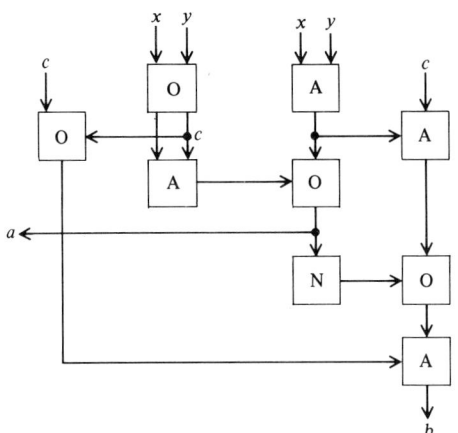

The outputs a and b are given by

$$a = (x + y)c + xy = xc + yc + xy$$
$$= xc(\bar{y} + y) + yc(\bar{x} + x) + xy(\bar{c} + c)$$
$$= xc\bar{y} + yc\bar{x} + xy\bar{c} + xyc = \text{carry},$$
$$b = [c + (x + y)][\overline{(x + y)c + xy} + (xy)c]$$
$$= x\bar{y}\bar{c} + \bar{x}y\bar{c} + \bar{x}\bar{y}c + xyc = \text{sum}.$$

Thus, this is a network for a full adder, but the roles of the half-adders are lost.

3.3.3 Parallel and Serial Addition

We can form the sum of any number of binary digits by means of appropriate combinations of half-adders and full adders connected. We consider first the open-ended system in which no carry need be accepted in the zero-order place. The connections are then:

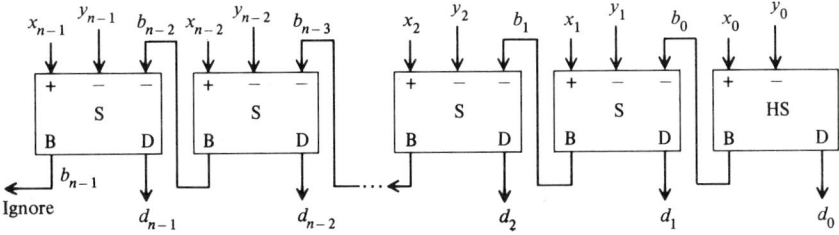

This diagram represents the logic of addition of residue classes modulo 2^n. The device will form the sum of two n-bit integers, ignoring any carry into the $(n + 1)$st position, so that it is capable of two's-complement addition. It forms a sum in the sense that as long as X is represented by the presence or absence of the x_i as its digits in accordance with their one or zero values and the same is true for Y, then the output digits s_i will be present or absent to give the correct sequence of zeros and ones in the sum. If we add storage capability to the device in the form of an "accumulator register" we can think of X as a number in the X-register and the whole device as one which *accumulates* the number Y (already in the accumulator register) and the number X to form $S = X + Y$. The digits of S can be thought of as replacing those of Y in the accumulator register. The fact that the original content Y and the accumulated result S are simultaneously present need not concern us; we can imagine that the digits of Y are taken from the accumulator register and brought to the input of the adder to make room for the sum. We will refer to the device as an *n*-digit open-ended binary *accumulator*. We note that in this kind of accumulator the accumulator register is purely a storage device with the logical operations that produce the sum occurring in the adders. An accumulator in which storage and addition are carried out in a single register will be considered later.

The necessary alterations to produce a closed *n*-digit accumulator are obvious. We replace the zero-order half-adder by a full adder and provide end-around carry as illustrated below.

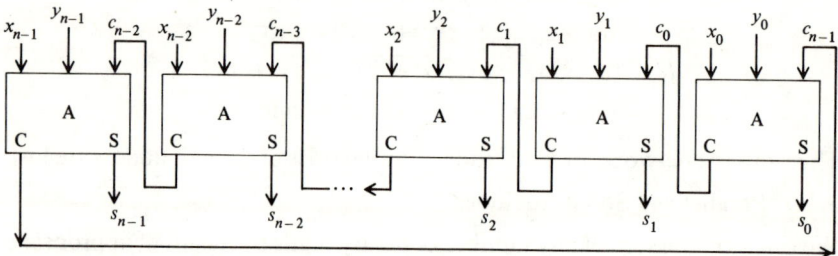

This diagram illustrates the logic of addition of residue classes modulo $2^n - 1$. We see that all stages of the adder register are alike. Since stage zero follows stage n in exactly the same manner as stage i follows stage $i - 1$, the device can be thought of as circular.

The adder arrangements above are based on the assumption that all digits are handled simultaneously; that is, in each triplet the digits (x_i, y_i, c_{i-1}) are combined to produce (s_i, c_i) at the same time (x_j, y_j, c_{j-1}) are combined to produce (s_j, c_j) for all admissible indices. All n combinations of digits are given parallel treatment in time; that is, digits are distinguished by where they appear, not when. As a consequence, this kind of operation is called *parallel addition*. It corresponds to a mechanical desk calculator in which all digits of Y are set into the accumulator, all digits of X into the keyboard, and a single stroke of the add key to all intents and purposes forms all digits of S at once.

As opposed to this parallel mode of operation, we can, as in normal long-hand computation, deal with the pairs of digits in order from lowest to highest, remembering the carry each time. Some accumulators are designed this way. That is, they combine (x_i, y_i) digits in pairs in order from $i = 0$ to $n - 1$. To achieve the analog of remembering the carry, a suitable delay device ⟶ D ⟶ is used in the circuit. Then an open-ended accumulator could be achieved by:

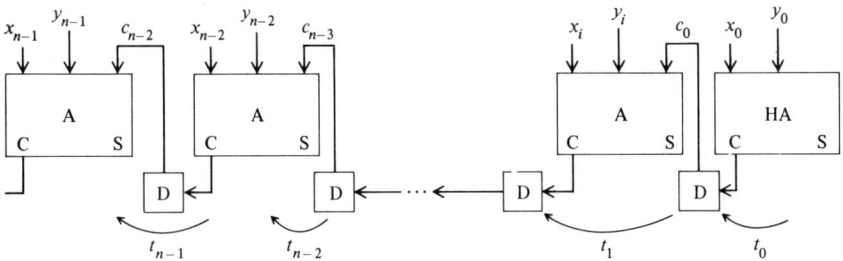

The digits are treated at discrete intervals of time, $t_0, \ldots, t_{n-1}$. Thus, each such time t_k corresponds to the kth position or to 2^k in binary representation. The delay device retains the carry c_0 formed at time t_0 until time t_1, when it can be combined with the pair (x_1, y_1) in a full adder and so forth. The process of computation, as indicated by the arrows, is thus serial in time and is called *serial addition*. This means digits are identified by when they appear and not where. Clearly the operation can be carried out so that the *same* adder is used in each interval of time t_i, effecting a considerable saving in equipment over the parallel accumulator. It is apparent that parallel addition holds time advantages over serial, since all digits are treated at once, but that serial addition holds economic advantages in that less equipment is needed. The choice of one or the other has usually been based on engineering and economic criteria, and we shall not concern ourselves with them here. However, we note that there is no logical distinction in the circuitry of the adders used in either case.

3.4 THE LOGIC OF BINARY SUBTRACTION

As we showed in Chapter 2, the subtraction operation can be carried out as a sequence of applications of the division algorithm in the form

$$x_i - y_i + b_{i-1} = b_i r + d_i, \qquad 0 \le d_i < r,$$

where $b_i = 0$ or -1. Here x_i and y_i represent correspondingly positioned digits of the minuend and subtrahend and b_{i-1} a possible borrow from the next lower-order position. We retain the remainder d_i as the difference digit and propagate a borrow when the quotient $b_i = -1$. For $r = 2$, the x_i, y_i, $|b_i|$, and d_i are all binary digits. The analogy of Boolean variable and binary digit will be retained and, as in the discussion of adders, exploited.

Binary subtraction parallels binary addition closely in terms of its logical structure. As in addition, the ultimate goal in a single digital position is a Boolean network or device with three logical input variables representing the minuend, subtrahend, and the lower-order borrow digits, and two logical output variables are to represent the difference digit and borrow to the next higher order. A device which will accept these inputs and produce the specified outputs will be called a *full subtractor*. Again we can develop this directly or by means of simpler devices, called *half-subtractors*, which can then be combined to provide the full result.

3.4.1 The Binary Half-Subtractor

A binary half-subtractor is a device which will accept as input the minuend and subtrahend (or lower-order borrow) digits and produce as output the partial higher-order borrow and difference digits. The function table is given by the following, where the Boolean variable b represents the absolute value of the borrow.

x	0	0	1	1	(+) Minuend
y	0	1	0	1	(−) Subtrahend
d	0	1	1	0	Difference digit
b	0	1	0	0	Absolute value of borrow

From this table we readily see that d is logically the same as a sum digit, for example,

$$d = (x + y)(\overline{xy}) = x\bar{y} + \bar{x}y.$$

However, $b = \bar{x}y$ is no longer symmetric in the two digits. We can obtain many versions of equivalent networks to produce half-subtraction. We will illustrate two of these.

Example 3.9 From the two expressions for d and b we could obtain half subtractors by either of the following circuits.

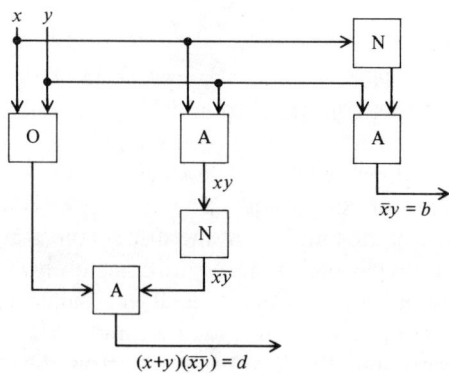

Alternatively,

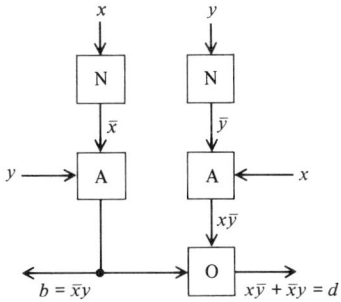

Whatever the internal circuitry, we shall symbolize a half-subtractor by

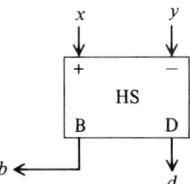

We emphasize that half-subtractors are not symmetric with respect to the input variables.

3.4.2 The Binary Full Subtractor

With subtraction, as with addition, we may utilize a direct approach from the function table of a full subtractor, or we may build up a subtractor by means of half-adders and half-subtractors. In either case we require three input variables and two output variables.

Example 3.10 The basic problem for full subtraction is to combine three bits x, y, and b (borrow from lower order) in the form $x - y - b$. One way of doing this with half-subtractors is by using the order $(x - y) - b$. Thus, we might do the following:

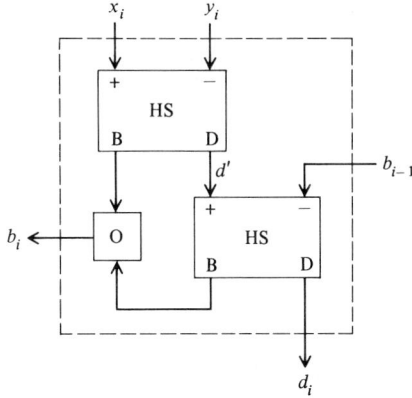

It is immediately apparent that we can interchange y_i and b_{i-1} but neither of these with x_i.

Example 3.11 We can also obtain full subtraction by combining a half-subtractor with a half-adder. This is achieved by considering $x - y - b$ as $x - (y + b)$.

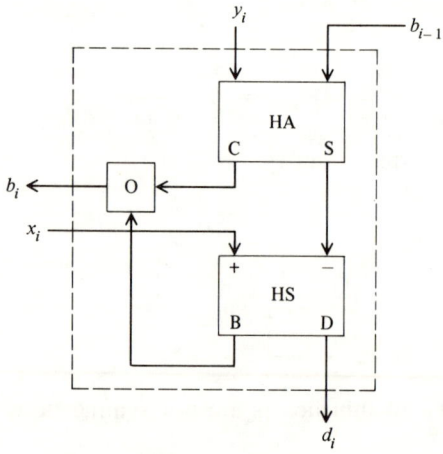

Because of the symmetry of the half-adder, y_i and b_{i-1} can be interchanged.

We symbolize a full subtractor in a manner similar to that of a full adder, but we must label the inputs:

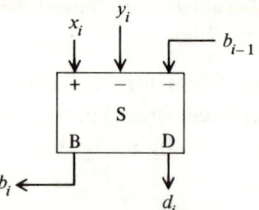

From the combinations of half-subtractors and/or half-adders illustrated above we can obtain a Boolean expression for a full subtractor. We will only sketch this direct approach to full subtractors, since it can obviously be modeled on the direct approach to full adders. We start with the function table for a full subtractor.

x	0	0	1	1	0	0	1	1	(+)
y	0	1	0	1	0	1	0	1	(−)
b	0	0	0	0	1	1	1	1	(−)
d	0	1	1	0	1	0	0	1	
b'	0	1	0	0	1	1	0	1	

Inspection of the table shows that one form for d and one form for b' are given by:

$$d = x\bar{y}\bar{b} + \bar{x}y\bar{b} + \bar{x}\bar{y}b + xyb,$$
$$b' = \bar{x}y\bar{b} + \bar{x}\bar{y}b + \bar{x}yb + xyb.$$

We note that if we replace b by the carry c, the expression for d is the same as that for the sum digit in full addition. This fact can be used to conserve components in machines which have both addition and subtraction. Many variations of the Boolean expressions given above have been derived. For example,

$$b' = \bar{x}y\bar{b} + \bar{x}\bar{y}b + xyb + \bar{x}yb = d(b + y) + \bar{x}yb.$$

We can also obtain this directly from the function table. We leave it to the reader to verify this.

3.4.3 Parallel and Serial Subtraction

We can provide subtraction of one n-digit binary number from another by providing the necessary full or half-subtractors for each digit. Thus, we can form an open *subtractive accumulator* by the following connections.

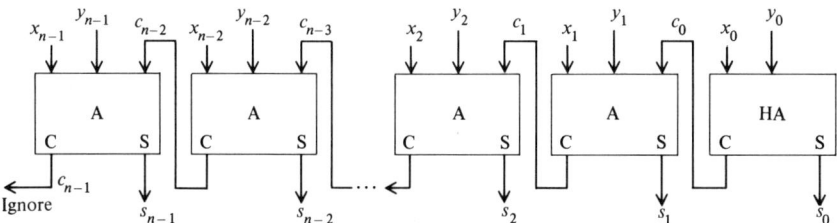

With this device we can perform subtraction modulo 2^n, that is, two's complement. By the addition of a storage register we can form a subtractive accumulator. The analogy with the additive type is obvious.

For one's-complement arithmetic we need only replace the zero-order half-subtractor by a full subtractor and connect the $(n - 1)$'s-order borrow output to it. For either the open or closed subtractive accumulator in the above arrangement, all digits are handled simultaneously to give *parallel subtraction*. If minuend and subtrahend digits are handled two at a time, with a delay device to retain each borrow long enough to be present at the appropriate instant, we have *serial subtraction*. The connections are equivalent to those for serial addition, with the implication that a digital position is determined by a specific instant in time.

3.5 FLIP-FLOPS

A *bistable element* is a physical entity or device which can be in either one of two stable states, by which we mean that the device, once in a state, remains there until some external influence causes it to change to its other state. Examples of bistable elements are two-position switches, electric lights on or off, coins heads or tails, and the like. We shall characterize each of the stable states with a Boolean value of zero or one, and this choice is arbitrary. Thus, we may characterize a light off as zero and on as one, or we may reverse this and assign the on state a value of zero and the off state value of one.

By a *flip-flop* we will mean a bistable element characterized by either of two properties:

1. It is capable of being arbitrarily set to its zero state upon receipt of one of two Boolean input variables and to its one state upon receipt of the other Boolean input variable.
2. It is capable of being changed to its alternate state upon receipt of a single Boolean input variable.

We symbolize these two kinds of flip-flops as follows.

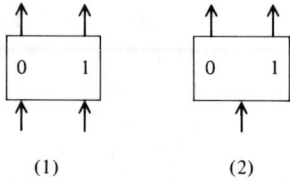

In each case the Boolean value of the output is characterized by either the zero-state present or the one-state present, but not both. Thus, the output variable corresponding to the zero side of the flip-flop will have the value one, or present, when the flip-flop is in the zero state and the value zero, or absent, otherwise. A similar statement holds for the output variable corresponding to the one side. Both output variables cannot be present simultaneously. We use the values zero and one in two ways, as defining the state of a bistable element, and as the values of Boolean output variables describing the presence or absence of each state. This should not cause confusion.

Example 3.12 The one-sides of two flip-flops are connected to an or-element as follows:

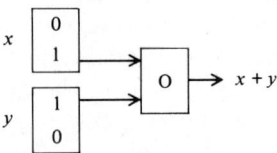

where x and y are Boolean variables representing the states of their associated flip-flops. We see that the output is still $x + y$. That is, it is $x + y$ in the sense that it is one if and only if x is present or y is present. In this sense we also note that the zero side of the flip-flop gives $\bar{x}$ or $\bar{y}$.

While one or the other of the output variables of a flip-flop will always be present (and remain present unless altered by an input), the input variables on both sides of flip-flop (1) above may both be zero. Indeed, while we have thought of Boolean values as remaining constant as long as a logical circuit is functioning, we will now consider that the input to a flip-flop may retain a Boolean value of one for an arbitrarily short duration. Thus, for a two-input flip-flop,

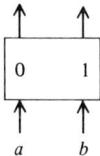

if $a = 1$ for even a short time, the state will be zero regardless of the original state and will remain there until, similarly, $b = 1$ sets the device to one. When $a = b = 0$, the element remains in whatever state it is. The case $a = b = 1$ is undefined, and we assume that it does not occur. For a single input device

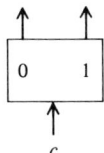

a value $c = 1$ will always throw the flip-flop to its *other* state or value $(0 \rightarrow 1, 1 \rightarrow 0)$ regardless of the time duration for $c = 1$. The value $c = 0$ leaves the Boolean output variables unchanged.

In practice many flip-flops will have all three inputs a, b, and c available. To avoid confusion, however, we will utilize whichever characteristic is logically required and will present only a two-input or one-input kind. Inputs of a, b, or $c = 1$ of short duration are frequently referred to as pulsed signals or simply pulses.

The property of a flip-flop which allows it to retain either of the two Boolean values indefinitely is a very useful one, since it provides *storage* of the value. In the same way it also stores a binary digit of zero or one. A set of n flip-flops can thus store an n-bit integer, and the two kinds of flip-flop input permit setting the value of the integer or complementing it. In the next section we will see how flip-flops can also provide for simultaneous addition and storage.

Example 3.13 A set of n flip-flops with dual input can be set to an arbitrary integral value in the range zero to $2^n - 1$ in the binary representation. Thus, for example with $n = 4$, we can use inputs $a_3 = 1, b_2 = 1, a_1 = 1, b_0 = 1$ to set 0101, the number five in binary.

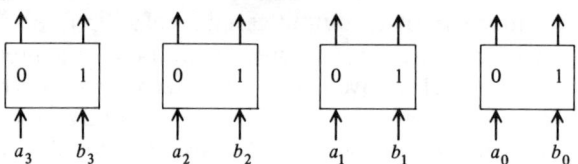

Similarly, with a single input we can form the one's complement. Thus, if 0101 is the number in the above register, a value $c_3 = c_2 = c_1 = c_0 = 1$ will cause each element to go to its other state to form 1010.

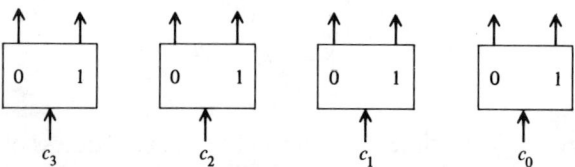

We note again that the input variables may be of arbitrarily short duration, and that when a given integer is set it will remain until replaced by new input pulses.

The functions of a flip-flop, like those of the half-adder and adder, can also be described in terms of the fundamental logical elements. We consider each kind with two examples.

Example 3.14 A two-input flip-flop can be characterized by the following logical circuit:

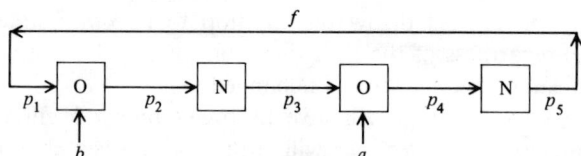

We consider the value of f to represent the state of the flip-flop, zero or one. If there is no input ($a = b = 0$) and $f = 1$, then the points of the circuit will be, in order, from left to right: $p_1 = f = 1$, $p_2 = p_1 + b = 1$, $p_3 = \bar{p}_2 = 0$, $p_4 = p_3 + a = 0$, and finally $p_5 = \bar{p}_4 = 1 = f$. Similarly, if $f = 0$, $p_2 = p_1 + b = 0$, $p_3 = \bar{p}_2 = 1$, $p_4 = p_3 + a = 1$, $p_5 = \bar{p}_4 = 0 = f$, and we have, for $a = b = 0$, $p_1 = p_5 = f$ remaining unchanged. Thus, each state of the flip-flop is stable. Now no matter what the current value of f is, a pulse at a, that is, $a = 1$ for any length of time, gives:

$$p_4 = p_3 + a = p_3 + 1 = 1 \quad \text{and} \quad p_5 = \bar{p}_4 = 0 = f = p_1,$$

and this state will then remain. Similarly, $b = 1$ gives, no matter what f is,

$$p_1 + b = p_1 + 1 = 1 = p_2 \quad \text{and} \quad p_3 = \bar{p}_2 = 0 = 0 + a = p_4,$$

so that $p_5 = \bar{p}_4 = 1 = f$. Thus, $a = 1$ sets the device to zero and $b = 1$ sets it to one, so it is equivalent to the following schematic:

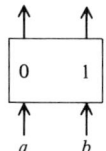

We can split f by means of an inverter to obtain the two output variables of the diagram:

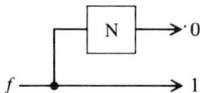

Each side yields a positive indication, that is, a value *one* for its corresponding state.

Example 3.15 If we superimpose the following network on the circuit of Example 3.14, we have a single-input flip-flop.

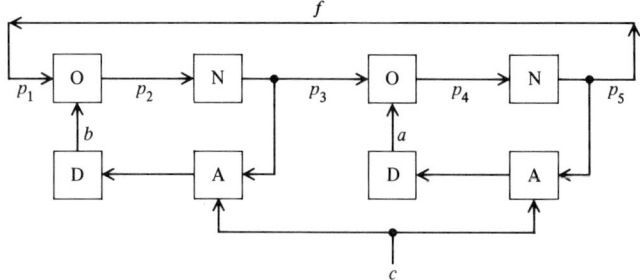

We analyse only one case. Suppose $f = 1$ and a momentary signal assigns to c the value $c = 1$. Since $p_3 = 0$, $p_3 \cdot c = 0$, and there is no input to b; but with $p_5 = f = 1$, $p_5 \cdot c = 1$, which gives $a = 1$ and sets the flip-flop to the zero state. The delay device is necessary so that the initial signal $c = 1$ is terminated prior to setting the value of $a = 1$ and so the state goes to zero. Otherwise a signal would be present on the b line and the state would be reset to *one*; in that case, oscillation between states would occur. With only the c input available, the circuit is equivalent to the single-input flip-flop of the following schematic diagram,

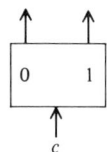

but it is apparent that with possible input signals a, b, and c, it can also perform the function of a dual-input flip-flop.

3.5.1 The Flip-Flop Accumulator

To consider a combination of full adders as forming an accumulator, we had to adjoin to the logical circuitry a dual-purpose storage register. We thought of the Y input and S output as being stored in the register, with Y representing the initial value in the accumulator and S the final sum. We shall now show how both accumulation and storage can be attained simultaneously by means of flip-flops. Many versions of such accumulators are possible, but since we wish only to present the idea, we give just one. We can think of Y as being *replaced* as the number in the accumulator by the sum $S = Y + X$. It is immediately apparent that if S is to replace Y, the replacement must occur at some point in time, that is, on receipt of an appropriate signal. We shall assume that this replacement takes place in two stages, corresponding to the fact that at each stage there will be two pairs of digits to handle. Thus, at the proper instant, there will be an *add signal* to combine the digits of X and Y, and this will be followed by a *carry signal* which will form the final digits of $S = X + Y$.

We first deal with the logical arrangement for the add signal which essentially gives half-addition. We start with the value Y in the n flip-flops of the accumulator, while X is in the n flip-flops of an auxiliary register. On receipt of the add signal each stage of the accumulator is to be set to the appropriate intermediate sum digit of half-addition and the carry of half-addition is to be generated. The carry signal will then be used to generate the ultimate carry and sum digit of full addition. If we consider the function table for a typical digital position,

x	0	0	1	1
y	0	1	0	1
s'	0	1	1	0
c	0	0	0	1

we see that the accumulator representing the digital value at the position should remain unchanged if $x = 0$ but should be switched to the alternate state if $x = 1$. We can achieve this by the following arrangement.

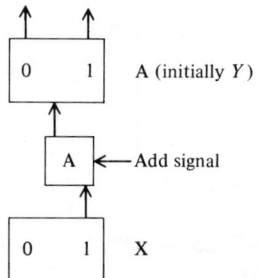

Duplicating this arrangement n times, with a common add signal, we see that the digits of Y in the accumulator A are altered by the add signal to the intermediate s' digits of half-addition. We must also generate a possible carry. This carry will occur if and only if the digit in the flip-flop at the A level is initially one and the corresponding digit in the X level is one. After partial addition has occurred by receipt of the add signal, a carry will be present if and only if the A-level digit is zero and the X-level digit is one. Thus, the following diagram gives a logical device which performs the functions of half-addition with storage of intermediate digits.

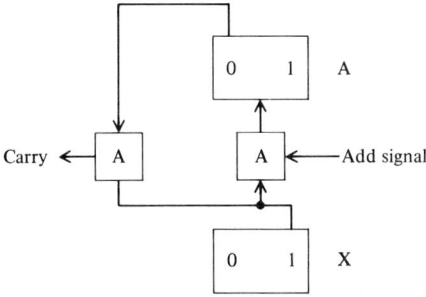

The function of half-addition takes place on receipt of the add signal.

The carry from the half-adder is sent to the next higher-order stage under control of the carry signal. This may in turn generate a carry, which will occur after the add signal has been issued, if and only if the intermediate value s' in A is one and the carry is one.

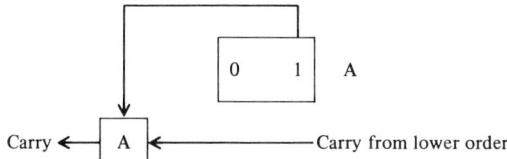

If a carry equal to one is generated either in partial addition from the next lower stage or in carry from below, the value of the digit in A must be switched on receipt of the carry signal. In case the flip-flop in A is *one* prior to the carry signal, we must provide a delay to retain the *one* long enough to generate the possible carry to the next higher order before the state is switched to zero. Thus, a complete, full-adder stage with add signal followed by carry signal might be as shown in the diagram below.

Example 3.16 Suppose we have a three-stage accumulator with each stage connected to the flip-flops on an X-register, as above, with carry and add signals in common. The number initially in the accumulator is three, 011, and $X = 010$, or two. On receipt of the add signal, the middle digit of X is one, so the corresponding digit one in A goes to zero, giving the intermediate

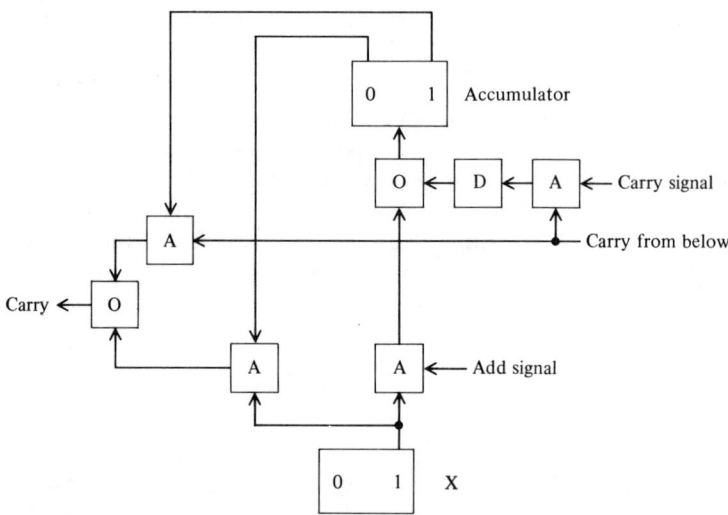

value 001. Since no end-around carry is assumed, no carry is generated to the lowest order. The combined value of $x = 1$ and $s' = 0$ in the middle digit, however, generates a value *one* from their common and-element. On receipt of the carry signal this sets the highest order flip-flop of A to its other state to give a final result of 101.

The logical relations for a flip-flop accumulator are the same as those of a full adder except that instead of accommodating carry and addition simultaneously, they are carried out in order. It is possible to provide a single signal which will perform both functions, but since the logical relations are more clearly seen in stages, we will not consider such systems here. It is immediately apparent that either open-ended or closed, end-around accumulators can be fashioned. For closed accumulators, the digits are all treated alike in a circular connection. An open-ended one needs to provide no input carry to the lowest-order stage and no output carry from the highest. If add-carry signals are generated sequentially for each pair of digits, we have serial addition. For the discussion above we have assumed a single add and single carry signal applied simultaneously to all digits, which provides parallel addition.

In a manner parallel to the development of a full subtractor we can devise a subtracting flip-flop accumulator. The reasoning is similar and so we will not take it up in detail. We leave it as an exercise for the reader to show that he can make the accumulator subtractive by reversing the zero state and one state of the A-level flip-flops. He may see this intuitively if he thinks of each flip-flop as a two-position wheel, as it might be in a mechanical binary machine. Reversing the zero and one states then corresponds to changing the direction of rotation.

3.6 ARITHMETIC IN ANY BASE

In developing the logic of binary addition and subtraction we have exploited the direct correspondence between the two possible values, one and zero, of binary digits and the two possible values, true and false, of logical variables. In developing the logic of base-r addition and subtraction, we will proceed in an analogous manner by the introduction of a correspondence between the r possible values of base-r digits and the true-false values of logical variables. This correspondence is usually referred to as a *binary code*. For example, an ordinary mechanical desk calculator represents each digit by means of a geared wheel or ratchet, which can assume and maintain any one of ten positions corresponding to the ten decimal digits. Such a device, however, can be looked at from another point of view. We can think of each of the ten ratchet positions as being in the up location or not. If it is in the up location, the wheel corresponds to that particular digit. Since we can think of each of the ten possibilities as either up or not, we can assign each a Boolean value of one for up and zero for not. Thus, we have the equivalent of ten Boolean variables (or binary digits), only one of which can be true (have the value one). The digit 5 could thus correspond to a sequence of Boolean values or binary digits of 0000100000 and 9 to 1000000000. Thus any element having r stable states can be considered as a set of r bistable elements in which at any time exactly $r - 1$ of the possible states have value zero, while the remaining one has the value one. By assigning each Boolean element to a different digit, we get a binary code for digits, base r. By analogy to the geared wheel of a desk computer, such a code is called a *ring code*.

A decimal ring code consists of ten binary digits, only one of which can have the value one. Thus, the decimal digits in order can be symbolized by

0000000001 = 0	0000100000 = 5
0000000010 = 1	0001000000 = 6
0000000100 = 2	0010000000 = 7
0000001000 = 3	0100000000 = 8
0000010000 = 4	1000000000 = 9

We have already given the example of the geared wheel on a desk computer. A punched card produces another example in which the ring code is utilized. Here we assume that in any column a hole may be punched in any one of ten locations but in not more than one. Thus, a hole represents one and the lack of it or blank a zero. We are ignoring the extensions utilizing twelve positions per column and possible multiple punching, since they represent a system with more than ten possible states including possible codes for alphabetical characters and other symbols.

The ring code is a simple way to translate base-r digits into either-or binary digits, but it is very wasteful. To construct an element with r states from binary elements we need at least k binary elements, where k is the smallest integer such that $2^k \geq r$. For example, with r equal to ten, k equal to three is too small, since $2^3 = 8$ implies that only eight different integers (states) can be represented by three binary digits or elements. However, $k = 4$ will do since $2^4 = 16$, and the sixteen different integers are more than sufficient to represent ten different states. Thus, a certain inefficiency will occur in using ten binary elements and the ring code to represent decimal digits since there are six unused configurations. A similar inefficiency occurs for any base r which is not a power of two. Hence, of the totality of $2^k \geq r$ configurations available, a subset of r must be selected, and a correspondence between these r selected configurations and the radix-r digits must be introduced. It is obvious that an unbounded number of different codes could be defined. The form and the details of the base-r adder which forms the fundamental logical unit of base-r arithmetic would, of course, depend on the code selected. From the multitude of possibilities, we will consider only two codes using the minimum number of binary digits. Discussions on the relative merits of various codes will not be entered into here.

3.6.1 Binary Coded Decimal

To define a code for each of the ten decimal digits, we need at least four binary digits, from which we can form sixteen patterns, of which six will necessarily be wasted. Even four bits provide the possibility of a very large number of distinct codes. The totality of possible choices is $(16!)/(6!)$, which is in the vicinity of 3×10^{10}. To qualify from this selection as a useful code, a particular one must have some advantageous aspect. We shall consider only two in detail, one of which has the advantage in complement arithmetic. We will give some examples of other useful codes.

3.6.2 The 8-4-2-1 Code

The most obvious binary code for decimal digits is their straightforward digital representation in binary. Thus, with four bits, each decimal digit takes the form $b_8 b_4 b_2 b_1$ where

$$d = b_8 \cdot 2^3 + b_4 \cdot 2^2 + b_2 \cdot 2^1 + b_1 \cdot 2^0.$$

We have used subscripts for each bit which correspond to the value of the power of two associated with it. This also suggests the name 8-4-2-1 code. This code is a special case of a *weighted* code, in which each position is assigned a weight, so that if the code for the decimal digit d is $b_3 b_2 b_1 b_0$, the value of the digit is

$$d = w_3 b_3 + w_2 b_2 + w_1 b_1 + w_0 b_0.$$

With $w_j = 2^j$, we have the 8-4-2-1 code:

Decimal	Binary 8-4-2-1
0	0000
1	0001
2	0010
3	0011
4	0100
5	0101
6	0110
7	0111
8	1000
9	1001

The combinations 1010, 1011, 1101, 1110, 1111, which correspond to decimal 10, 11, 12, 13, 14, 15, are not used.

This code has a simplicity in the logic of its arithmetic, since it permits digit-by-digit operations in decimal to be carried out to a large extent by binary arithmetic. In particular, if the sum of two decimal digit does not exceed nine, a four-bit accumulator provides decimal half-addition, with the variations from binary arithmetic coming because of the six unused codes and the necessity of providing a *decimal* carry.

Ordinary binary addition can be used as a basis for decimal addition, with corrections made where necessary. To create the decimal analog of the binary half-adder, we must build up a logical element which will accept eight input elements (four from each coded decimal digit). The output for the decimal half-adder must provide four correctly coded binary digits for the decimal sum digit and a carry. Since carry for any base r is zero or one, it can still be provided by a single Boolean variable. If an ordinary four-position binary accumulator is used to perform the first stage of decimal addition, we must reconcile the fact that it does arithmetic modulo $2^4 = 16$, with the fact that we wish arithmetic modulo 10. Three cases arise. Let the decimal digits be

$$x = x_8 x_4 x_2 x_1,$$
$$y = y_8 y_4 y_2 y_1,$$

in the 8-4-2-1 code. The possibilities are:

CASE 1. $0 \leq x + y \leq 9$.
CASE 2. $10 \leq x + y \leq 15$.
CASE 3. $16 \leq x + y \leq 18$ (maximal).

Only Cases (2) and (3) generate a carry. In Case (1) the sum is itself a digit and already in properly coded form. Thus, addition of the decimal digits 3 and 4, coded respectively as 0011 and 0100, produces 0111, which is the proper code for 7. No carry needs to be propagated and no correction to the binary addition is required.

In Case (2) a decimal carry is generated. It can be detected by the fact that one of the unused codes occurs as the result of the preliminary binary addition. The results are given in the following table.

Decimal sum	Code generated			
	8	4	2	1
10	1	0	1	0
11	1	0	1	1
12	1	1	0	0
13	1	1	0	1
14	1	1	1	0
15	1	1	1	1

It is easily seen that a *one* is present in the eight's position in all cases, together with a one in either the four position, the two position or both. This does not occur in the legal codes. Thus, the necessity of decimal carry can be detected by simultaneous ones in the (8, 4), (8, 2), or (8, 4, 2) positions. However, no binary carry which can be used for this purpose is generated from the adder in the eight's position. The code generated is for a sum which exceeds ten and is therefore not a decimal digit. It is too large by ten, so that ten must be subtracted. This is then balanced by the decimal carry. We have, however,

$$-10 \equiv 6 \pmod{16}$$

so the subtraction of ten can be accomplished by adding six and ignoring the carry which is generated in the eight's position. Ignoring the latter carry is equivalent to subtracting sixteen, so that $-10 = -16 + 6$ produces the subtraction of ten by addition of six.

Example 3.16 The sum of $5 + 7 = 12$ in the 8-4-2-1 code is started by

$$\begin{array}{r} 0101 \\ 0111 \\ \hline 1100 \end{array}$$

We detect and propagate a decimal carry by one in (8, 4) and we add six to give

$$\begin{array}{r} 1100 \\ 0110 \\ \hline \boxed{1}\ 0010 \end{array}$$

and ignore the binary carry $\boxed{1}$. This gives the correct sum digit of two.

The carry generated in Case (3) is easily detected. It occurs if and only if

$x + y > 15_{10} = 1111_2$ which also generates a binary carry from the eight's position. This happens for $8 + 8 = 16$, $8 + 9 = 17$, and $9 + 9 = 18$. The arithmetic is done modulo 16 by the accumulator, so that, including the carry the sum s' formed is

$$s' = 16 + c,$$

where c is the residual code digit generated by binary addition. Since $s' = 16 + c \equiv 6 + c \pmod{10}$, the decimal sum digit required is $s = 6 + c$, so we again need to correct by adding 6.

Example 3.17 The sum $9 + 9 = 18$ in the 8–4–2–1 code is given at the first step of binary addition by

$$\begin{array}{r} 1001 \\ 1001 \\ \hline \boxed{1} \; 0010 \end{array}$$

where the carry $\boxed{1} = 16$. This is equal to $10 + 6$ where 10 is used as the decimal carry and 6 is added back in for correction:

$$\begin{array}{r} 0010 \\ +0110 \quad \text{(correction)}. \\ \hline 1000 \end{array}$$

We note that the binary carry leaves a zero in the eight's position so the conditions of Case (2) cannot occur.

To summarize, binary addition gives the correct sum digit with no carry if the sum is no greater than nine. A decimal carry is detected by either binary sum digits $s'_8 = 1$ and $s'_4 = 1$ or $s'_2 = 1$ or both, or else it is detected by a binary carry in the eight's position. In any case where a carry occurs the code generated by binary addition must be corrected by adding six (mod 16). A possible decimal full adder is given in the following example. Note the simplicity of converting the above arrangement for a half adder to a full adder.

Example 3.18 One arrangement for a decimal full adder in the 8–4–2–1 code is given in the diagram below.

If the adder in the one's position at the lowest order is replaced by a half-adder, the configuration is a decimal half-adder. We leave it to the reader to satisfy himself that the discussion above is not changed in any essential way when a decimal carry is included.

Just as in the binary case, we can combine decimal adders and half-adders with appropriate storage devices to form open or closed decimal accumulators. We have shown for binary addition that we can combine the storage and addition functions by use of flip-flops. Since we have built up a decimal adder based on binary accumulators, it is clear that flip-flop techniques can be

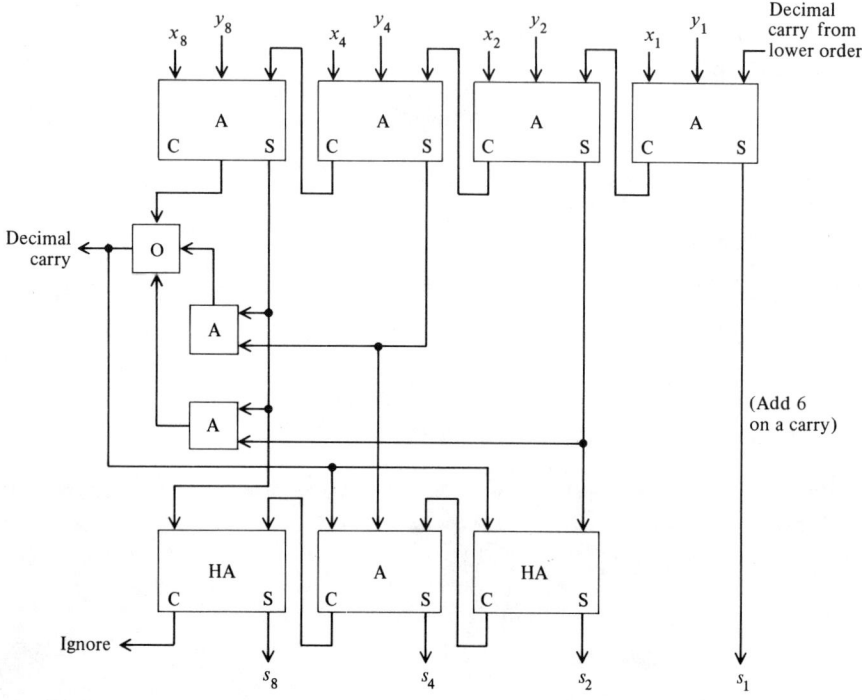

used to provide simultaneous decimal addition and storage. We leave it to the reader to elaborate on these ideas.

The formation of complements is a complicating factor in doing decimal arithmetic in the 8–4–2–1 code. In binary, we simply reverse every digit for one's complement or do this and then add one for two's complement arithmetic. If we form the one's complement of a decimal digit expressed in the 8–4–2–1 code, however, we actually form $d^* = 2^4 - 1 - d = 15 - d$, while for nine's complement we require $d' = 9 - d = 15 - (6 + d)$. Thus, we can form the nine's complement by first adding six and then forming the one's complement of the resulting code.

Example 3.19 To form the nine's complement of the digits 3, 7, and 8 we need to generate 6, 2, and 1. We do this by

	0011	0111	1000
Add six	0110	0110	0110
	1001	1101	1110
One's complement	0110	0010	0001
Decimal	6	2	1

3.6.3 Self-Complementing Codes; The Excess-Three Code

Of the many possible codes some have the property of being *self-complementing*; that is, formation of the one's complement of the code automatically gives the coded form of the nine's complement of the decimal digit. This is true of many weighted codes.

Example 3.20 The weighted 4–2–2–1 code can be made self-complementing.

	4221		
0	0000		
1	0001		
2	0010	or	0100
3	0011	or	0101
4	1000	or	0110
5	1001	or	0111
6	1010	or	1100
7	1011	or	1101
8	1110		
9	1111		

Note that the codes are not unique, but by appropriate choice of code the listings are self-complementary.

$$0 = 0000 \qquad 1111 = 9$$
$$1 = 0001 \qquad 1110 = 8$$
$$2 = 0100 \qquad 1011 = 7$$
$$3 = 0011 \qquad 1100 = 6$$
$$4 = 0110 \qquad 1001 = 5$$

The use of self-complementing weighted codes materially simplifies complement formation, but it may increase the difficulties of addition and subtraction. A self-complementing code which is not weighted, but which keeps arithmetic simple, is the *excess-three* code. In this code each decimal digit is arbitrarily represented by the 8–4–2–1 code for its true value plus three. Thus, the code for the decimal digit d is given by $d_3 d_2 d_1 d_0$ where

$$d_3 \cdot 8 + d_2 \cdot 4 + d_1 \cdot 2 + d_0 = d + 3.$$

Example 3.21 The excess-three code for the decimal digits and its self-complementing feature are illustrated in the table below.

The advantages of the excess-three code are its self-complementing feature and the fact that binary addition can be used easily as the basis for decimal addition. For two decimal digits d and c we use $D = d + 3$ and $C = c + 3$ expressed as four-digit binary integers. Ordinary addition gives

$$D + C = d + c + 6 = (d + c + 3) + 3.$$

If $d + c$ is a decimal digit, its excess-three code is too large by three and must be corrected by subtraction. Assuming we are using a four-place open binary accumulator, we may achieve the subtraction of three by adding $13 = 1101$ (true binary representation) and ignoring the carry of sixteen. We note that direct addition of the excess-three codes in this case will never result in either a decimal carry or a carry modulo 2^4, since if $0 \leq d + c \leq 9$, then $6 \leq d + c + 6 \leq 15$.

Decimal		Excess-Three		One's Complement		Decimal
0	=	0011		1100	=	9
1	=	0100		1011	=	8
2	=	0101		1010	=	7
3	=	0110		1001	=	6
4	=	0111		1000	=	5
5	=	1000		0111	=	4
6	=	1001		0110	=	3
7	=	1010		0101	=	2
8	=	1011		0100	=	1
9	=	1100		0011	=	0

Example 3.22 The sum of $3 + 4 = 7$ requires no carry. We use

$$\begin{array}{r} 0110 \\ 0111 \\ \hline 1101 \\ 1101 \quad \text{(Add true binary 13)} \\ \hline \boxed{1}\ 1010 \end{array}$$

and ignore the binary carry $\boxed{1}$ to obtain 1010, the correctly coded form of the sum digit.

If two decimal digits d and c have a sum $d + c > 9$, a decimal carry must be generated. If $d + c > 9$,

$$D + C = d + c + 6 > 15$$

is used to detect and propagate the decimal carry. Thus, another advantage of the excess-three code is that it generates the decimal carry naturally in the highest order. Just as with the 8-4-2-1 code, this binary carry is 16, which is too large by 6. We get $d + c + 6 - 16$ which we must restore to the excess-three code $d + c - 10 + 3$, so we can correct by adding three.

Example 3.23 The decimal sum $5 + 8 = 13$ requires a carry and a sum digit 3 or 0110 in excess-three form. We have

$$\begin{array}{r} 1000 \\ 1011 \\ \hline ① \leftarrow \boxed{1}\ 0011 \end{array}$$

where the binary carry $\boxed{1}$ gives the decimal carry $\textcircled{1}$. We add 3 or 0011 (true binary):

$$\begin{array}{r} 0011 \\ \underline{0011} \\ 0110 \end{array}$$

Thus, we may summarize addition in excess-three by the steps:

1. Ordinary binary addition (mod 2^4) of the coded form of the decimal digits.
2. If no carry is generated in the highest order, send no decimal carry and subtract three (add thirteen (mod 2^4)).
3. If a carry is generated in the highest order, send a decimal carry and add three (mod 2^4).

We leave it to the reader to devise circuitry to achieve this result and to develop the steps for subtraction.

EXERCISES

SECTION 3.2.1

3.1 By means of mathematical induction, extend the associative law for Boolean multiplication. That is, show that all groupings of $a_1 \cdot a_2 \cdot a_3 \cdots a_n$ are the same and unambiguous without parentheses.

3.2 Show that the order for the elements in Exercise 3.1 is immaterial.

3.3 Do the equivalent of Exercises 3.1 and 3.2 for Boolean addition.

3.4 Does the law of cancellation extend to Boolean algebra? That is, does
$$ab = ac$$
imply
$$b = c?$$

3.5 Devise a rule for "transposing terms" in Boolean algebra. For example if
$$a + b = c,$$
in what form can we transpose b to the right side of the equation?

3.6 Extend deMorgan's laws to any finite number of elements by mathematical induction.

SECTION 3.3.1

3.7 Establish the validity of the various forms of the sum digit s in Example 3.5.

SECTION 3.3.2

3.8 Establish the validity of the assertion that the carries c' and c'' of Section 3.3.2 cannot both be *one*.

3.9 Establish the validity of the alternate forms for s and c' in Example 3.6.

3.10 Devise a scheme for providing end-around carry for modulo $2^n - 1$ serial addition.

SECTION 3.4.2

3.11 Show that simultaneous borrows cannot be generated on the two half-subtractors of Example 3.10.

3.12 Show that a carry on the half-adder and a borrow on the half-subtractor of Example 3.11 cannot occur simultaneously.

3.13 Deduce the expressions for difference and borrow digits given for the function table for a full subtractor by using the Boolean expressions for two half-subtractors.

3.14 Redo Exercise 3.13 using one half-adder and one half-subtractor.

SECTION 3.5

3.15 For the logical circuit of Example 3.15, analyse the case in which $f = 0$ initially and c is set to $c = 1$.

SECTION 3.5.1

3.16 Devise a flip-flop subtractor with successive subtract and borrow signals.

3.17 Use the result of Exercise 3.16 to show that an additive flip-flop accumulator can be made subtractive by reversing the zero and one state of each of the A-level flip-flops.

SECTION 3.6.2

3.18 Various weighted codes for binary representation of decimal digits in the form $b_3 b_2 b_1 b_0$ with

$$d = w_3 b_3 + w_2 b_2 + w_1 b_1 + w_0 b_0$$

have been devised. For each of the following values of the weights w_i, write out the ten decimal codes.

	w_3	w_2	w_1	w_0
a)	5	2	1	1
b)	3	3	2	1
c)	7	5	3	-6

Note that the weights need not be positive. Do the weights determine a unique code for each digit?

3.19 For each of the following decimal additions, analyse each step if the decimal digits are represented in the 8–4–2–1 code.

```
   a)        b)        c)
   231       752       976
  +406      +318      +897
```

(Assume four positions where necessary.)

3.20 Use the Boolean expressions for the components of the 8–4–2–1 decimal adder given and derive Boolean expressions for the sum and carry digits. Experiment with simplifying these.

3.21* Use whatever full and half-adders and subtractors are needed and make a combination to produce one stage of an 8–4–2–1 decimal subtractive accumulator.

3.22* Devise one stage of an additive 8–4–2–1 decimal accumulator using flip-flops and add and carry signals.

SECTION 3.6.3

3.23 Show that the three codes given in Exercise 3.18 are self-complementing if the appropriate choice is made for each code.

3.24* Utilize any full and half-adders and Boolean elements needed and form one stage of an excess-three decimal additive accumulator.

3.25* Analyse subtraction for the excess-three code and redo Exercise 3.24 for a subtractive accumulator, utilizing any binary subtractors needed.

3.26* Devise one stage of an excess-three, decimal, additive accumulator using flip-flops.

CHAPTER 4

SHIFTING

4.1 INTRODUCTION

In the extension of the digital representation of integers to include all real numbers, considered in Chapter 1, we introduced the idea of the base, or radix, point. The primary function of this base point in a positional notation is to separate those positions corresponding to nonnegative powers of the base from those that correspond to negative powers. Alternatively, in any sequence of digital positions, base r, we can think of the base point as a reference point which uniquely defines the power of r associated with each position. Relocation of its place in the sequence, or "moving" the base point, thus redefines the exponent of the base assigned to any position. It is apparent that the change in the power of the base which such a relocation produces is the same in all places. Since the increase or decrease in the exponent is the same in all positions, the net effect is multiplication of the entire sequence by some fixed power (positive or negative) of r. Thus, multiplication and division by a power of the base are extremely simple operations which we accomplish in pencil-and-paper arithmetic by moving the radix point, to the right for multiplication, and to the left for division. We can also adopt the point of view that the position of the radix point is kept fixed and the digits are moved or *shifted* to the left or to the right relative to the point. Since operations in a computer consist essentially of digital manipulations, it is the latter view we adopt.

Example 4.1 For the decimal number 123.456 we can multiply by 100 by moving the decimal point two places to the right to give 12345.6. Alternately, we can think of the decimal point as fixed and shift the digits two places left.

$$123.456$$
$$12345.6$$

Similarly, we can divide by 10 by shifting the digits one place to the right

$$123.456$$
$$12.3456$$

which is equivalent to moving the decimal point one place to the left.

The shifting of digits in a register can roughly be defined to be the replacement of each digit by the digit immediately adjacent to it on its right or on its left. Intuitively, we may feel that if the original sequence of digits represented an operand N, the shifted sequence should represent Nr or Nr^{-1}, depending on the direction of shift. Because we are really dealing with residue classes, however, more careful investigation will reveal that this need not necessarily be the case. Multiplication and division operations more general than by powers of the base can be expressed in terms which depend on a sequence of multiplications and divisions by the base. Therefore, it is of importance for the later study of multiplication and division to determine precisely what we mean by a shifting operation. It is equally important to determine under what conditions a shifting operation can be interpreted as either a multiplication or a division by a power of the base.

While the shifting of digits is a necessary component operation of multiplication and division, it is also a useful arithmetic operation in its own right. Thus, the control of a general purpose digital computer will have the ability to execute shifting instructions as well as instructions involving the usual operations of arithmetic. We will find the execution of shifting instructions to be particularly useful in our discussion of scaling in Chapter 7. The shifting of digits in a register is also of importance in connection with certain nonarithmetic aspects of computer operation, but we will not consider these here.

4.2 DEFINITION OF SHIFTING OPERATIONS

In an n-digit register, with the usual ordering from right to left, a *left-shift of one position* is defined to be the replacement of the digit originally in the ith position by the digit originally in the $(i - 1)$st position for $i = 1, 2, \ldots, n - 1$. This definition does not state the digit which is to be placed in the position vacated by the lowest-order digit, or what is to be done with the highest-order digit that has no higher-order position to shift into. Two useful kinds of left-shift operations can be constructed in accordance with the manner in which the lowest- and highest-order positions are to be treated. We consider these cases in the definitions below.

Definition 4.1 A *closed* or *circular* left shift of one position is the replacement of the digit originally in the ith position by the digit originally in the $(i - 1)$st position for $i = 1, \ldots, n - 1$. The digit originally in the zero position is replaced by the digit originally in the $(n - 1)$st-order position.

Definition 4.2 An *open* or *end-off* left shift of one position is the replacement of the digit originally in the ith position by the digit originally in the $(i - 1)$st position for $i = 1, \ldots, n - 1$. The digit originally in the zero position is replaced by zero and the digit originally in the $(n - 1)$st position is discarded.

4.2 DEFINITION OF SHIFTING OPERATIONS

We emphasize that for a circular left shift the highest order digit is shifted into the position vacated by the lowest-order digit, while for an end-off left shift, the highest-order digit is discarded and the lowest-order digit position filled with zero.

Example 4.2 Let $n = 4$ and $r = 2$, and let the initial content of the register be

$$1011.$$

Then, after a circular left shift of one, the content will be

$$0111,$$

and, after an open left shift of one, it will be

$$0110.$$

In Example 4.2, if the digit configuration represents the absolute value of an integer, the original integer is eleven and the shifted integers are seven for the circular shift and six for the end-off shift. Thus, in neither case has the shift had the effect of a multiplication by the base two. However, we observe that after the circular shift we obtain digits which represent an integer congruent to twice the original number modulo $2^4 - 1$. After the open shift, we obtain a digital configuration which represents an integer congruent to twice the original integer modulo 2^4. That is

$$7 \equiv 22 \ (\mathrm{mod} \ 2^4 - 1)$$
$$6 \equiv 22 \ (\mathrm{mod} \ 2^4)$$

In the following section we will show this to be true in general.

In an n-digit register a *right shift of one position* is the replacement of the digit originally in the ith position by the digit originally in the $(i + 1)$st position for $i = 0, 1, \ldots, n - 2$. The definition leaves open the question of what will be done with the digit shifted from the lowest-order position and what digit will be shifted into the position vacated by the highest-order digit. Again two kinds of shifting operations can be defined in accordance with the way in which the lowest- and highest-order digital positions are treated.

Definition 4.3 A *closed* or *circular* right shift of one position is the replacement of the digit originally in the ith position by the digit originally in the $(i + 1)$st position for $i = 0, 1, \ldots, n - 2$. The digit originally in the $(n - 1)$st-order position is replaced by the digit originally in the zero position.

Definition 4.4 An *open* or *end-off* right shift of one position is the replacement of the digit originally in the ith position by the digit originally in the $(i + 1)$st position for $i = 0, 1, \ldots, n - 2$. The digit originally in the zero-order position is discarded, and the digit originally in the $(n - 1)$st position is replaced by zero.

108   SHIFTING   4.3

In the manner of a circular left shift, a circular right shift brings the zero-order digit into the highest-order position. In either case the effect is exactly the same as though all positions were connected in a circle with the highest and lowest orders contiguous. This is completely analogous to the end-around carry and is sometimes called an end-around shift. We shall see later that both concepts are intimately connected with residue-class arithmetic modulo $r^n - 1$.

The end-off right shift has the same effect, for integers, as moving a digit to the right of the base point and discarding it. In the meantime the vacated high-order position is filled with a zero, to keep all digital positions defined.

Example 4.3 With the original sequence of digits 1011 of Example 4.2 the result of a right circular shift of one position is 1101, while the result of an open shift would be 0101. Interpreted as integers, the results are, respectively, eleven, thirteen, and five. Clearly the shift has not accomplished a division by the base two.

We have observed from the above examples that the operations of shifting cannot always be expected to accomplish either division or multiplication by the base. In the following sections, however, we will determine conditions under which shifting can be interpreted as such elementary multiplication and division.

4.3 LEFT SHIFTING

Let the nonnegative integer $N = d_{n-1} d_{n-2} \ldots d_1 d_0$ occupy an additive accumulator. N will be the machine representation of a number X with

$$N \equiv X \pmod{r^n}$$

for an open accumulator, or

$$N \equiv X \pmod{r^n - 1}$$

for a closed one. Now consider the result of transmitting these digits to the accumulator $r - 1$ times. This additive transmission of N to an accumulator already containing N is equivalent to transmitting to a cleared accumulator the digits of N a total of r times and the true result will be $R = Nr$; then, as before, we will denote the integer defined by the resultant accumulator digits to be R^*. The digit d_0 will be added to zero in the zero-order position r times. By the definition of accumulator operation given in Chapter 2, the amount rd_0 thus transmitted will give a sum digit zero in the zero position and d_0 carries to the first position. Thus, the total amount added to an original zero in the first position is $rd_1 + d_0$. This results in a sum digit equal to d_0 in the first position with a total carry of d_1 to the second position. In general, in the ith position ($i > 1$), the amount accumulated will equal

$rd_i + d_{i-1}$, and this will give a sum digit in the ith position equal to d_{i-1}. It will also propagate a total carry equal to d_i to the $(i + 1)$st position. If the accumulator functions modulo r^n, there will be no carry to the lowest-order position, which will then remain at zero, and any carry from the $(n - 1)$st position will be discarded. On the other hand, if the accumulator is closed, the end-around carry generated from position $n - 1$ will cause the lowest-order position to advance to the value d_{n-1}. Thus, in an open accumulator, R^* can be determined by an open left shift of the original digits in one position, and in a circular accumulator, R^* can be determined by a circular left shift of the original digits by one position. Clearly, open shifting is appropriate to the open accumulator, and circular or closed shifting to the closed accumulator. Unless otherwise stated, this will henceforth be assumed to be the case.

Example 4.4 Suppose that in a five-position decimal accumulator the number 12345 is initially set and is then transmitted additively nine times. For an open and closed accumulator we have

Open	Closed
12345	12345
12345	12345
12345	12345
12345	12345
12345	12345
12345	12345
12345	12345
12345	12345
12345	12345
12345	12345
Discard ← [1]23450	[1]23450 End-around
	→ 1
	23451

The net effect is to give, respectively, open and closed left shifts. For the open case, $R = 123450$ and $R^* = 23450$, with

$$R = 123450 \equiv 23450 = R^* \pmod{10^5}.$$

For the closed shift, $R = 123450$ and $R^* = 23451$, with

$$R = 123450 \equiv 23451 = R^* \pmod{10^5 - 1}.$$

We have so far defined a left shift of one position. By a left shift of $s \geq 1$ positions we will mean the equivalent of s consecutive left shifts of one position, either open or closed. Since we have seen that a left shift (open or closed) of one position is equivalent to additive transmission to a cleared accumulator (open or closed) r times, it follows that the result R^* of a left shift of s positions is the same as the result of an additive transmission of the original digits into the initially cleared accumulator r^s times. Such a transmission of digits into the accumulator is a residue-class addition, and we may thus make the following statements:

$$R^* \equiv Nr^s \pmod{r^n}$$

if the accumulator is open, and

$$R^* \equiv Nr^s \pmod{r^n - 1}$$

if the accumulator is closed. Here, R^* is the integer represented by the digits as they finally appear in the accumulator after the shift and N is the nonnegative integer given by the original digits. Obviously, $R = Nr^s$ is the true result of an equivalent pencil-and-paper computation.

To determine when the shifted result R^* actually represents Nr^s, we have only to apply the criteria which were derived in Chapter 2. In case

$$Nr^s < M, \tag{4.1}$$

where M is the appropriate modulus for the accumulator, it is apparent that $R = R^* = Nr^s$. To see what this implies relative to the actual number X represented by N, we consider first the effect of shifting zero. Since we are doing arithmetic modulo M, the representation of M and zero are the same. Thus, for $M = r^n$, M is represented by a true zero and for $M = r^n - 1$, M is represented by a negative zero consisting of all digits equal to $r - 1$. In the end-off left shift in the first case and the circular left shift in the second, the representation of Mr^s is identical to the representation of M. That is, in arithmetic modulo M, M and zero are equivalent. Now if $|X| < r^n$ and $X > 0$, $N = X$ and from (4.1), $R^* = Nr^s = Xr^s$. If $X < 0$, it is represented by its complement $N = M - |X|$, and $Nr^s = Mr^s - |X|r^s$, which is the same as $M - |X|r^s$. This follows from the above discussion of the representation of M, and tells us that the result of shifting the complement is the same as the complement of the shifted number.

Example 4.5 Suppose that in a three-position accumulator we have $X = 6$ with $N = 006$ or $X = -6$ and $N = 994$ for $M = 10^3$ or $N = 993$ for $M = 10^3 - 1$. For the two cases of open and closed left shifts of two we have:

```
        Open                Closed
    006     994         006     993
     ↓       ↓           ↓       ↓
    600     400         600     399
```

In all cases $Nr^s = N \cdot 10^2 < M$. For $X = 6 > 0$ we have

$$R^* = 6 \times 10^2 = 600 = X \cdot 10^2$$

4.3 LEFT SHIFTING

in either open or closed shifting. For $X = -6 < 0$ and $M = 10^3 = 1000$ we have

$$R^* = 400 = 1000 - 600 = M - 6 \times 10^2 = M - |X|10^2,$$

and for $M = 10^3 - 1 = 999$,

$$R^* = 399 = 999 - 600 = M - 6 \times 10^2 = M - |X|10^2.$$

If we want the leading digit of R^* to determine uniquely the sign of Xr^s, we can impose the more stringent condition

$$|Xr^s| < r^{n-1}. \tag{4.2}$$

If condition (4.1) holds, $|X|$ must have at least s leading zeros in its machine representation and if $|X|$ has at least s such leading zeros, condition (4.1) will hold. If, instead of $|X|$, the register digits represent $-|X|$ by means of its complement, the s leading zeros of the absolute value become s leading digits equal to $r - 1$. In case we shift only absolute-value operands with a separately held sign indicator, this second situation will not arise and condition (4.1) will always hold if the representation of X has at least s leading zeros.

If condition (4.2) holds, $|X|$ must have at least $s + 1$ leading zeros, and, conversely, if $|X|$ has at least $s + 1$ leading zeros, condition (4.2) will hold. Therefore, R^* resulting from a left shift of s positions can be interpreted as presenting the product of the original operand X and r^s or the complement of Xr^s. Furthermore the leading digit of R^* determines the true algebraic sign of R^*, provided that the original digits representing X contain at least $s + 1$ leading "sign" digits, that is, digits all equal to zero or to $r - 1$.

If $s \geq n$, condition (4.1) can be rewritten as $|X| \leq 0$, and hence $|X| = 0$. Thus, unless $X = 0$, we must restrict ourselves to shifts of less than n positions if we wish to interpret the result as Xr^s or its complement. Similarly, if $s \geq n - 1$, condition (4.2) can also be rewritten as $|X| \leq 0$ or $X = 0$. Accordingly, in applying (4.2), we must restrict ourselves to shifts of less than $n - 1$ positions, unless $X = 0$. Thus, for left shifts $s \geq n$ or, in some cases, $s \geq n - 1$, and $X \neq 0$, the only conclusion is that the shifted result R^* is congruent to Xr^s in the appropriate modulus.

In a left shift of $s \geq n$ positions, the division algorithm gives

$$s = qn + t, \quad 0 \leq t < n.$$

In general, for open shifting we have

$$R^* \equiv Nr^s \pmod{r^n}.$$

We can write

$$Nr^s = Nr^{qn} \cdot r^t = (Nr^{(q-1)n}r^t)r^n.$$

112   SHIFTING   4.3

This last term is an integral multiple of the modulus and so is congruent to zero in the modulus. Thus, for an open left shift of $s \geq n$ positions,

$$R^* \equiv 0 \pmod{r^n}.$$

For a closed left shift of $s \geq n$ positions, R^* must satisfy

$$R^* \equiv Nr^s \pmod{r^n - 1}.$$

We can express Nr^s as

$$Nr^s = Nr^{qn} \cdot r^t + Nr^t - Nr^t = Nr^t + Nr^t(r^{qn} - 1).$$

Since the last term on the right is an integral multiple of $r^n - 1$, it is congruent to zero in this modulus. Therefore, for a circular left shift of $s \geq n$ positions,

$$R^* \equiv Nr^t \pmod{r^n - 1},$$

where t is the remainder obtained on division of s by n. The amount t is the *effective* left shift in this case.

Example 4.6 Let $n = 10$ and $r = 10$. If the accumulator operates modulo 10^{10}, open shifting will be utilized, but if the accumulator operates modulo $10^{10} - 1$, closed shifting will be employed. We will take as the original configuration of digits 9999543210, a complement number. In the open accumulator, these digits represent the operand

$$X = -456790,$$

and in the closed accumulator, the operand

$$X = -456789.$$

Since there are four leading sign digits in the accumulator, the result of a left shift of $1 \leq s \leq 3$ positions should represent $X10^s$. We check that this is so in the following table.

s	Open shift	Interpretation		Closed shift	Interpretation	
1	9995432100	-4567900	$= X10$	9995432109	-4567890	$= X10$
2	9954321000	-45679000	$= X10^2$	9954321099	-45678900	$= X10^2$
3	9543210000	-456790000	$= X10^3$	9543210999	-456789000	$= X10^3$
4	5432100000		$\equiv X10^4$	5432109999		$\equiv X10^4$
5	4321000000		$\equiv X10^5$	4321099995		$\equiv X10^5$
6	3210000000		$\equiv X10^6$	3210999954		$\equiv X10^6$
7	2100000000		$\equiv X10^7$	2109999543		$\equiv X10^7$
8	1000000000		$\equiv X10^8$	1099995432		$\equiv X10^8$
9	0000000000		$\equiv X10^9$	0999954321		$\equiv X10^9$
10	0000000000		$\equiv X10^{10}$	9999543210		$\equiv X10^{10}$

If we shift left more than three positions, we get results only congruent to the products. We check this for a shift of four in both cases. In the open case we compute

$$R^* - 10^4 X = (54321)10^5 - (-456790)10^4$$
$$= (543210 + 456790)10^4$$
$$= 10^6 \cdot 10^4 = 10^{10},$$
$$R^* \equiv 10^4 X \pmod{10^{10}}.$$

In the circular case a similar computation yields

$$R^* - 10^4 X = 5432109999 - (-456789)10^4$$
$$= 9999999999 = 10^{10} - 1;$$

and it follows that

$$R^* \equiv 10^4 X \pmod{10^{10} - 1}.$$

We note in Example 4.6 that an open shift of $s = 10 = n$ has set the digits all equal to zero, while a similar circular shift has returned the digits to their original configuration.

4.4 CIRCULAR RIGHT SHIFTING

Consider a circular left shift of $n - 1$ positions in an n-digit register whose original configuration was specified by $d_{n-1} \ldots d_0$. The digit d_0 moves $n - 1$ positions to the left and finally occupies the position vacated by d_{n-1}, while d_{n-1} moves around to position zero and then through all positions until it finally occupies the position vacated by d_{n-2}. It follows that the final configuration will be $d_0 d_{n-1} d_{n-2} \ldots d_1$. This is the same result we would have obtained by shifting the original configuration circularly one position to the right. Therefore, we may state that a left circular shift of $n - 1$ positions is equivalent to a right circular shift of one position.

Just as we did in the case of left shifting, we extend the definition of right shifting by more than one position. By either a circular or end-off right shift of $s \geq 1$ positions, we mean the equivalent of s consecutive shifts by one position. For completeness we define either a right or left shift of $s = 0$ positions to mean no shift. This, of course, leaves the sequence of digits intact. By an extension of the above argument for a circular left shift of $n - 1$ positions, we may conclude that a right circular shift of amount $0 \leq s \leq n$ can be obtained by an equivalent left shift of $n - s$ positions.

Example 4.7 In a five-position shifting register, consider the result of successive circular left shifts and the equivalent circular right shifts. The original content is 12345 (see table below).

Suppose we perform a circular right shift of s positions where s is any positive integer. Then by the division algorithm we can write

$$s = qn + t, \quad q \geq 0, \quad 0 \leq t \leq n - 1.$$

	Number of shifts		
←Left	L	R	Right→
12345	$0 \equiv -5$ (mod 5)	5	12345
23451	$1 \equiv -4$ (mod 5)	4	23451
34512	$2 \equiv -3$ (mod 5)	3	34512
45123	$3 \equiv -2$ (mod 5)	2	45123
51234	$4 \equiv -1$ (mod 5)	1	51234
12345	$5 \equiv 0$ (mod 5)	0	12345

Thus, a circular right shift of any amount s can be accomplished by q circular right shifts of an amount n, followed by a final circular right shift of amount t. Each circular right shift of n positions returns all digits to their original locations and can be ignored; that is, circular shifting in an n-digit register is modulo n. Therefore, the *effective* right shift is of amount t, and the result of any circular right shift can be accomplished by performing the effective shift. We have seen that the circular right shift can be carried out by performing an appropriate circular left shift. For a circular right shift of $s = qn + t$ positions, the q right shifts of n positions are equivalent to q left shifts, each of $n - n = 0$ positions. Thus, we do not need a separate operation for circular right shift but can accomplish it by a circular left shift of amount $n - t$. Circular left shifting as an arithmetic operation is appropriate only in an accumulator which functions modulo $r^n - 1$, and circular right shifting as a special kind of circular left shifting is also appropriate to this type of accumulator.

To investigate the interpretation of a circular right shift on an operand X, let R^* be the integer resulting from an effective circular right shift of t positions, that is, from an equivalent circular left shift of $n - t$ positions. By the properties of circular left shifting, if X is the operand represented by the original digits, then

$$R^* \equiv Xr^{n-t} \pmod{r^n - 1}.$$

If it can be shown that

$$Xr^{n-t} \equiv Xr^{-t} \pmod{r^n - 1}, \qquad (4.3)$$

then it will follow that

$$R^* \equiv Xr^{-t} \pmod{r^n - 1}. \qquad (4.4)$$

Condition (4.3) is true, and so also is condition (4.4), if and only if Xr^{-t} is an integer. If Xr^{-t} is not an integer (and it may well have a fractional part different from zero), it cannot be congruent to anything, since congruence is a relationship defined only among integers. Thus, if (4.3) holds, Xr^{-t} must be an integer. On the other hand, if Xr^{-t} is an integer, (4.3) must hold, for then

$$Xr^{n-t} - Xr^{-t} = (r^n - 1)Xr^{-t}$$

is an integral multiple of the modulus.

If X satisfies $|X| < r^n - 1$, then the original digits represent X or its complement, and $|Xr^{-t}| < r^n - 1$. For $t \neq 0$, $|Xr^{-t}| < r^{n-1}$. It follows from the results of Chapter 2 that if (4.4) is true, R^* either equals $|X|r^{-t}$ or its complement modulo $r^n - 1$. In either case, for $t \neq 0$, the algebraic sign can be determined from the leading digit of R^*. To see what this means in terms of the original digits representing X, let

$$|X| = \sum_{i=0}^{n-1} x_i r^i.$$

Then

$$|X|r^{-t} = \sum_{i=0}^{n-1} x_i r^{i-t}.$$

This last number is an integer, if and only if the coefficients of all negative powers of r in the summation are zero, that is,

$$x_i = 0, \quad i = 0, 1, \ldots, t - 1.$$

If X is nonnegative, the digits x_i will be in the shifting register, and if X is negative, the complement digits equal to $r - 1 - x_i$ will be in the shifting register. Therefore, when $|X| < r^n - 1$, we can summarize the results in the following statement:

The result of an effective circular right shift of t positions can be interpreted as representing Xr^{-t} if and only if, in the original digits representing X, at least the low-order t were all equal to sign digits. That is, they are equal to zero if X is nonnegative and to $(r - 1)$ if X is negative.

If we shift only absolute values, the shifted digits represent Xr^{-t} if and only if at least the low-order t of the original set are zeros.

If the original operand in the register does not actually equal X but is merely congruent to it, condition (4.4) will still hold if and only if Xr^{-t} is an integer. If the true result is sufficiently small in magnitude, that is, if

$$|Xr^{-t}| < r^n - 1,$$

R^* will represent Xr^{-t} or its complement. However, in this case, we cannot generally determine this in advance by observation of the t low-order digits of the original operand.

Example 4.8 Consider a three-digit decimal shifting register which shifts circularly, or modulo $10^3 - 1 = 999$. Let $X = 12300$. Then $X10^{-2} = 123$ is an integer and satisfies $|X10^{-2}| < 999$. By the reduction scheme of Chapter 2, the operand in the register which represents X is 312. Two successive circular right shifts give, in order, 231 and 123. The final result is correct, but the original operand 312 does not have two low-order zeros.

Even when Xr^{-t} is not an integer, we may interpret the result of an effective right shift of t in a manner which is useful. We confine our attention *only* to the low-order $n - t$ digits after the shift, and we can then say that they represent an integer Xr^{-t} to within an error due to round-off. We will have more to say about this type of shifting in the chapters on scaling and round-off error.

Example 4.9 Let $n = r = 10$ with an initial configuration

$$9999543219.$$

These digits represent the operand $X = -456780$ as the complement (mod $10^{10} - 1$). According to our criterion, a circular right shift of one position would yield digits representing $X10^{-1}$, but a circular right shift of two positions would not represent $X10^{-2}$. Shifting right one position circularly, we obtain the digital configuration

$$9999954321.$$

This sequence is $(10^{10} - 1) - 45678$ and so represents the integer $-45678 = X10^{-1}$, as expected. A circular right shift of two positions results in

$$1999995432,$$

which bears no obviously useful relationship to $X10^{-2}$. The low-order $10 - 2$ digits, however, are

$$99995432,$$

which can be interpreted as representing the integer -4567. This is a truncated approximation to $-4567.8 = X10^{-2}$. We note that this interpretation does not utilize a rounding technique.

4.5 OPEN OR END-OFF RIGHT SHIFTING

A circular right shift, followed by a circular left shift of the same number of positions, always returns the accumulator to its initial digital configuration, and vice versa. Hence, for closed shifting, left shifts and right shifts are the inverse operations of each other. Indeed, in circular shifting we may interpret a *negative* shift as one in the opposite direction. For open shifting, however, we will find that left- and right-shift operations are not necessarily the inverse of each other. Moreover, open right shifts cannot necessarily be expressed in terms of left shifts. The end-off right shift does not have the same properties as the right circular shift, since the properties of the latter operation were derived from those of the left shift which always yields a congruent result.

Let the initial configuration of digits in a shifting accumulator equal $|X|$. An open right shift of $s < n$ positions will introduce s high-order zeros, and so the digit configuration resulting from the shift will have at least s leading zeros. If the resultant configuration equals the integer $|X|r^{-s}$, a left open shift of s would yield a configuration equal to the original $|X|$. Such a left shift, however, would introduce s low-order zeros into the accumulator. Thus, if the right shift of s is to provide a configuration equal to $|X|r^{-s}$, the original representation of the $|X|$ should have at least s low-order zeros. Conversely, if $|X|$ has s low-order zeros, an open right shift of s positions will yield the integer $|X|r^{-s}$. In other words, the open right shift of s will yield $|X|r^{-s}$ if and only if the $|X|$ has at least s low-order zeros. In this case, the results of an open right shift of s (or fewer) positions and a circular right shift of the same number of positions are indistinguishable.

Example 4.10 Suppose a five-position decimal shifting register contains the digits $01200 = |X|$. The result of two successive open right shifts of one would yield

$$00120 = |X|10^{-1}, \quad 00012 = |X|10^{-2},$$

since there are two low-order zeros. A shift of three, however, would give

$$00001,$$

which is only a truncated version of $|X|10^{-3}$. For a right shift of one or two, the equivalent left shift would restore the original number. For a shift of three, however, an attempt to restore the original by a left shift would give 01000.

If the original configuration of digits is the complement of $|X|$, that is, equals $r^n - |X|$, the shifted digits do not represent $-|X|r^{-s}$, since the open right shift introduces zeros on the left. These digits should be complemented digits $(r - 1)$ for a correct complement representation. If we complement before shifting, shift the complemented digits, and then recomplement, the digital configuration will represent $-|X|r^{-s}$ to within truncation error. Complementation of the original digits gives the $|X|$ and the previous discussion applies. If $|X|$ has s low-order zeros, the result of an open right shift of s positions will be an n-digit integer equaling $|X|r^{-s}$. The recomplementation will yield the correct complement representation of $-|X|r^{-s}$.

When r's complements are taken, as is appropriate in open shifting operations, any low-order zero digits of the absolute value remain zero digits in the complement. We can thus summarize our discussion in the following manner:

An open right shift of s positions yields a configuration which can be interpreted as the result of dividing the originally represented operand by r^s if and only if there are at least s low-order zeros in the original sequence of digits. If the original digits are the complement of the number represented, we must complement before the shift and the shifted digits must be recomplemented after the shift.

If we know that we deal only with digital configurations representing nonnegative integers, that is, with absolute values, our conditions can be modified. We can then ignore the initial complementation and the recomplementation operations before and after the shift.

Complementation before and after shifting can be avoided if a modification in the definition of open right shifting is made. When the number to be shifted is a complement, instead of filling the position vacated by the highest-order digit with zero, we will fill it with a digit equal to $(r - 1)$. This is easy to implement if we assume that the original operand X satisfies $|X| < r^{n-1}$ so that the leading digit is either zero or $(r - 1)$ in accordance with the sign of the number represented. In this case we fill each vacated high-order position with a digit equal to the original leading or sign digit. We will call such a shift an open right shift with *sign extension*. If the leading digit is zero, the open right shift with sign extension yields the same result as the previously defined open right shift. If the leading digit is $(r - 1)$, the sign extension property will cause digits $(r - 1)$ to be filled in on the left as the shift proceeds. The result will then be the same as if the original digits had been complemented, shifted in the ordinary way, and then recomplemented.

Example 4.11 Consider a five-position decimal register with open right shifting capability. If $|X| = 01200$, it satisfies the condition $|X| < 10^4$, and the leading digit determines the sign. For $X = 01200 > 0$, an open right shift of two yields 00012, which is $X \cdot 10^{-2}$. If $X = -01200 < 0$, it will be represented by 98800, where the leading nine indicates a complement representation of a negative number. If we right-shift by two with sign extension, we have 99988, which represents $-00012 = X \cdot 10^{-2}$. This is the same as if we:

a) Complement 98800 to obtain 01200.

b) Open right shift by two to obtain 00012.

c) Recomplement the shifted result to obtain 99988.

If the condition for correct right-shifting does not apply, we may still interpret the shifted digits as representing the integer Xr^{-s} with truncation and loss of all digits shifted out of the register. If we have sign extension or the equivalent, we need not, as in the circular case, confine our attention to the low-order $n - s$ digits only. If we do not have sign extension, we can interpret only the low-order $n - s$ digits as approximating Xr^{-s} if $X < 0$. If $X \geq 0$, we may assign the interpretation to all n digits.

For open right shifts of $s \geq n$ positions, the final configuration will be all zeros or all $(r - 1)$'s. These can be considered to represent the value obtained in dividing by a large power of the base and truncating the result. For the modulus r^n, however, the case of all $(r - 1)$'s is more or less meaningless since this cannot be interpreted as a negative **form of zero**.

Example 4.12 For $n = r = 10$ consider the configuration

$$9999543210,$$

employed in Example 4.6. Since there is one low-order zero, we may shift right once and be sure to obtain $X10^{-1}$ since we extend the sign digit. An open right shift of two or more positions will yield a result which only approximates $X10^{-s}$. A right shift of one yields

$$9999954321,$$

and a second right shift of one gives

$$9999995432.$$

The first configuration represents the integer

$$-45679 = X10^{-1},$$

while the second represents the integer -4568, which approximates $X10^{-2} = -4567.9$. We note that for the complement number we get a properly rounded result. If we had shifted the absolute value twice, however, 0000456790 would become 0000004567, a less accurate approximation to 4567.9.

EXERCISES

SECTION 4.2

4.1 Under what circumstances will a closed left shift of one position be the equivalent of multiplication by r?

4.2 Under what circumstances will an open left shift of one position be the equivalent of multiplication by r?

4.3 Under what circumstances will a closed right shift of one position be the equivalent of division by r?

4.4 Under what circumstances will an open right shift of one position be the equivalent of division by r?

4.5 For a six-digit, decimal shifting register, what is the relation between each of the following numbers and that obtained by open or closed left or right shifts of one position?
   a) 034560   b) 345600   c) 003456   d) 345607

SECTION 4.3

4.6 By explicit addition in a three-digit decimal accumulator initially containing 456, compute the final value if 456 is transmitted additively nine times in
   a) An open accumulator   b) A closed accumulator

   What is the relation between the final result and 456?

4.7 Do the equivalent of Exercise 4.6 if the accumulator initially contains 054.

4.8 For the numbers given in Exercises 4.6 and 4.7, what is the effect of open and closed shifts of two or three positions? What is the relation between the original number and the final result?

4.9 A circular left shift of n positions gives the original configuration of digits. What is the implication of this in either modulus r^n or $r^n - 1$?

4.10 In a six-bit accumulator whose content is represented in octal, deduce the effect on the octal digits of circular left shifts of one, two, and three binary positions.

SECTION 4.4

4.11 Interpret the fact that a left circular shift of $n - s$ positions is the same as a right circular shift of s positions in terms of modular arithmetic.

4.12 Under what circumstances is a right circular shift of s positions the same as an end-off shift?

SECTION 4.5

4.13 Under what circumstances will an open left shift of s positions produce $|X|r^s$ if the initial content is $|X|$?

4.14 If the content of the register is X in complement form, what modification in open left shifting is needed to produce correct results?

4.15 With the modified shifting of Exercise 4.14, under what circumstances may we interpret an open left shift as yielding Xr^s?

4.16 If we consider open right or left shifts with sign extension, under what circumstances will they be the equivalent of circular right or left shifts?

4.17 An open right shift of more than the number of low-order sign digits, with sign extension, produces correct results in the sense that the final configuration is at least a truncated version of Xr^{-s}. Is the same true for a left shift of more than the number of leading sign digits?

4.18 Interpret the final result of an open left or right shift of s positions in terms of modular arithmetic. Is there always a congruence relationship?

4.19 Under what circumstances will the truncation effect of an end-off right shift give a correctly rounded value if $X \geq 0$?

4.20 Under what circumstances will the truncation effect of an end-off right shift give a correctly rounded version of $|X|$ if $X < 0$, represented in complement form?

CHAPTER 5

MULTIPLICATION

5.1 INTRODUCTION

In Chapter 1 we assumed that the operation of multiplication is well defined for integers (and, indeed, for all real numbers). In the case of nonnegative integers the product of two numbers can be achieved by repeated addition. We can, for example, compute 3×4 as $4 + 4 + 4$. Thus, if we let one of two integral machine factors in multiplication be $X^* \geq 0$ and the other $Y^* \geq 0$, the most naive way to form the product $P^* = X^*Y^*$ would be to add X^* to an initially clear accumulator Y^* times. Utilizing the commutative law we could alternatively reverse the process and add Y^* into an initially clear accumulator X^* times. In this process the integer added into the accumulator is called the multiplicand and the other factor which determines how many multiplicands are added is called the multiplier. While in practice we do not perform multiplication in this manner, it is instructive to investigate the process, since any algorithm we employ must somehow be equivalent to it.

It is possible to carry out multiplication through repetitive addition by use of paraphernalia for addition such as we have considered in Chapters 2 and 3. This approach is not very practical, but we can at least use it to determine modifications and extensions so that the arithmetic facility can more effectively perform multiplication. With this in view, suppose we place the multiplicand in the X-register. We must now provide for transmission of the digits of the multiplicand from the X-register additively to the accumulator Y^* times. In terms of those properties which we assumed for an additive accumulator in Chapter 2, we see that there was no provision for counting the number of transmissions from the X-register to the accumulator, that is, for counting to a value equal to the multiplier. Although the multiplier's digits could be held in one of the general storage registers during multiplication, provision of the necessary counting facility would complicate storage in a costly way. Instead, the arithmetic unit will be provided with its own registers for performing such special operations as well as for storing the operands and results of arithmetic. Hence, we will assume that there is available, in the arithmetic unit, a multiplier register with the necessary counting properties.

In general, we may not operate in the arithmetic unit with the true factors x and y and, moreover, these may not be nonnegative. Instead we will use their machine representations X^* and Y^*, properly reduced to n-digit form so that

$$x \equiv X^*, \qquad y \equiv Y^* \qquad (\text{mod } M_1),$$

where the modulus M_1 is either r^n or $r^n - 1$, whichever is appropriate. In all cases $X^* \geq 0$, $Y^* \geq 0$. Let

$$P^* = X^*Y^*$$

be the true product of the machine representations. It follows from the rules of residue-class arithmetic that for $p = xy$,

$$p \equiv P^*(\text{mod } M_1).$$

From the discussion of addition in Chapter 2, it is clear that on forming P^* by repetitive additions in the accumulator, we do not necessarily achieve a result R^* equal to P^*. However, we can assert that

$$R^* \equiv P^*(\text{mod } M_1).$$

Thus it follows that

$$p \equiv P^* \equiv R^*.$$

Example 5.1 Consider a three-position, decimal open accumulator. Let $x = 4567$ and $y = -997$ be the original operands. For the modulus $M_1 = 10^3$, the proper reduction to machine form yields $X^* = 567$ and $Y^* = 003$. The true product $p = xy$ is -4553299, and $P^* = 1701$. To form R^* we add the multiplicand ($X^* = 567$) additively to a clear accumulator a number of times equal to the multiplier ($Y^* = 003$) to obtain

$$\begin{array}{r} 567 \\ 567 \\ \underline{567} \\ \boxed{1}\,701 \end{array}$$

Ignore ⟵┘

with $R^* = 701$. We have

$$p = -4553299 \equiv 1701 = P^* \ (\text{mod } 10^3),$$

since $1701 - (-4553299) = 4555000 = (4555)10^3$. Similarly,

$$R^* = 701 \equiv 1701 = P^* \equiv p \ (\text{mod } 10^3).$$

By direct computation, $R^* - p = 701 - (-4553299) = (4554)10^3$.

In Example 5.1 the only relation between the true product p and the final machine result R^* is one of congruence. It would require more informa-

5.1 INTRODUCTION

tion than the machine gives directly to reconstruct the negative integer $p = -4553299$ from $R^* = 701$. In some instances, however, if p satisfies appropriate bounds, we may get true results even though P^* is too large for an n-digit accumulator. We will explore this idea in more depth later.

Example 5.2 Suppose that, on the machine of the preceding Example 5.1, the operands are $x = -002$ and $y = -003$. The true product is $p = xy = 006$. Since $X^* = 998$ and $Y^* = 997$, we have $P^* = X^*Y^* = 995006$, which is much too large for the three-position accumulator. The result R^* of adding 998 to a clear accumulator 997 times and ignoring carry to the fourth position yields $R^* = 006$. In this case $R^* = 006$ gives the true result p, even though intermediate results represent only equivalent numbers.

The factor x may be signed, that is,

$$x = |x| \quad \text{or} \quad x = -|x|.$$

The same is true of y, and, furthermore, both $|x|$ and $|y|$ may exceed the allotted n digits, base r. The combination of algebraic sign and excessive magnitude may require a double reduction to obtain the machine representation of the number. To avoid unnecessary complication in notation we will assume in what follows that X and Y are signed representatives of the true operands which require no more than n digits, base r, that is,

$$|X| < r^n, \quad |Y| < r^n. \tag{5.1}$$

That such representations are available follows from the division algorithm, for if M_1 is the modulus and the true operands are x and y, then

$$x = rM_1 + X, \quad y = sM_1 + Y,$$

where X and Y satisfy (5.1) and are of the same sign as x and y. If x and y satisfy (5.1) also, then $r = s = 0$ and $x = X$ and $y = Y$. It is obvious that $p \equiv P \equiv XY \pmod{M_1}$.

Example 5.3 For the three-position decimal machine of Examples 5.1 and 5.2, the operands $x = 1234$ and $y = -2234$ are signed and require more than three-digit representation. Since

$$x = 1234 = (1)(10^3) + 234,$$

and

$$y = -2234 = (-2)(10^3) + (-234),$$

we take $X = 234$ and $Y = -234$, so that

$$x \equiv X \pmod{10^3}, \quad y \equiv Y \pmod{10^3}.$$

The true machine representation of $Y = -234$ requires the second reduction to $Y^* = 1000 - 234 = 766$. For a closed accumulator, $M_1 = 10^3 - 1 = 999$, and we would use $X = 235$ and $Y = -236$. Again, the negative $Y = -236$ requires $Y^* = 999 - 236 = 763$.

The simplifying assumption that X and Y satisfy (5.1) produces no loss of generality, for the transitivity property of congruences tells us that any modular results we obtain for X and Y are valid for the initial operands x and y. Thus, in the following discussion X and Y satisfy (5.1) but are assumed to have the appropriate algebraic sign.

Let X^* be the n-digit machine representation of X. Then, if $X \geq 0$, $X^* = X$, and if $X < 0$, $X^* = M_1 + X = M_1 - |X|$. We may write

$$X^* = kM_1 + X,$$

where $k = 0$ if $X \geq 0$ and $k = 1$ if $X < 0$. Similarly, we may write

$$Y^* = lM_1 + Y.$$

We can interpret the congruence between the accumulator result R^* and the true product P to imply that R^* equals P or its complement, provided that the magnitude of P is sufficiently small. Similar statements hold with respect to P^*. Unfortunately, since both P^* and P are the products of n-digit factors neither of these products can be expected, in general, to be sufficiently small in magnitude with respect to the modulus M_1. However, it is true that

$$0 \leq P^* < r^{2n} - 1,$$

and that

$$0 \leq |P| < r^{2n} - 1.$$

Thus, if we extend the accumulator to $2n$ positions and perform consistent residue-class arithmetic modulo M_2, where $M_2 = r^{2n}$ or $M_2 = r^{2n} - 1$, whichever is appropriate, the congruence relationship will be interpretable as equality in the case of P^* and also for P provided that its sign is known in advance.

Consistent arithmetic modulo M_2 requires that the n-digit machine operands X^* and Y^* be replaced by $2n$-digit operands X^{**} and Y^{**} such that

$$X^{**} \equiv X \pmod{M_2} \quad \text{and} \quad Y^{**} \equiv Y \pmod{M_2}.$$

Let E be the necessary extension to X^* to produce X^{**}. Thus

$$X^{**} = kM_2 + X = X^* + E = kM_1 + X + E.$$

These equalities are achieved if $E = k(M_2 - M_1)$. If k is zero,

$$X^* = X \quad \text{and} \quad E = 0,$$

so the extension requires merely that we adjoin n leading zeros to X^*. In case $k = 1$,

$$X^* = M_1 - X \quad \text{and} \quad E = M_2 - M_1 = r^{2n} - r^n.$$

5.1 INTRODUCTION

Since $r^{2n} - r^n = r^n(r^n - 1)$ is represented digitally by n leading digits equal to $(r - 1)$, followed by n zeros, the extension requires that we adjoin n maximal digits to X^*, each equal to $r - 1$. Thus, if $X \geq 0$, we extend by adding zeros, and if $X < 0$, we extend by appending digits equal to $(r - 1)$. A similar statement holds for Y^*.

Example 5.4 In a three-digit, decimal machine, $X = 15$ and $Y = -25$. The operands X^* and Y^* are $X^* = X = 015$ and $Y^* = 975$ for $M_1 = 10^3$ and $Y^* = 974$ for $M_1 = 10^3 - 1 = 999$. For either modulus, the correct extensions to six digits are

$$000015 \quad \text{for } X$$

and

$$999975 \quad \text{or} \quad 999974 \quad \text{for } Y.$$

With either modulus, the negative operand is extended by adding three leading 9's where $9 = 10 - 1 = r - 1$.

For correct representation of the true operands X and Y, the extended versions X^{**} and Y^{**} must satisfy

$$X^{**} \equiv X \pmod{M_2} \quad \text{and} \quad Y^{**} \equiv Y \pmod{M_2}$$

for the larger modulus. We note that congruence in the modulus M_2 implies congruence (mod M_1).

Since

$$X^{**} = k(M_2 - M_1) + X^* = kr^n(r^n - 1) + X^*$$

and

$$Y^{**} = l(M_2 - M_1) + Y^* = lr^n(r^n - 1) + Y^*,$$

we always have

$$X^{**} \equiv X^* \equiv X \pmod{M_1} \quad \text{and} \quad Y^{**} \equiv Y^* \equiv Y \pmod{M_1}$$

from the transitivity property of congruences. However, X^{**} and Y^{**} are not necessarily congruent to X^* and Y^* in the modulus M_2; but in case $k = 0$ and $l = 0$, we will also have congruence in the larger modulus M_2 for both extensions, since equality implies congruence in any modulus.

Example 5.5 For the operands of Example 5.4, the n- and $2n$-digit representatives are

$$X = 15, \quad X^* = 015, \quad \text{and} \quad X^{**} = 000015$$

so that $X = X^* = X^{**}$. In this case, as usual,

$$X^{**} \equiv X \pmod{M_2},$$

and also

$$X^{**} \equiv X^* \pmod{M_2}.$$

126   MULTIPLICATION   5.1

For $Y = -25$,

$$Y^* = 975 \text{ or } 974 \quad \text{and} \quad Y^{**} = 999975 \text{ or } 999974.$$

Although

$$Y^{**} \equiv Y \pmod{M_2}$$

from

$$999975 \equiv -25 \pmod{10^6} \quad \text{or} \quad 999974 \equiv -25 \pmod{10^6 - 1},$$

we see that

$$999975 \not\equiv 975 \pmod{10^6} \quad \text{and} \quad 999974 \not\equiv 974 \pmod{10^6 - 1}.$$

Thus

$$Y^{**} \not\equiv Y^* \pmod{M_2}.$$

If the products are formed by the equivalent of repeated addition, we have

$$P = XY, \quad P^* = X^*Y^*,$$
$$P^{**} = X^{**}Y^{**},$$

and R^{**} is the net machine result in $2n$ digits. Because of the way in which the accumulator does residue-class arithmetic we will always have

$$R^{**} \equiv P^{**} \pmod{M_2}$$

by the discussions of Chapter 1. We also have

$$P^{**} = X^{**}Y^{**} = [kM_2 + X][lM_2 + Y]$$
$$= [klM_2 + lX + kY]M_2 + XY,$$

so that

$$R^{**} \equiv P^{**} \equiv P \pmod{M_2}.$$

Example 5.6 For the operands of Example 5.4 with $M_1 = 10^3$ and $M_2 = 10^6$, we have

$$\begin{aligned} P &= XY &&= (15)(-25) &&= -375, \\ P^* &= X^*Y^* &&= (015)(975) &&= 014625, \\ P^{**} &= X^{**}Y^{**} &&= (000015)(999975) &&= 14999625, \end{aligned}$$

so that

$$R^{**} = 999625 \equiv -375 = P \pmod{10^6}.$$

However, while

$$P^{**} = 14999625 \equiv 014625 = P^* \pmod{10^3},$$

$P^{**} \not\equiv P^* \pmod{10^6}$, and thus

$$R^{**} \not\equiv P^* \pmod{10^6},$$

5.2 MULTIPLICATION BY SHIFTING THE MULTIPLICAND

although $R^{**} \equiv P^*$ (mod 10^3). A similar result holds for $M_1 = 10^3 - 1$ and $M_2 = 10^6 - 1$.

The inequalities (5.1) guarantee that $|XY| < r^{2n}$. Thus, R^{**} will correctly represent XY or the complement of XY depending on algebraic sign. We may not, however, be able to determine whether R^{**} represents a true value or a complement. If we impose the restrictions

$$|X| < r^{n-1} \quad \text{and} \quad |Y| < r^{n-1}, \tag{5.2}$$

R^{**} will have a leading digit which determines the correct sign of XY. This is easily seen from the fact Eqs. (5.2) imply $|XY| < r^{2n-2} < r^{2n-1}$, so that the leading digit of R^{**} (in fact, the two leading digits) will be zero or $r - 1$ according to sign. Conditions (5.2) provide that the leading digits of X^* and Y^* determine algebraic sign and these are extended to give X^{**} and Y^{**}.

Example 5.7 Suppose we have

$$n = 1, \quad r = 10$$

and, for $M_1 = 10$,

$$X = 4, \quad Y = -9.$$

Since $X^* = 4$ and $Y^* = 1$, the extensions to $2n$ are $X^{**} = 04$ and $Y^{**} = 91$. We have $P^{**} = 364$ so $R^{**} = 64$. Thus, $R^{**} \equiv P = -36$ (mod 10^2) but the leading digit 6 is not a 9 for sign detection. We cannot tell whether 64 represents, say, $08 \times 08 = 64$, or the correct result. If we assume $n = 2$, however, $X^* = 04$ and $Y^* = 91$ with leading sign digits. The extensions to $2n$ are $X^{**} = 0004$ and $Y^{**} = 9991$. This yields $R^{**} = 9964$ with two leading 9's to indicate the sign.

In the remainder of this chapter, we will consider the problem of forming the product of factors represented by two sequences of digits taken from storage registers. We will assume that the factors satisfy either (5.1) or (5.2), whichever is appropriate. We will develop various algorithms based on counting and shifting for forming the product. Algorithms will first be derived for the case of nonnegative factors. We will determine the changes in the organization of the arithmetic unit so that these algorithms can be performed without disturbing the ability to do addition and subtraction. Finally, we will investigate the changes necessary to the algorithms to make them correct for signed factors.

5.2 MULTIPLICATION BY SHIFTING THE MULTIPLICAND

Let $X \geq 0$ and $Y \geq 0$ be defined by digits X_i and Y_i, $i = 0, \ldots, n - 1$, be such that

$$X = \sum_{i=0}^{n-1} X_i r^i, \quad Y = \sum_{i=0}^{n-1} Y_i r^i.$$

Thus, $X = X^*$ and $Y = Y^*$. We may express the product P as

$$P = XY = X \sum_{i=0}^{n-1} Y_i r^i = \sum_{i=0}^{n-1} Y_i (Xr^i).$$

In this way of writing the product, Y is the multiplier and X the multiplicand. Consider a typical term $Y_i(Xr^i)$ of this last sum. Imagine that the X-register has also been extended to a length of $2n$ digits. In accordance with the method of extending nonnegative operands to $2n$ digits, the digits of X will occupy the low-order n positions and digits equal to zero will occupy the high-order n positions. The reader should not be confused by the fact that the variable X and the register which holds its digits have the same name. The $2n$-digit integer in the X-register represents the multiplicand X. Since this integer represents an absolute value and has n leading zeros, it may be shifted left i places, $i = 1, 2, \ldots, n$, in either an open or circular manner, and the resultant digits will represent Xr^i. Hence, to form $Y_i(Xr^i)$, we can shift the digits in the X-register left i positions and transmit them to the accumulator Y_i times. Thus, the process of forming P consists of successive additive transmissions, to the accumulator, of properly shifted multiplicand digits. The number of transmissions and the amount of each shift are controlled by the values and positions of the individual multiplier digits. The introduction of shifting simplifies counting to the point that we need only count to the value of a digit rather than to the value of the multiplier.

Example 5.8 We illustrate the above procedure for four-digit factors.

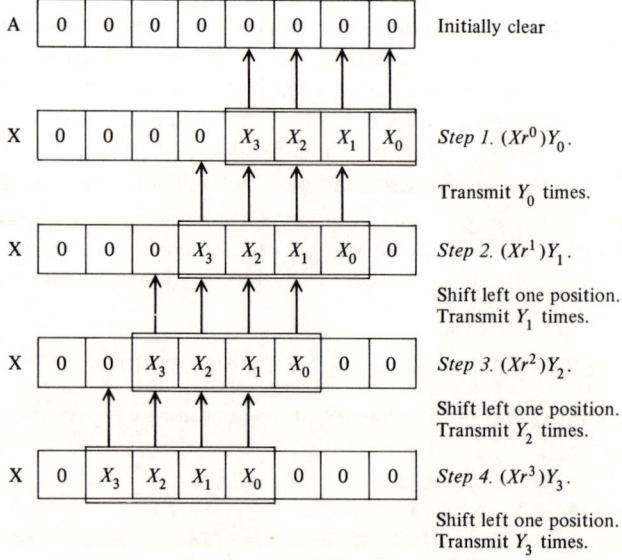

5.2 MULTIPLICATION BY SHIFTING THE MULTIPLICAND

Several points of interest are immediately apparent from the example. First, the kind of left shift makes no difference because the result of either open or closed left shifting will be the same in this case. Second, the zero digits need not be transmitted. This has been emphasized by the placement of the arrows. Finally, since the zeros need not be transmitted, we need not actually have a 2n-digit X-register. Instead, an n-digit X-register can be used if, in effect, it can be picked up and its units position shifted relative to the units position of the accumulator. This last point has been emphasized by drawing the actual part of the X-register with double lines.

Example 5.9 Suppose the four-digit factors of Example 5.8 are decimal numbers with $X = 2345$ and $Y = 1321$. Consider a four-digit X-register which can be shifted left. The successive steps and intermediate results are as follows.

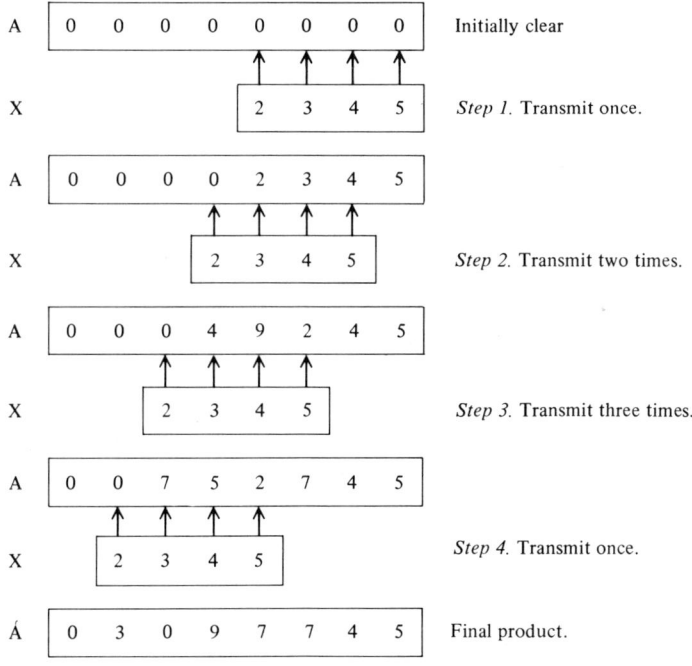

If $r = 2$, the counting procedure based on the individual digits Y_i is rather simple. We either transmit or suppress the transmission of digits to the accumulator corresponding to $Y_i = 1$, or $Y_i = 0$, respectively.

In this multiplication algorithm and the succeeding ones to be described, the intermediate values appearing in the accumulator will be referred to as *partial products*. Thus, in Example 5.10 some partial products are 2345, 49245, and 752745.

5.3 SHIFTING PARTIAL PRODUCTS RELATIVE TO THE MULTIPLICAND

The multiplication technique of Section 5.2 is frequently used with mechanical hand or desk-top calculators. In such a machine the keyboard plays the role of the X-register and the moving carriage provides the shift of the units position of the X-register relative to that of the accumulator. Shifting the X-register relative to A in this manner requires that all but the highest order of the positions of A accept the additive transmission of a digit from the X-register. The connections between the two registers would be much simpler if it were only necessary to provide for the transmission of digits from the X-register into *fixed orders* of A. Then, to achieve the effect of shifting the X-register relative to A, we can instead shift the digits of the partial product within A while holding the multiplicand digits fixed in the X-register.

There are, of course, many ways in which to select n fixed accumulator positions into which digits from the X-register will be transmitted. It is reasonable to make this selection so that the interconnection will also be suitable for addition and subtraction operations. In the previous discussions we have considered the operands and results to be integers. Thus, we can imagine that a radix point lies just to the right of every register. If we are to add or subtract the content of one register from the content of another, we must line up the radix points. This can be accomplished for the $2n$-position accumulator and n-digit X-register by transmitting from the ith position of the X-register into the ith position of A, $i = 0, 1, \ldots, n - 1$. We will call an accumulator *integral* if it is connected to an X-register in this manner. This kind of interconnection is illustrated in the following diagram, with $n = 4$.

$$
\begin{array}{c|c|c|c|c|c|c|c|c|}
\cline{2-9}
A & A_7 & A_6 & A_5 & A_4 & A_3 & A_2 & A_1 & A_0 \\
\cline{2-9}
& & & & & \updownarrow & \updownarrow & \updownarrow & \updownarrow \\
\cline{6-9}
X & & & & & X_3 & X_2 & X_1 & X_0 \\
\cline{6-9}
\end{array}
$$

The two-headed arrows connecting the two registers indicate an additive transmission operation into the accumulator and a copying operation into the X-register. The arrow above A indicates the dual shifting property of digits within A. The implicit radix points, although not really existent, are shown on the right. In Section 5.3.1 we will develop a multiplication algorithm suitable for an integral accumulator.

Our preceding discussions of addition, subtraction, and multiplication have centered around operations on residue classes in which the particular operands used were necessarily integral representatives of each class. We pointed out in Chapter 1, however, that the ultimate location of the base point will be determined by a separate algorithm and that dealing with finite sequences of digits is only equivalent to interpreting them as integers with an implicit base point at the right-hand end. Thus, the decimal number 123.45

uniquely determines the integer 12345, which in turn determines the appropriate residue class. The specific *interpretation* of the location of the base point in any sequence is immaterial so long as it gives a fixed reference point which allows proper alignment of digits relative to each other. The ultimate algorithm for location of the true position of the base point must, of course, correspond to this interpretation. From some points of view there are advantages to an intermediate interpretation of the base point as being in a different position from the right-hand end of the sequence. In Chapter 7 we will consider the specific relationship between the various interpretations in detail. In particular, many computer designers prefer to consider that the machine operands are always fractional. In this case each sequence of digits in a register can be thought of as representing a fraction, and the radix point can be imagined to lie just to the left of each register. Thus, in order to line up the radix points of A and the X-register, we must transmit from the X-register into the n high-order digits of A. The resulting configuration, which will then be suitable for addition and subtraction of fractional operands, is illustrated below for the case $n = 4$.

A	A_7	A_6	A_5	A_4	A_3	A_2	A_1	A_0
X	X_3	X_2	X_1	X_0				

With an X-register effectively joined to A in this manner, we will refer to an accumulator as *fractional*. Multiplication algorithms suitable for a fractional accumulator will be developed in Section 5.3.2.

The fact that we use either an integral or fractional accumulator does not prevent us in practice from dealing with operations on mixed numbers. We merely multiply each operand by a power of the base, chosen so that the product will be an integer (or a fraction as the case may be). At the conclusion of the digital operations, we can correct for the introduction of these powers. This technique, which determines the algorithms for true base-point location, is called scaling, and it will be discussed at greater length in Chapter 7. For the balance of this chapter, we will retain the assumption that all sequences of digits represent integers even though we may at times be dealing with the fractional accumulator.

5.3.1 The Integral Accumulator

In a $2n$-digit integral accumulator, a transmission of the multiplicand digits from the n-digit X-register to A yields a result that is congruent modulo M_2 to the sum of the multiplicand X and the operand represented by the initial content of A. If this transmission is repeated Y_i times, a result congruent to the sum of Y_iX and the original operand is obtained. We can multiply or divide these partial products by powers of r by means of either left or right shifting, and, since only nonnegative powers of $Y_i X$ appear in the product P, only left shifting will be necessary. Thus, to form a result congruent to $P = XY$ in an integral accumulator, we must express the product so that it can be evaluated by a combination of additive transmissions from the X-register to A and left-shifting of the partial products. The following formula is appropriate.

$$P = XY = \sum_{i=0}^{n-1} Y_i(Xr^i)$$

$$= [\ldots((0 + Y_{n-1}X)r + Y_{n-2}X)r + \cdots + Y_1X]r + Y_0X.$$

The quantities in parentheses give the values of the various partial products $Y_i X$, each to be multiplied by r in shifting left one position. The leading zero indicates that we will start with an initially cleared accumulator.

For a multiplication operation we will assume the accumulator initially cleared to zero, the multiplicand in the X-register, and the value of Y in the multiplier register. We then transmit digits from the X-register to A a number of times equal to Y_{n-1}. This produces an initial partial product that is congruent to $0 + Y_{n-1}X$. The digits in A are then shifted left one position. After the shift, the digits from X are transmitted Y_{n-2} times. This forms the partial product which is congruent to $(0 + Y_{n-1}X)r + Y_{n-2}X$. The process continues in this manner until finally the digits from X are transmitted Y_0 times to add $Y_0 X$ to the last shifted partial product, thus forming a result congruent to P. Note that according to the discussion in Section 5.1, all congruences are modulo M_2. The final result must be congruent to P, since only left-shift operations and transmissions from the X-register to A have been employed. We know from the assumptions on the factors X and Y that $0 \le P \le r^{2n} - 1$, and since the congruence is with respect to either r^{2n} or $r^{2n} - 1$, this also ensures that the final accumulator result equals P.

Example 5.10 We illustrate an integral-accumulator multiplication process for the case $n = 4$, $r = 2$, and $X = 1011$, $Y = 1101$. Since we are dealing with the case of nonnegative factors, the fact that both X and Y start with the digit one has no significance here for sign detection in terms of complements. Consequently, the multiplicand represents the integer eleven and the multiplier the integer thirteen. The digits of the product should then represent the integer one-hundred forty-three (see table below).

In the preceding example we could have eliminated a separate multiplier register by utilizing the high-order digits of A. Thus, we initially place the multiplier digits in these positions and employ an open left shift. With one preliminary shift after detecting Y_3, the multiplier digits can be shifted out of the high-order end of the register. The appropriate order from high to low is

5.3 SHIFTING PARTIAL PRODUCTS

Y_3	Y_2	Y_1	Y_0
1	1	0	1

Y

X_3	X_2	X_1	X_0
1	0	1	1

X

Operation	Symbolic Content of A	Numerical Content of A								A
		A_7	A_6	A_5	A_4	A_3	A_2	A_1	A_0	
Clear A	Zero	0	0	0	0	0	0	0	0	
Transmit X → A	$0 + Y_3X$	0	0	0	0	1	0	1	1	
$Y_3 = 1$ times										
Shift left	$(0 + Y_3X)r$	0	0	0	1	0	1	1	0	
Transmit X → A	$(0 + Y_3X)r + Y_2X$	0	0	1	0	0	0	0	1	
$Y_2 = 1$ times										
Shift left	$((0 + Y_3X)r + Y_2X)r$	0	1	0	0	0	0	1	0	
Transmit X → A	$((0 + Y_3X)r + Y_2X)r + Y_1X$	0	1	0	0	0	0	1	0	
$Y_1 = 0$ times										
Shift left	$[((0 + Y_3X)r + Y_2X)r + Y_1X]r$	1	0	0	0	0	1	0	0	
Transmit X → A	$[((0 + Y_3X)r + Y_2X)r + Y_1X]r$ $+ Y_0X = XY = P$	1	0	0	0	1	1	1	1	
$Y_0 = 0$ times										

maintained without interfering with the partial products generated during the formation of P.

Example 5.11 Suppose that in Example 5.10 we initially place $Y = 1101$ in the high-order end of A. The steps are then as follows:

Recognize $Y_3 = 1$	1	1	0	1	0	0	0	0
Shift left	1	0	1	0	0	0	0	0
Transmit X, $Y_3 = 1$ times	1	0	1	0	1	0	1	1
Recognize $Y_2 = 1$	1	0	1	0	1	0	1	1
Shift left	0	1	0	1	0	1	1	0
Transmit X, $Y_2 = 1$ times	0	1	1	0	0	0	0	1
Recognize $Y_1 = 0$	0	1	1	0	0	0	0	1
Shift left	1	1	0	0	0	0	1	0
Transmit X, $Y_1 = 0$ times	1	1	0	0	0	0	1	0
Recognize $Y_0 = 1$	1	1	0	0	0	0	1	0
Shift left	1	0	0	0	0	1	0	0
Transmit X, $Y_0 = 1$ times	1	0	0	0	1	1	1	1

Note that the magnitudes of X and Y preclude interference between the shifted digits of Y with the shifted partial products.

5.3.2 The Fractional Accumulator

For a fractional accumulator we can think of the n multiplicand digits in the X-register as having n low-order zero digits adjoined to them. Thus, in effect we may consider that the transmission of digits from the X-register to A as the equivalent of additive transmission from one $2n$-digit register to another. The result is the same as adding an amount Xr^n to the content of A. If this is repeated Y_i times, the original content will have been increased by an amount congruent to $Y_i X r^n$ modulo M_2. As before, intermediate results will be called partial products. By shifting, we can multiply or divide a partial product by the base r. Since r^n is the highest power of X to appear in the product, and additive transmissions from the X-register to A always introduce Xr^n into the accumulator, we can confine ourselves to divisions by r, that is, to right shifting. Thus, for a fractional accumulator, we seek a formula for P which can be evaluated in terms of forming partial products by means of successively adding to the initial content of A terms of the type $Y_i X r^n$ followed by divisions by r, or right shifting. An appropriate formula is

$$P = XY = \sum_{i=0}^{n-1} Y_i X r^i = \sum_{i=0}^{n-1} Y_i X r^n r^{i-n}$$
$$= [\ldots ((0 + Y_0 X r^n) r^{-1} + Y_1 X r^n) r^{-1} + \cdots + Y_{n-1} X r^n] r^{-1}.$$

The factor r^{-n} in the summation is split into n successive right shifts, and quantities in parentheses represent the fractional partial products which are to be shifted to the right.

For a closed accumulator which employs right circular shifting, we can use the previously developed theory of residue-class arithmetic to establish that the accumulator result obtained from the above formula for P actually equals P. The initial transmission of Xr^n a total Y_0 times clearly gives a partial product congruent to $0 + Y_0 X r^n$ modulo M_2. We can think of the right circular shift of one position as a left circular shift of $2n - 1$ positions. After such a shift, the accumulator will contain a partial product congruent to $(0 + Y_0 X r^n) r^{2n-1}$. We know from Section 4.4 that this, in turn, will be congruent to $(0 + Y_0 X r^n) r^{-1}$, if and only if this last expression is an integer. Since the factor in parentheses contains r^n, which provides n low-order zeros, it follows that this and all subsequent partial products are integers; thus the congruence of the partial product to $(0 + Y_0 X r^n) r^{-1}$ is established. We may repeat the argument for the succeeding partial products. Assume that congruence has been established for the ith partial product of the form $(\sum_{k=0}^{i} Y_k X r^{n-(i-k)}) r^{-1} = P_i$. Then transmission of Xr^n a total of Y_{i+1} times gives a result congruent to $P_i + Y_{i+1} X r^n$ and the right shift in turn yields a new partial product congruent to $(P_i + Y_{i+1} X r^n) r^{2n-1}$. This, in turn, is congruent to the term $((P_i + Y_{i+1} X r^n) r^{-1}$, if and only if this term is an integer. This will be the case, since each term of P_i contains factors r^{n-j}, $j \leq i \leq n$, and thus, a sufficiently high power of r to compensate for the successive multiplications by r^{-1}. Therefore, the final result obtained in the accumulator from the sequence of transmissions from the X-register and successive right shifts corresponding to the fractional formula for P will be congruent to P modulo M_2. Since for circular shifting $M_2 = r^{2n} - 1$ and P satisfies $0 \leq P < r^{2n} - 1$, the congruence implies equality, that is, the accumulator result must be equal to P.

Example 5.12 We repeat Example 5.10 for a closed fractional accumulator.

Y_3	Y_2	Y_1	Y_0
1	1	0	1

X_3	X_2	X_1	X_0
1	0	1	1
↓	↓	↓	↓

Operation	Symbolic Content of A	Numerical Content of A							
Clear A	Zero	0	0	0	0	0	0	0	0
Transmit X → A $Y_0 = 1$ times	$0 + Y_0 X r^3$	1	0	1	1	0	0	0	0
Shift right	$(0 + Y_0 X r^3) r^{-1}$	0	1	0	1	1	0	0	0
Transmit X → A $Y_1 = 0$ times	$(0 + Y_0 X r^3) r^{-1} + Y_1 X r^3$	0	1	0	1	1	0	0	0
Shift right	$((0 + Y_0 X r^3) r^{-1} + Y_1 X r^3) r^{-1}$	0	0	1	0	1	1	0	0
Transmit X → A $Y_2 = 1$ times	$((0 + Y_0 X r^3) r^{-1} + Y_1 X r^3) r^{-1}$ $+ Y_2 X r^3$	1	1	0	1	1	1	0	0
Shift right	$[((0 + Y_0 X r^3) r^{-1} + Y_1 X r^3) r^{-1}$ $+ Y_2 X r^3] r^{-1}$	0	1	1	0	1	1	1	0
Transmit X → A $Y_3 = 1$ times	$[((0 + Y_0 X r^3) r^{-1} + Y_1 X r^3) r^{-1}$ $+ Y_2 X r^3] r^{-1} + Y_3 X r^3 = Pr$	0	0	0	1	1	1	1	1
Shift right	P	1	0	0	0	1	1	1	1

Note that the closed accumulator makes the necessary end-around carry and circular shifts.

For an open fractional accumulator with open right shifting, we can no longer rely on the theory of congruence, since open right shifting in the event that nonzero digits are lost does not yield a congruent result. For a $2n$-digit accumulator, however, there are n low-order zeros in the initial partial product. There are no digital transmissions or end-carries into these positions, and just n right shifts. Thus, each right shift can be interpreted as a division by r of the quantity equal to the partial product. Therefore, if the partial products are equal to the quantities appearing between parentheses in the fractional formula for P, the final result of the sequence of transmissions from the X-register to A and the right shifts corresponding to this formula will equal P. Since we are dealing only with absolute values in an open accumulator, the criterion for equality is that the magnitude of any partial product be less than r^{2n}. Unfortunately, this criterion may not always be met. For example, let X and Y be as large as possible, that is, $X = r^n - 1$ and all $Y_i = r - 1$. The largest possible partial product, which appears just before the final right shift, has the value XYr. With maximum values of X and Y this has a magnitude

$(r^n - 1)^2 r$, which for $n > 1$, satisfies

$$r^{2n} < (r^n - 1)^2 r < r^{2n} + 1.$$

Thus, a partial product may not satisfy the condition for equality in a $2n$-digit register, but it will always satisfy it in a $(2n + 1)$-digit register. Therefore, to obtain correct results we will increase the length of the accumulator register by one position. The position will be of order $2n$ and will be referred to as the *overflow position*. We will not change the orders of the accumulator into which digits from X are transmitted.

Example 5.13 We repeat Examples 5.10 and 5.12 to illustrate fractional multiplication in an open accumulator.

Y_3	Y_2	Y_1	Y_0
1	1	0	1

Y

X_3	X_2	X_1	X_0
1	0	1	1

X

Operation	Symbolic Content of A	Numerical Content of A								
		O.F.	A_7	A_6	A_5	A_4	A_3	A_2	A_1	A_0
Clear A	Zero	0	0	0	0	0	0	0	0	0
Transmit X → A, $Y_0 = 1$ times	$0 + Y_0 X r^3$	0	1	0	1	1	0	0	0	0
Shift right	$(0 + Y_0 X r^3) r^{-1}$	0	0	1	0	1	1	0	0	0
Transmit X → A, $Y_1 = 0$ times	$(0 + Y_0 X r^3) r^{-1} + Y_1 X r^3$	0	0	1	0	1	1	0	0	0
Shift right	$((Y_0 X r^3) r^{-1} + Y_1 X r^3) r^{-1}$	0	0	0	1	0	1	1	0	0
Transmit X → A, $Y_2 = 1$ times	$((Y_0 X r^3) r^{-1} + Y_1 X r^3) r^{-1} + Y_2 X r^3$	0	1	1	0	1	1	1	0	0
Shift right	$[((Y_0 X r^3) r^{-1} + Y_1 X r^3) r^{-1} + Y_2 X r^3] r^{-1}$	0	0	1	1	0	1	1	1	0
Transmit X → A, $Y_3 = 1$ times	Pr	1	0	0	0	1	1	1	1	0
Shift right	P	0	1	0	0	0	1	1	1	1

This result checks those obtained in the integral accumulator and closed fractional. We note that the overflow position is unnecessary in the closed case because the overflow is accommodated by the end-around carry.

We observe that an open fractional accumulator is simpler than an integral one since the n low-order digital positions are merely storage positions. They receive no carries and have no digits transmitted to them except by shifts. In an integral accumulator, although the n high-order positions receive no direct transmissions from the X-register they must be able to propagate carries.

Again, as in the integral case, the multiplier can be initially placed in an open fractional accumulator. We store the multiplier in the low-order n positions. Then, after an initial right shift, all multiplier digits will appear at the low-order end of the accumulator at the time and in the correct order required for each step of the algorithm. In this situation circular shifting is not appropriate because the digits of Y would be successively shifted into high order digits in A.

5.4 NEGATIVE FACTORS

In Section 5.2 and 5.3 we have assumed that we were operating with nonnegative factors. Thus, any of the algorithms which we developed perform multiplication in a manner which is equivalent to the long-hand operation illustrated in the following binary example:

```
         1011      Multiplicand
         1101      Multiplier
         ────
         1011
        0000
       1011
      1011
      ────────
      10001111    Product
```

This calculation is equivalent to all of the previous algorithms in the sense that if the multiplier digits are put into the multiplier register and the multiplicand digits into the X-register, any of the algorithms would cause the same sequence of product digits to appear in A. In the following pages we will refer to these algorithms for nonnegative factors as *basic algorithms*.

If we employ absolute-value and sign representation for positive and negative factors, the basic algorithms solve the multiplication problem almost completely. We simply segregate the sign indicators and apply any appropriate basic algorithm to the digits of the absolute values to obtain the absolute value of P in the accumulator. The sign of the product is determined, as it is with long-hand multiplication, by examination of the sign indicators of the factors. The resultant sign is then appended to the absolute value of the product. If the digit zero is used for a plus sign and the digit one for a minus sign, the sign digit of the product can be determined by addition of the sign digits of the factors modulo two, that is,

$$0 + 0 \equiv 0 \pmod{2} \Rightarrow (+) \cdot (+) = +,$$
$$0 + 1 \equiv 1 \pmod{2} \Rightarrow (+) \cdot (-) = -,$$
$$1 + 0 \equiv 1 \pmod{2} \Rightarrow (-) \cdot (+) = -,$$
$$1 + 1 \equiv 0 \pmod{2} \Rightarrow (-) \cdot (-) = +.$$

In dealing with absolute values we have interpreted the result of the binary multiplications in Examples 5.10, 5.11, 5.12, and 5.13 as eleven times thirteen equals one hundred forty-three. If we restrict n-digit operands to be less than r^{n-1} in absolute value, and give them a complement interpretation, a sequence of digits starting with zero represents a nonnegative operand and a sequence starting with $r - 1$ represents the complement of a negative operand. Using this interpretation of operands, the result of the basic algorithm in these examples is, in decimal,

$$(-4)(-2) = -112 \quad \text{(one's complement interpretation)},$$
$$(-5)(-3) = -113 \quad \text{(two's complement interpretation)}.$$

Evidently, these are not correct. We must adjust the basic algorithms if the results are to be correctly interpreted for operands in complement form.

A direct approach to this problem is by means of a check of the leading digit of each factor. In any factor with leading digit $r - 1$, we replace the operand by its complement and retain a separate indication of the sign. This procedure is equivalent to the absolute-value and sign case, and the basic algorithms apply, as before. If the product is negative, the value formed in the accumulator must be complemented. When this reduction to the absolute-value and sign case is used, the leading digit of each factor in the basic algorithm will be zero, which eliminates any need for an overflow position in a fractional $2n$-digit accumulator.

When we introduce complement factors into the arithmetic unit, we will require a correct extension of the n-digit operands to $2n$-digit form for consistent arithmetic with respect to the residue classes modulo M_2. The correct extension is accomplished by adjoining n leading zeros to a factor if it begins with zero and adjoining n leading digits $r - 1$ to one which begins with $r - 1$. We first consider the extension of the multiplicand X. For an integral accumulator we can imagine that the X-register has n high-order digital positions added to it. These positions are always filled with digits equal to the (sign) digit in position $n - 1$. Thus, both absolute values and complements will be correctly transmitted from the X-register to A. For nonnegative operands we have the following picture.

		A
00..................0	$0X_{n-2}........X_0$	X

For negative operands represented by complements which have a leading digit $(r - 1)$, the zeros of the configuration above are replaced by $(r - 1)$'s.

		A
$r - 1\, r - 1 \ldots r - 1$	$(r - 1)X_{n-2} \ldots X_0$	X

Thus, for the multiplicand of the examples given above, the effect would be to transmit 11111011, rather than 1011, from the X-register. When the X-register and accumulator operate in the manner just described, we say that A

5.4 NEGATIVE FACTORS

has *extension facilities*. Extension facilities are also required to maintain consistency of residue-class arithmetic for addition and subtraction in double-length integral accumulators. We can avoid this requirement by ignoring the high-order digits and returning to a single-length accumulator. In the case of binary arithmetic, we can imagine that the extension facilities can be introduced without actually making a $2n$-digit register out of the X-register, for example, by forcing carries into the n high-order positions. In general, we will assume that the X-register is of n-digit length and that some such technique provides the extension in A. Extension facilities in an integral accumulator precludes storing multiplier digits in its high-order positions, since the multiplier digits and the extended complement digits will interfere with each other.

Example 5.14 We form a product of 2 and -3 in an open decimal accumulator with $n = 2$. The representations are $Y = 02$ and $X = 97$. For the double-length accumulator, we must extend X to 9997. The effective computations are illustrated below.

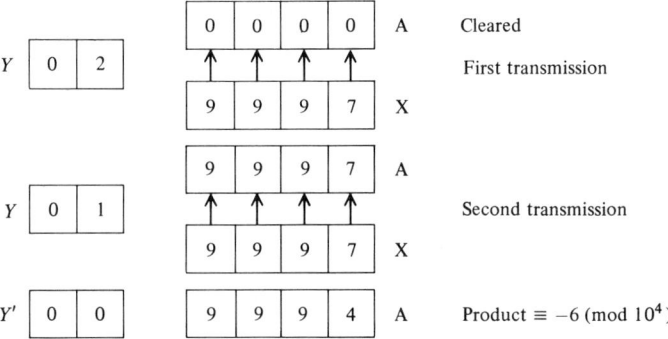

In an open binary accumulator ($n = 3$) the computations could be achieved as follows:

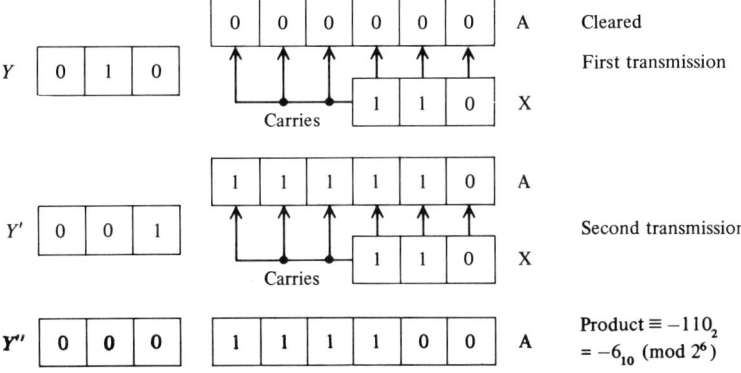

In neither case could we have placed the digits of Y in the high-order positions of A.

In a fractional accumulator we transmit Xr^n from the X-register to A. Thus, for nonnegative X, the extension of the multiplicand to $2n$ digits merely consists of adjoining n low-order zeros to the n-digit representation. For negative values of X, however, $X^* = M_1 - |X|$ and so transmission in a fractional accumulator sends $(M_1 - X)r^n$ from the X-register into A. In this case consistent residue-class arithmetic modulo M_2 requires the transmission of $M_2 - |X|r^n$. In order to produce this result we need a correction term C such that $C + (M_1 - X) = M_2 - |X|$. Thus, $C = M_2 - M_1 r^n$. For the open accumulator $M_2 - M_1 r^n = 0$ and for the closed accumulator $M_2 - M_1 r^n = r^n - 1$. Therefore, in an open fractional accumulator the extension for both negative and nonnegative multiplicands consists in adjoining n low-order zeros to the n-digit representation. In effect, no extension is necessary. In a closed accumulator we must append n low-order digits equal to $r - 1$ to the n-digit representation of the multiplicand. In view of this, the open fractional accumulator is preferable for complement multiplication, since it requires no special extension facilities. For this reason, we will consider only open fractional accumulators. This will permit the low-order n positions of the accumulator to comprise an n-digit register which can shift in conjunction with the high-order n positions. The low-order n positions receive no direct digital transmissions from the X-register and do not propagate or receive carries. In addition and subtraction in an open fractional accumulator, the low-order n positions can be ignored.

In both integral and fractional accumulators the extension of Y to correct $2n$-digit form requires addition of n leading zeros if Y is nonnegative and n leading digits equal to $r - 1$ if Y is negative. That is, we must adjoin n leading digits equal to the sign digit of the single-length representation. Extension of Y is not required in order to preserve the addition and subtraction operation, and so we will not extend Y to $2n$-digit form. Rather, we will utilize the basic multiplication algorithms as if Y were nonnegative and then determine corrections necessary to convert the product produced to a correct complement product. These correction techniques will be considered in the following sections.

5.4.1 Integral Accumulator with Complement Factors

If we consider changes which must be made to the integral basic algorithm for multiplication when the factors and product are given a complement interpretation, there are four cases to examine.

Case	Multiplier	Multiplicand
I	+	+
II	−	+
III	+	−
IV	−	−

5.4 NEGATIVE FACTORS

We note immediately that for Case I no change to the basic algorithm is required.

In Case II the multiplicand digits, stored in the X-register, start with zero and represent $|X|$. The multiplier digits, stored in the multiplier register, start with $r - 1$ and give an integer with absolute value $M_1 - |Y|$. The basic algorithm treats these digital sequences as the absolute value of two integers and yields a product in the accumulator

$$(r^n - |Y|)|X| \quad \text{or} \quad (r^n - 1 - |Y|)|X|.$$

In Case II the true product is equal to $-|X||Y|$, and its correct $2n$-digit complement will be

$$P = M_2 - |X||Y|.$$

To determine the required correction C we compute the difference between the correct representation of the true product and the result of the basic algorithm. This gives

$$C = [M_2 - |X||Y|] - [(M_1 - |Y|)|X|]$$
$$= M_2 - M_1|X|.$$

Since $M_2 \equiv 0 \pmod{M_2}$, the first term in C can be ignored. Thus, the effective correction is $-r^n|X|$ or $-r^n|X| + |X|$, depending on $M_1 = r^n$ or $r^n - 1$.

In Case III, the digits of the multiplier represent an integer with $Y = |Y|$. Those of the multiplicand give an integer with absolute value equal to $M_1 - |X|$. The extension facilities of the accumulator, however, will give the same effect as if the digits of $M_2 - |X|$ were transmitted. Thus, the application of the basic algorithm will be to form in A the product $(M_2 - |X|)|Y|$. In this case, the true product is equal to $-|X||Y|$ and should be represented in the accumulator by $M_2 - |X||Y|$. The correction is then

$$C = M_2 - |X||Y| - (M_2 - |X|)|Y|$$
$$= M_2(1 - |Y|).$$

Since the correction term is an integral multiple of the modulus, it is congruent to zero. The net effect of the correction term is zero, and no correction is needed. In effect the correction has been supplied by the extension facility.

In Case IV the multiplier will be represented by a nonnegative integer equal to $M_1 - |Y|$ and the multiplicand by a similar integer equal to $M_1 - |X|$. Because of the extension facilities the latter is equivalent to $M_2 - |X|$, so that $[M_1 - |Y|][M_2 - |X|]$ is formed. In either modulus the product that should appear in A is $P = |X||Y|$. Thus, the correction term is

$$C = |X||Y| - [M_1 - |Y|][M_2 - |X|]$$
$$= -M_2(M_1 - |Y|) + M_1|X|.$$

After dropping out terms congruent to zero, this correction is $M_1|X|$. We summarize the corrections in the following table.

Case	Correction modulo r^{2n}	Correction modulo $r^{2n} - 1$
I	None	None
II	$-r^n|X|$	$-r^n|X| + |X|$
III	None	None
IV	$r^n|X|$	$r^n|X| - |X|$

The correction term $-r^n|X|$ is easily achieved. Since for Case II, $X = |X|$ represents the initial content of the X-register, the extended complement of X can be introduced into A, either by prior complementation in the X-register (if available) or by complementation in A. A single left shift in the accumulator then yields a result congruent to $-r|X|$ modulo M_2, and the successive $n - 1$ left shifts of the basic algorithm change this to $-r^n|X|$. If complementation is in the X-register, the result must be recomplemented prior to initiating multiplication. If the modulus $M_2 = r^{2n} - 1$, the additional correction of $|X|$ is made by one additional transmission from the X-register to A at the conclusion of the basic algorithm. The same steps can be used to obtain the corrections $r^n|X|$ and $(r^n - 1)X$ when they are required.

Since corrections are required if and only if the multiplier is negative, we need a sign test for the multiplier. If it is negative, the correction steps are the same for both Case II and Case IV, that is, the sign of the multiplicand is immaterial. For Cases I and III, then, we apply the basic algorithm, and for Cases II and IV we can amend the basic algorithm as follows:

1. Complement digits in X-register   Transmit to A
2. Transmit complemented digits to A   (or)   Complement in A
3. Recomplement digits in A   Omit
4. Shift left one position   (Same)
5. Apply the basic algorithm   (Same)
6. If $M_2 = r^{2n} - 1$, transmit to A   (Same)

Example 5.15 We compute in the table opposite the products of the operands of the preceding examples, using a one's-complement interpretation.

We read the final result as $(-4)(-2) = 8$. This is seen to be the correct interpretation. In carrying out the arithmetic we must extend complements and propagate end-around carries.

5.4.2 Fractional Accumulator with Complement Factors

As we have noted above, no extension properties are necessary for a fractional accumulator, since we assume that it is open. Thus, when the multiplier and multiplicand are negative, they are respectively represented in n-digit storage

5.4 NEGATIVE FACTORS

Y_3	Y_2	Y_1	Y_0
1	1	0	1

Y

Extension facility

X_3	X_2	X_1	X_0
1	0	1	1

X

	Operation	A_7	A_6	A_5	A_4	A_3	A_2	A_1	A_0
	Initially clear	0	0	0	0	0	0	0	0
Correction	Transmit digits of complement of X	0	0	0	0	0	1	0	0
	Left circular shift of one position	0	0	0	0	1	0	0	0
Basic algorithm	Transmit X → A, Y_3 times	0	0	0	0	0	1	0	0
	Left-circular-shift one position	0	0	0	0	1	0	0	0
	Transmit X → A, Y_2 times	0	0	0	0	0	1	0	0
	Left-circular-shift one position	0	0	0	0	1	0	0	0
	Transmit X → A, Y_1 times	0	0	0	0	1	0	0	0
	Left-circular-shift one position	0	0	0	1	0	0	0	0
	Transmit X → A, Y_0 times	0	0	0	0	1	1	0	0
Final correction	Transmit X → A	0	0	0	0	1	0	0	0

registers by integers having absolute-value $r^n - |Y|$ and $r^n - |X|$. By an analysis similar to that made for the integral accumulator, we derive the following table of corrections.

Case	Multiplier	Multiplicand	Correction
I	+	+	None
II	−	+	$-|X|r^n$
III	+	−	$-|Y|r^n$
IV	−	−	$|X|r^n + |Y|r^n$

From the table of corrections we observe that when the multiplier is negative a correction term $-|X|r^n$ or $|X|r^n$ must be added to the result of the basic algorithm. Although this term has a different sign for positive and negative multiplicands, it is not necessary to test this sign, since the digital manipulations are precisely the same in each case. These consist of applying the basic algorithm, complementing the content of the X-register, and transmitting the complemented digits to A. If the multiplicand is negative, correction terms $-|Y|r^n$ or $|Y|r^n$, depending on the sign of the multiplier, must be included. We do not have to test this sign, since the digital manipulations are the same in each case.

A correction term involving the multiplicand requires no alteration of the content of the X-register, since the operand X is already there. Unfortunately, this is not true of the multiplier. In fact, the multiplier digits, if they were initially stored in the low order of A, may be lost and have to be brought back into the arithmetic unit from general storage. We can avoid this by using the multiplier digits during the basic algorithm to construct partial corrections as well as partial products. Whatever scheme is used, it must be the equivalent of transmitting the complement of the multiplicand to A on completion of the basic algorithm. We illustrate the general technique in the following example.

Example 5.16 We repeat Example 5.15 for a fractional accumulator.

Y_3	Y_2	Y_1	Y_0			X_3	X_2	X_1	X_0	
1	1	0	1	Y		1	0	1	1	X

	Operation	O.F.	A_7	A_6	A_5	A_4	A_3	A_2	A_1	A_0
	Initially clear	0	0	0	0	0	0	0	0	0
	X → A, Y_0 times	0	1	0	1	1	0	0	0	0
	Shift right	0	0	1	0	1	1	0	0	0
Basic	X → A, Y_1 times	0	0	1	0	1	1	0	0	0
algorithm	Shift right	0	0	0	1	0	1	1	0	0
	X → A, Y_2 times	0	1	1	0	1	1	1	0	0
	Shift right	0	0	1	1	0	1	1	1	0
	X → A, Y_3 times	1	0	0	0	1	1	1	1	0
	Shift right	0	1	0	0	0	1	1	1	1
Corrections	Comp X → A	0	1	1	0	1	1	1	1	1
	Comp Y → A	1	0	0	0	0	1	1	1	1

This is to be interpreted as $(-5) \times (-3) = 15$. The one in the overflow position does not, of course, serve as a sign digit in the eight-position accumulator.

5.5 MULTIPLY-ADD

In many computations a multiplication will be followed by an addition. For this reason it would often be convenient to combine the multiplication and the addition into a single operation. We can do this if we can add to the integer already in the double-length accumulator the product of the factors represented by the multiplier and multiplicand digits. For the first basic multiplication algorithm in which the multiplicand digits are shifted relative to the partial product digits, this multiply-add operation comes free and is, in fact, simpler than multiplication by itself, since we need *not* clear the accumulator. For the fractional and integral basic multiply algorithm, however, this does not prove to be true. Indeed, in the open fractional accumulator, the modifications required make it generally unfeasible to incorporate a multiply-

5.5 MULTIPLY-ADD

add operation. For integral accumulators, the use of open shifting would destroy the high-order digits of the initial content of A in the process of applying the basic algorithm. This is not true in the case of a closed accumulator with circular shifting, and we present below the steps necessary to achieve a multiply-add. We leave it to the reader in the exercises to justify these steps. We note that the multiply-add operation would prevent using the accumulator for storage of the multiplier digits.

For a multiply-add operation using complement factors, we change the algorithm for an integral accumulator to the following pattern:

1. Do *not* initially clear A.
2. Shift left circularly n positions in A.
3. If the multiplier is negative, transmit the digits of the complement of X to A.
4. Shift the digits in A left one position.
5. Apply the basic algorithm.
6. If multiplier is negative, transmit digits of X to A.

Example 5.17 As an illustration we apply this algorithm to the operands of the preceding examples, with the added assumption that the number -7 is initially in the accumulator. The final result is then sufficiently small in magnitude so that the accumulator result will equal it.

Y_3	Y_2	Y_1	Y_0
1	1	0	1

Y

X_3	X_2	X_1	X_0
1	0	1	1

X

	Operation	A_7	A_6	A_5	A_4	A_3	A_2	A_1	A_0
Set up multiply-add	Initial content	1	1	1	1	1	0	0	0
	Shift left 4 positions	1	0	0	0	1	1	1	1
Correction term	Transmit complement of X to A	1	0	0	1	0	0	1	1
	Shift left one position	0	0	1	0	0	1	1	1
Basic algorithm	Transmit X → A, Y_3 times	0	0	1	0	0	0	1	1
	Shift left one position	0	1	0	0	0	1	1	0
	Transmit X → A, Y_2 times	0	1	0	0	0	0	1	0
	Shift left one position	1	0	0	0	0	1	0	0
	Transmit X → A, Y_1 times	1	0	0	0	0	1	0	0
	Shift left one position	0	0	0	0	1	0	0	1
	Transmit X → A, Y_0 times	0	0	0	0	0	1	0	1
Correction term	Transmit X → A	0	0	0	0	0	0	0	1

5.6 SERIAL MULTIPLICATION

We have more or less tacitly assumed that the digits in a register are all available simultaneously, and that a given digit is determined to be associated with a specific power of r by its spatial position in the register. Thus, we have assumed that in shifting digits we change their spatial position within a register. Registers which operate on this basis are called *parallel* registers, and the arithmetic carried out by use of such registers is called parallel arithmetic. We have discussed this idea for addition and subtraction in Chapter 3.

There is another type of storage device which makes its digits available one at a time, usually with the low-order digit coming first and then the remaining digits in order. These registers are called *serial* registers, and the arithmetic carried out with these registers is called *serial arithmetic*. We can employ the same algorithms for multiplication with serial registers but we must change our point of view. Instead of thinking of the simultaneous transmission of digits we think in terms of sequential transmission. We must think of shifting as a change in the time of availability of digits, instead of a change in physical location of digits. In serial operation we match digits against a clock, and any digit arriving at a given time is considered to be the coefficient of a fixed power of r, while the digit arriving at the next greater unit of time is considered to be the coefficient of the next greater power of r.

If serial operations are employed, assuming similar components, it will take longer to carry out a given algorithm than if parallel ones are used. However, serial operations may be considered more economical and reliable, since there is less equipment needed. There are certain inconveniences in implementing algorithms in serial arithmetic, however. In particular, operations depending upon a knowledge of the sign of an operand may be excessively time consuming, since the sign indicator or digit may be the last in the sequence to be available. With this in mind, we can develop algorithms which minimize the need to test the sign digits in advance. We will not consider such algorithms here.

EXERCISES

SECTION 5.1

5.1 Establish by induction the fact that the product of two positive integers $P = XY$ can be formed by adding to zero the value X a number of times equal to Y. What properties of addition and multiplication do you utilize?

5.2* Suppose we define the product of an integer X and an integer Y in the form XY, as the result of adding Y to zero X times. Prove that $XY = YX$ (commutative law). Use whatever properties of addition are needed.

5.3 Assume the definition and validity of the commutative law of Exercise 5.2 and establish the associative law of multiplication (Chapter 1).

5.4 Assume the definition and validity of the commutative law of Exercise 5.3 and establish the distributive law (Chapter 1).

5.5 Extend the results of Exercises 5.1 through 5.4 to include a factor zero.

5.6 Establish directly that if
$$x \equiv X^* \pmod{M},$$
$$y \equiv Y^* \pmod{M},$$
and
$$X^* Y^* \equiv R^* \pmod{M},$$
then $xy \equiv R^* \pmod{M}$.

5.7 In effect the use of a double-length accumulator for multiplication establishes n overflow positions for the product. Suppose X^* and Y^* are n-digit operands satisfying (5.1), and let X^{**} and Y^{**} be their $2n$-digit extended versions. If $X^{**} + Y^{**}$ is formed in a $2n$-digit accumulator, under what circumstances is it a properly extended version of $X^* + Y^*$?

5.8 If the n-digit operands X^* and Y^* satisfy (5.2) and have a sign digit, and X^{**} and Y^{**} are the $2n$-digit versions with a sign extension, establish that $X^{**} + Y^{**}$ is a properly sign-extended version of $X^* + Y^*$.

5.9 The $2n$-digit accumulator in Exercise 5.8 produces more sign digits than needed. If X^* and Y^* are n-digit operands satisfying (5.2), how many digits in the accumulator are required to guarantee that $X^{**} + Y^{**}$ is a properly extended version of $X^* + Y^*$ with at least one sign digit?

SECTION 5.2

5.10 If an n-digit, base r, integer X is transmitted to a $2n$-digit cleared accumulator r times, according to the definition of Exercise 5.2, the result is rX. Establish the fact that the final digit is zero and that the digits of X are shifted left one position.

5.11 In forming the product of the decimal numbers 32×95, we may add 32 to zero 95 times, or we may add 32 to zero five times, shift left once and add 32 nine more times. If one shift is counted the same as one addition, what is the ratio of additions in the first method to that in the second?

5.12 If we form the product XY by brute force by adding X to zero Y times or Y to zero X times, is there any way to minimize the total number of additions?

5.13* Suppose the product of X and Y is formed by adding and shifting as, for example, in Example 5.10. Does the number of additions and shifts in forming XY or YX differ? In other words, computationally, does the commutative law hold?

SECTION 5.3

5.14 Suppose that an n-digit register has an even number of digits, say $n = 2k$. Thus, the $2n$-digit accumulator has $2n = 4k$ digits. If the n digits are added to the central portion of the accumulator

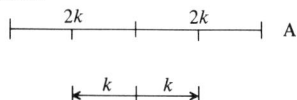

148   MULTIPLICATION

and the accumulator is a shifting register, can a reasonable form of addition and multiplication be attained? What kind of extension facilities do we need to assume?

SECTION 5.3.1

5.15 The formula given for forming the product of X and Y is equivalent to
$$XY_{n-1}r^{n-1} + XY_{n-2}r^{n-2} + \cdots + XY_1 r + XY_0.$$
Can we use an algorithm which transmits X a total of Y_{n-1} times, shifts left $n-1$ times, then transmits X a total of Y_{n-2} times, shifts left $n-2$ times, and so on?

SECTION 5.3.2

5.16 Although the results given for a fractional accumulator are interpreted as integers, is the algorithm described valid if we literally interpret the base point at the left-hand end of the sequence?

5.17 With a base-point interpretation as given in Exercise 5.16, what is the significance of the overflow positions, if any?

SECTION 5.4

5.18 Show that for correct complement arithmetic the extension of a number requires appending all sign digits.

SECTION 5.4.1

5.19 Apply the basic algorithm to the binary product of 1101 and 1011 in binary, and justify the corrections, if any.

5.20 What modification is needed, if any, in Example 5.16 if we use two's-complement interpretation?

SECTION 5.4.2

5.21* What extension facilities are required if a fractional accumulator is considered closed for complement factors?

5.22 Do Example 5.17 for a closed fractional accumulator.

5.23 If overflow positions in a fractional accumulator are used with a literal fractional interpretation, what is the interpretation?

SECTION 5.5

5.24 The multiply-add operation is a combination of multiplication and addition. Multiplication (Exercise 5.2) is a repetition of addition. Formulate this interpretation as a multiply-add operation.

5.25 Justify the steps of the multiply-add operation for an integral accumulator.

5.26* Derive an algorithm equivalent to that of Exercise 5.25 for a fractional accumulator and indicate why such a formulation is difficult.

5.27* We have described a method of multiplying by combinations of addition and shifting. Refer to Chapter 3 and devise a logical method of forming a multiplier. Use base two and determine what components, including adders, half-adders, and the like, would be needed.

CHAPTER 6

DIVISION

6.1 INTRODUCTION

Of the four operations of arithmetic, division is the most difficult for both men and computing machines. One method of forming $N/D = ND^{-1}$ is to take the reciprocal D^{-1} of D and then multiply by N. The formation of the inverse of D can be accomplished by an iterative technique which employs only additions and multiplications. While we will not discuss this method here, the idea serves to remind us that division and multiplication are inverse operations of each other. Hence, we will investigate the possibility of making a division algorithm by reversing the steps of a multiplication algorithm.

Here, as in the discussion of multiplication, we will initially deal exclusively with nonnegative integers. If division were always exact for integers, that is, if the result of dividing a dividend N by a divisor D in the form

$$\frac{N}{D} = Q,$$

always produced an integral quotient Q, the reversal of the algorithmic steps of multiplication would be simplified. In this case, the equivalent relation $N = QD$ gives the obvious correspondence

$$\text{product} \longleftrightarrow \text{dividend,}$$
$$\text{multiplicand} \longleftrightarrow \text{divisor,}$$
$$\text{multiplier} \longleftrightarrow \text{quotient.}$$

Unfortunately the integer N will not always be an integral multiple of the second integer D. Thus, a more realistic definition of division is based on the division algorithm of Chapter 1 in the form

$$N = QD + R.$$

Here R is the remainder chosen so that $0 \leq R < D$ and has no analog in ordinary multiplication except as an understood zero value. We see that, in fact, the full generality of division corresponds to an inversion of a multiply-add operation, wherein R plays the role of the integer initially in the accumulator to which the product is to be added.

150 DIVISION 6.1

Given Q, D, and R, we can form $N = QD + R$ by placing R in the accumulator and then additively transmitting the digits of D a number of times equal to Q. Conversely, we can determine Q from N and D by placing N initially in the accumulator and counting the number of times we need to transmit the complement digits of D (subtracting) until we have a nonnegative content of the accumulator less than D. The number of times we must subtract will be equal to Q and the final content of the accumulator will be R.

Example 6.1 For a four-digit open decimal integral accumulator with extension facilities, we could form the quotient for 17/5 as follows. We place the extended value of 17 in A and the complement of 5 in the X-register and transmit additively until the net result in A is less than 5.

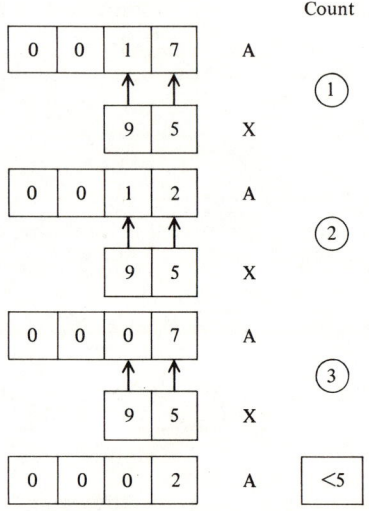

This gives the quotient 3 and the remainder 2.

The technique just discussed is the equivalent of forming the product XY by adding X a total number of times equal to Y in a cleared accumulator. We saw in Chapter 5 that this is inefficient. However, by replacing a count equal to a power of r by a shift, we were able to reduce the count to the value of a multiplier digit. In division, too, we want to use shifting to improve the efficiency of the counting procedure. Thus, by analogy, we might expect to produce the quotient and remainder from a given dividend and divisor by counting, one digit at a time, the number of times it is possible to subtract the divisor from the dividend, where counts equal to powers of r are obtained by shifting. The individual counts would then give the digits of the quotient. Unfortunately, difficulties arise in carrying out this program which are not encountered in carrying out its analog for multiplication.

6.1 INTRODUCTION

The dividend, as analogous to the product, will be $2n$ digits. The divisor, as analogous to the multiplicand, will be only n digits. There are a variety of ways of choosing the orders of the dividend from which to begin subtracting the divisor digits. Ordinarily, when we divide by hand, we arrange this initial subtraction so that the first counting operation will yield the most significant digit of the quotient.

Example 6.2 Consider the division of the four-digit dividend 0316 by the two-digit divisor 15. Ordinary long-hand division yields

```
           2 1              Quotient
    15)0 3 1 6
       -3 0                 (Equivalent of 2 subtractions)
       ─────
         1 6
        -1 5                (Equivalent of 1 subtraction)
        ────
           1                Remainder
```

In Example 6.2, we could have divided 15 initially into three and obtained a leading quotient digit of zero, or perhaps have divided 15 initially into 316. In this last case we would have had to provide for a two-digit quotient. Instead we "lined up" the dividend and divisor digits so that the initial operations produced a single, most significant, quotient digit. In a mechanized process it is more difficult to "line up" the dividend and divisor digits in this variable manner. The analog of locating the high-order digit by inspection is hard to automate. Thus, in machine division, the initial subtraction of the divisor will generally be from the same orders of the dividend. This may lead to insignificant high-order zero quotient digits. If the divisor is small relative to the dividend, the procedure can produce a loss of the most significant quotient digits by generating counts greater than the largest digit of the number system. In order to avoid this problem, the relative magnitudes of dividend and divisor must be restricted. It will be the responsibility of the user of a computing machine to observe these restrictions or face the need for additional hardware components. This problem will be discussed at greater length in Chapter 7.

In multiplication it is known in advance that the multiplicand must be transmitted to the accumulator a number of times equal to a given multiplier digit. Since the quotient digits are not known in advance, it is not a priori certain, in division, how many times it will be necessary to subtract the divisor from a given order of the dividend. Thus some kind of test must be introduced. This frequently turns out to be a trial-and-error process in which it is not known that the divisor has been subtracted a sufficient number of times until it has been subtracted once too often. In such a case a backtracking or correction procedure must be provided.

Another problem which must be dealt with in division arises from the fact that the operation, unlike the other three arithmetic operations, produces two results, the quotient and the remainder. We must provide for the disposition of the remainder as well as for a possible round-off of the quotient.

It is always possible to generate the correct $2n$-digit product from any two n-digit factors. It is not always possible to generate a quotient and remainder which can be stored in arithmetic registers from arbitrary dividend and divisor which are storable in arithmetic registers. For example the quotient 0333 generated by the dividend 0666 and divisor 02 cannot be stored in a two-digit register. Consequently, the division operation may not produce correct results for various combinations of dividend and divisor which can be introduced into the arithmetic unit. This problem is usually dealt with by means of either machine or hand scaling, a subject we consider in Chapter 7.

6.2 THE APPLICATION OF THE BASIC DIVISION ALGORITHM TO MACHINE DIVISION

The problem we consider is the following: given a nonnegative integer N and a positive integer D, find integers Q and R such that

$$N = QD + R, \quad 0 \le R < D. \tag{6.1}$$

The discussion of machine division will be based on a study of conditions (6.1). We know from the division algorithm, proved in Chapter 1, that there exist unique nonnegative integers Q and R satisfying (6.1). Therefore, once we have determined values of Q and R satisfying (6.1), no matter how we accomplish it, they are the desired and only answers to the problem posed.

Since the quotient Q is the analog of the multiplier, it is natural to assume that Q will be stored in the same-sized register as that used for multiplier storage during multiplication. In many machines the same register is used for the multiplier and quotient. In the context of division, we will denote the digits of the multiplier-quotient register by q_i, $i = 0, 1, 2, \ldots, n - 1$. We assume then that Q can be represented as

$$Q = \sum_{i=0}^{n-1} q_i r^i. \tag{6.2}$$

If this is true, we may substitute (6.2) into (6.1) to obtain

$$N = \left(\sum_{i=0}^{n-1} q_i r^i \right) D + R, \quad 0 \le R < D. \tag{6.3}$$

Any means of finding a sequence of digits q_i satisfying (6.3) will solve the problem.

If we transpose the high-order product term to the left side of equality

6.2 BASIC DIVISION ALGORITHM APPLIED TO MACHINE DIVISION

(6.3), we have

$$N - q_{n-1}r^{n-1}D = \left(\sum_{i=0}^{n-2} q_i r^i\right)D + R. \tag{6.4}$$

Since all terms on the right in (6.4) are nonnegative, the left-hand term is nonnegative. Each q_i on the right has a maximum possible value of $r - 1$ and $0 \leq R < D$, and thus

$$\left(\sum_{i=0}^{n-2} q_i r^i\right)D + R < (r^{n-1} - 1)D + D = r^{n-1}D.$$

Therefore, we may write

$$0 \leq N - q_{n-1}r^{n-1}D < r^{n-1}D. \tag{6.5}$$

The inequality (6.5) characterizes the quotient digit q_{n-1}. Let q be any integer such that

$$0 \leq N - qr^{n-1}D < r^{n-1}D.$$

Substitution of N from (6.1) and Q from (6.2) converts this last inequality to

$$0 \leq \left(\sum_{i=0}^{n-1} q_i r^i\right)D + R - qr^{n-1}D < r^{n-1}D$$

or

$$0 \leq (q_{n-1} - q)r^{n-1}D + \left(\sum_{i=0}^{n-2} q_i r^i\right)D + R < r^{n-1}D.$$

Since $(\sum_{i=0}^{n-2} q_i r^i)D + R < r^{n-1}D$, we must have $q \leq q_{n-1}$ or the lower condition of this last inequality is not satisfied. On the other hand if $q < q_{n-1}$, the upper inequality is violated. Therefore $q = q_{n-1}$.

Suppose that q is an integer satisfying $0 \leq q < q_{n-1}$ (therefore q is a digit). Then, making use of (6.5), we obtain

$$N - qr^{n-1}D \geq N - q_{n-1}r^{n-1}D \geq 0.$$

Thus, any digit $q \leq q_{n-1}$ gives $N - qr^{n-1}D \geq 0$, but any $q > q_{n-1}$ yields, from (6.5),

$$N - qr^{n-1}D$$
$$= N - q_{n-1}r^{n-1}D - (q - q_{n-1})r^{n-1}D < [1 - (q - q_{n-1})]r^{n-1}D \leq 0.$$

If we consider systematically computing $N - qr^{n-1}D$ for values of $q = 0, 1, 2, \ldots, q_{n-1}$ (counting), we see that we can determine $q = q_{n-1}$ by counting the largest number of times that $r^{n-1}D$ can be subtracted without producing a negative result. Alternatively we can say that the first negative result occurs for $q = q_{n-1} + 1$, a digit one larger than the quotient digit.

We define

$$R_n = N,$$
$$R_i = R_{i+1} - q_i r^i D, \quad i = n-1, n-2, \ldots, 2, 1, 0. \qquad (6.6)$$

Then, if each q_i is the correct quotient digit,

$$R_0 = R_1 - q_0 r^0 D = R,$$

which gives the correct remainder. We will refer to the R_i defined by (6.6) as *partial remainders*. We have

$$R_i = R_{i+1} - q_i r^i D = N - \left(\sum_{j=i}^{n-1} q_j r^j\right) D$$

$$= \left(\sum_{j=0}^{i-1} q_j r^j\right) D + R.$$

It follows that each R_i satisfies

$$0 \leq R_i < r^i D. \qquad (6.7)$$

We can show, using (6.7) and the argument used above, that each q_i is equal to the largest number of times we can subtract $r^i D$ from R_{i+1} without achieving a negative result. This permits the following complete division algorithm.

1. Starting with $N = R_n$, we use the following procedures for $i = n-1, n-2, \ldots, 1, 0$, in that order.

2. Given the partial remainder R_{i+1}, we determine the quotient digit q_i by successively subtracting $r^i D$ from R_{i+1} until a negative result is reached. The digit q_i is then equal to one less than the number of subtractions, and we obtain q_i by subtracting one from the count.

3. At this point we have subtracted once too often and computed

$$(R_{i+1} - q_i r^i D) - r^i D = R_i - r^i D.$$

In order to obtain R_i and proceed to the evaluation of q_{i-1} we must correct by adding $r^i D$.

4. The remainder R is equal to R_0.

Example 6.3 We illustrate the division algorithm described above for $r = 10, n = 2$. We use $N = 0335$ and $D = 23$. We assume the computations

6.2 BASIC DIVISION ALGORITHM APPLIED TO MACHINE DIVISION

are carried out in a four-digit open accumulator.

Operation	Computation	Sign of remainder	Quantity represented	Count and/or quotient digit
	0335		$N = R_2$	
$-D(10)^1$	-230			
	0105	$+$	$R_2 - Dr^1$	$q = 1$
$-D(10)^1$	-230			
	9875	$-$	$R_2 - 2Dr^1$	$q = 2$
$+D(10)^1$	$+230$			$q - 1 = 1$
	0105	$+$	$R_2 - Dr^1 = R_1$	$q = q_1 = 1$
$-D(10)^0$	-23			
	0082	$+$	$R_1 - Dr^0$	$q = 1$
$-D(10)^0$	-23			
	0059	$+$	$R_1 - 2Dr^0$	$q = 2$
$-D(10)^0$	-23			
	0036	$+$	$R_1 - 3Dr^0$	$q = 3$
$-D(10)^0$	-23			
	0013	$+$	$R_1 - 4Dr^0$	$q = 4$
$-D(10)^0$	-23			
	9990	$-$	$R_1 - 5Dr^0$	$q = 5$
$+D(10)^0$	$+23$			$q - 1 = 4$
	0013	$+$	$R_1 - 4Dr^0 = R_0 = R$	$q = q_0 = 4$

In the form of the basic division algorithm this represents

$$0335 = (23)(14) + 13.$$

In the binary system the counting can be simplified. Since q_i is equal to zero or to one, we know immediately that

$$R_{i+1} - 2^i D < 0 \quad \text{implies} \quad q_i = 0$$

and

$$R_{i+1} - 2^i D \geq 0 \quad \text{implies} \quad q_i = 1.$$

In the latter event we need not continue subtracting until a negative result is achieved. Thus, we can avoid the necessity of a correction. In the former event, however, we have carried out the algorithm in the normal way by subtracting once too often, and a correction will be necessary.

Example 6.4 We illustrate a binary division with

$$N = 01000111 = 71_{10} \quad \text{and} \quad D = 0111 = 7.$$

We assume that computations are done in an 8-bit open accumulator.

156   DIVISION   6.2

Operation	Computation	Sign of remainder	Quantity represented	Count and/or quotient digit
	01000111		$N = R_4$	
$-D2^3$	-0111000			
	00001111	$+$	$R_3 = R_4 - D2^3$	$q = q_3 = 1$
$-D2^2$	$- \ 011100$			
	11110011	$-$	$R_3 - D2^2$	$q = 1$
$+D2^2$	$+ \ 011100$			$q - 1 = 0$
	00001111		$R_2 = R_3$	$q = q_2 = 0$
$-D2^1$	$- \ \ \ 01110$			
	00000001	$+$	$R_1 = R_2 - D2$	$q = q_1 = 1$
$-D2^0$	$- \ \ \ \ \ 0111$			
	11111010	$-$	$R_1 - D2^0$	$q = 1$
$+D2^0$	$+ \ \ \ \ \ 0111$			$q - 1 = 0$
	00000001		$R_0 = R_1 - 0D = R$	$q = q_0 = 0$

The result is

$$Q = 1010 = 10_{10},$$
$$R = 00000001 < D,$$
$$N = QD + R = (10)_{10} \cdot 7 + 1 = 71_{10} = N.$$

The algorithm was derived on the assumption that the quotient Q is an n-digit integer. This assumption was introduced because of the necessity of storing the digits of Q in the multiplier-quotient register. If Q is too large to be represented as an n-digit integer, this algorithm will break down. Even if we permit Q to be contained in a register with more than n digits, some similar assumption will have to be made, because we can only retain a finite number of digits. If Q is representable as an n-digit integer, it must be in the range $0 \leq Q \leq r^n - 1$. Therefore, we have

$$N = QD + R \leq (r^n - 1)D + R < (r^n - 1)D + D = r^n D.$$

This implies $N - r^n D < 0$. On the other hand, if $N - r^n D < 0$, we have

$$N = QD + R < r^n D,$$

from which it follows that

$$Q < r^n - \frac{R}{D} \leq r^n,$$

since

$$0 \leq R/D < 1.$$

Therefore, Q, being an integer, satisfies $Q \leq r^n - 1$. Thus Q is representable as an n-digit integer, if and only if

$$N - r^n D < 0. \tag{6.8}$$

6.2 BASIC DIVISION ALGORITHM APPLIED TO MACHINE DIVISION

In what follows, in order to ensure that this algorithm is applicable we will test condition (6.8) before initiating it. In other words, we will incorporate a test of (6.8) into the algorithm and terminate it if the test fails.

6.2.1 Fractional Mechanization of the Division Algorithm

The basic steps in the algorithm developed in Section 6.2 consist of forming terms $R_{i+1} - qDr^i$ and testing the sign of the result for values of q successively equal to $1, 2, \ldots$. When a negative result is obtained, we correct by adding Dr^i once. We start with $N = R_n$ in the accumulator and continue with steps in which R_{i+1} is in the accumulator and we wish to obtain q_i. The divisor D, as the analog of the multiplicand, can be assumed to be in the X-register. In an open fractional accumulator the operations that can be performed are:

1. add Dr^n,
2. subtract Dr^n,
3. multiply by r (left shift),
4. divide by r (right shift).

If we assume that a left shift of $(n - i)$ places on the partial remainder R_{i+1} corresponds to a multiplication by r^{n-i}, the result is $R_{i+1}r^{n-i}$. Thus, we can form $R_{i+1}r^{n-i} - qDr^n$ by repetitive transmission from X to A. These operations eventually yield

$$r^{n-i}(R_{i+1} - q_iDr^i) = r^{n-i}R_i.$$

Since r^{n-i} is positive, the sign of each of the products is the same as the sign of $R_{i+1} - q_iDr^i$, which is what we must know to control the process.

We start with $i = n - 1$ and proceed with sequentially decreasing values of i until $i = 0$. Thus, initially, we have, after a single left shift and subtraction,

$$r^{n-i}(R_{i+1} - q_iDr^i) = r^1(R_n - q_{n-1}Dr^{n-1})$$
$$= rR_{n-1}.$$

A second left shift yields $r^2R_{n-1} = r^{n-i}R_{i+1}$ for $i = n - 2$, and we are ready to determine the next digit q_{n-2}. Each subtraction of Dr^n requires the complement of D in the X-register. At each step we proceed until a negative partial remainder is attained. This demands a count reduction of one and a corrective addition of Dr^n. The latter will be achieved by complementing the X-register to obtain D, transmission to A, and recomplementation to go on to the next subtraction. Thus, the algorithm is mechanized by the following sequence, with the complement of D initially in the X-register.

1. Left shift in A.
2. Transmit from the X-register to A and count.
3. Test the sign of the partial remainder in A.

4a. If the partial remainder is nonnegative, repeat steps (2) and (3).

4b. If the partial remainder is negative, reduce the count by one. Complement the content of the X-register and transmit to A. Recomplement and repeat steps (1), (2), and (3).

Example 6.5 We illustrate the algorithm for a four-digit decimal accumulator with an overflow position. We assume the equivalent of extension into the overflow position. The operands are a dividend equal to 17 and a divisor equal to 5.

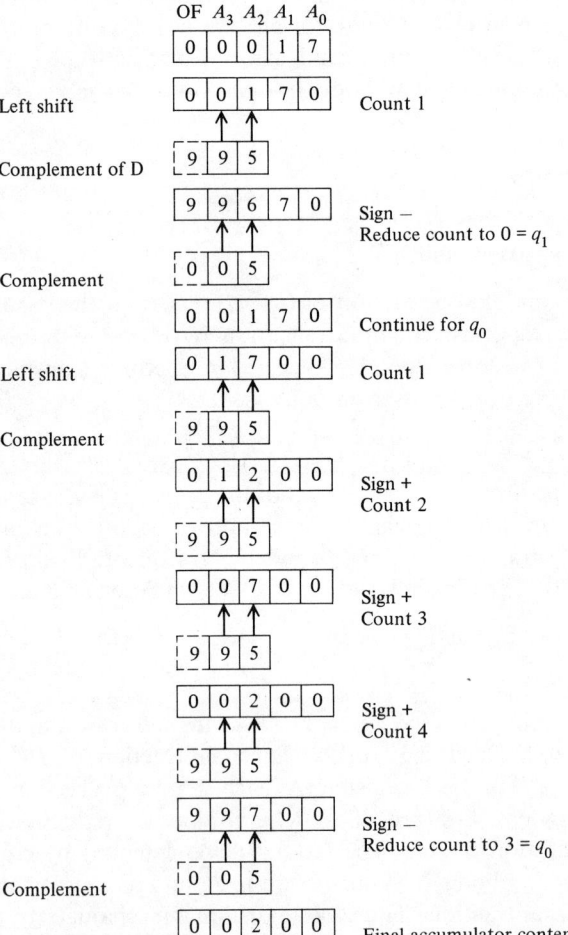

Thus, $Q = q_0 = 3$ and $R = 2$.

The algorithm requires additions, subtractions, and left shifts in the accumulator. The subtractions can be carried out by the addition of complements as they were in Example 6.5. All of these operations are consistent

6.2 BASIC DIVISION ALGORITHM APPLIED TO MACHINE DIVISION

residue-class operations and will always yield results congruent to $r^{n-i}R_{i+1} - qDr^n$. Since we are interested in the sign of this number, we cannot tolerate mere congruence. We must know that the accumulator result is either equal to the actual result or its complement, and we must be able to tell which case we have *a posteriori*. In Example 6.5, because of the small magnitudes involved, we were able to distinguish signs by examination of the leading digit. We can do this in general, as we know from Chapter 2, provided that all of the intermediate results are sufficiently small in magnitude. We have, from (6.7),

$$0 \le r^{n-i}R_{i+1} < r^{n+1}D,$$

and so

$$-qDr^n \le r^{n-i}R_{i+1} - qDr^n < r^{n+1}D - qDr^n = (r-q)r^nD$$

for $q \ge 1$. Since $D < r^n$, this last inequality implies

$$-qDr^n \le r^{n-i}R_{i+1} - qDr^n < (r-1)r^{2n} = r^{2n+1} - r^{2n} < r^{2n+1}.$$

This guarantees that the partial remainder can be contained in a $(2n + 1)$-digit register, but does not necessarily provide the sign designation.

Example 6.6 With $r = 10$, $n = 2$, $N = 9801$, and $D = 99$, we have, for a $(2n + 1) = 5$-digit accumulator, $R_n = R_2 = 09801$. Since

$$N - r^nD = 9801 - 9900 < 0,$$

the quotient is representable as a two-digit integer. For $i = n - i$ and $q = 1$,

$$r^{n-i}R_{i+1} - qDr^n = rR_2 - Dr^2 = 98010 - 9900 = 88110.$$

This number is representable in the five-digit register, but the leading digit 8 would give a false impression of the sign. We need a six-digit register with 088110 to indicate the nonnegative result.

To ensure that we have magnitudes sufficiently small to provide a leading sign digit, we will need in some instances $(2n + 2)$ digits in the accumulator, as seen in Example 6.6. Since

$$r^{n-i}R_{i+1} - qDr^n < r^{2n+1} = r^{(2n+2)-1},$$

an accumulator with $2n + 2$ digits, that is, two overflow positions, is always adequate for any base. However, for binary, $r - 1 = 1$ and, thus

$$2^{n-1}R_{i+1} - qD2^n < (2-1)2^{2n} = 2^{2n},$$

so that a $(2n + 1)$-digit accumulator, that is, one overflow position, will suffice.

Example 6.7 With $r = 2$, and $N = 01001$ in a five-digit register and $D = 11$, $N - r^nD = 1001 - 1100 < 0$, and Q can be represented by two digits. For $i = n - 1$, $N = R_2$, and $q = 1$,

$$r^{n-i}R_{i+1} - qDr^n = rR_2 - Dr^2 = 10010 - 1100 = 00110.$$

The leading zero bit correctly indicates the nonnegative result.

In Examples 6.6 and 6.7 the value of D was the maximum for the given values of r and n. Similarly, for these divisors, the dividend was the maximum which would still guarantee that $N - r^nD < 0$, so that a two-digit quotient would result. We see that for the binary case a $(2n + 1)$-digit accumulator suffices to determine sign, but for $r > 2$ we need to add a second overflow position. In either case, however, $N - r^nD < r^{2n+1}$, and the test of (6.8) for correctness of the algorithm presents no difficulty. We will incorporate it into the algorithm as a check.

Example 6.8 We illustrate fractional mechanization of the division algorithm in the binary system with $N = 211_{10}$, $D = 15_{10}$, and $n = 4$. In this example we also demonstrate how the low-order n digits of the accumulator can be used as a register to store the quotient. Note that for $r = 2$, no corrections are required if the partial remainder is positive.

					1	1	1	1			D
					↓	↓	↓	↓			

Operation	Symbolic content of A	Sign test	q	OF	A_7	A_6	A_5	A_4	A_3	A_2	A_1	A_0
	$N = R_4$			0	1	1	0	1	0	0	1	1
$-D2^4$	$N - D2^4$	Neg		1	1	1	1	0	0	0	1	1
$+D2^4$	R_4			0	1	1	0	1	0	0	1	1
Left shift	$2R_4$			1	1	0	1	0	0	1	1	0
$-D2^4$	$2(R_4 - D2^3) = 2R_3$	Pos	$q_3 = 1$	0	1	0	1	1	0	1	1	q_3
Left shift	2^2R_3			1	0	1	1	0	1	1	q_3	0
$-D2^4$	$2^2(R_3 - D2^2) = 2^2R_2$	Pos	$q_2 = 1$	0	0	1	1	1	1	1	q_3	q_2
Left shift	2^2R_2			0	1	1	1	1	1	q_3	q_2	0
$-D2^4$	$2^3(R_2 - D2) = 2^3R_1$	Pos	$q_1 = 1$	0	0	0	0	0	1	q_3	q_2	q_1
Left shift	2^4R_1			0	0	0	0	1	q_3	q_2	q_1	0
$-D2^4$	$2^4(R_1 - D)$	Neg	$q_0 = 0$	1	0	0	1	0	q_3	q_2	q_1	q_0
$+D2^4$	$2^4R_1 = 2^4R_0 = 2^4R$			0	0	0	0	1	q_3	q_2	q_1	q_0

The result in Example 6.8 is $Q = 1110 = 14_{10}$ and $R = 1$. Checking, we get correctly

$$N = QD + R = (14_{10})(15_{10}) + 1 = 211_{10}.$$

In particular, the steps for the determination of q_3 and q_0 illustrate the use of the overflow position. Note that in the computations the sign bit of D was extended into the overflow position.

6.2 BASIC DIVISION ALGORITHM APPLIED TO MACHINE DIVISION

Example 6.9 We show the first steps of the fractional mechanization of the division algorithm for the case $r = 10$, $n = 4$, $N = (10^4 - 1)^2$, $D = 9999$. The use of two overflow positions is illustrated.

Operation	Symbolic content of A	Sign test	q	OF	OF	A_7	A_6	A_5	A_4	A_3	A_2	A_1	A_0
									9	9	9	9	D
									↓	↓	↓	↓	
Initial state	$R_4 \equiv N$			0	0	9	9	9	8	0	0	0	1
Subtract overflow check	$R_4 - D10_4$	Neg		9	9	9	9	9	9	0	0	0	1
Add restore A	R_4			0	0	9	9	9	8	0	0	0	1
Shift	$10R_4$			0	9	9	9	8	0	0	0	1	0
Subtract	$10(R_4 - D10^3)$	Pos		0	8	9	9	8	1	0	0	1	0
Subtract	$10(R_4 - 2D10^3)$	Pos		0	7	9	9	8	2	0	0	1	0
Subtract	$10(R_4 - 3D10^3)$	Pos		0	6	9	9	8	3	0	0	1	0
Subtract	$10(R_4 - 4D10^3)$	Pos		0	5	9	9	8	4	0	0	1	0
Subtract	$10(R_4 - 5D10^3)$	Pos		0	4	9	9	8	5	0	0	1	0
Subtract	$10(R_4 - 6D10^3)$	Pos		0	3	9	9	8	6	0	0	1	0
Subtract	$10(R_4 - 7D10^3)$	Pos		0	2	9	9	8	7	0	0	1	0
Subtract	$10(R_4 - 8D10^3)$	Pos		0	1	9	9	8	8	0	0	1	0
Subtract	$10(R_4 - 9D10^3)$	Pos		0	0	9	9	8	9	0	0	1	0
Subtract	$10(R_4 - 10D10^3)$	Neg	$q_3 = 9$	9	9	9	9	9	0	0	0	1	$\frac{0}{q_3}$
Correct add	$10(R_4 - 9D10^3) = 10R_3$			0	0	9	9	8	9	0	0	1	$\frac{0}{q_3}$
Shift	$10^2 R_3$			0	9	9	8	9	0	0	1	$\frac{0}{q_3}$	0
Subtract etc.	$10^2(R_3 - D10^2)$	Pos		0	8	9	8	9	1	0	1	$\frac{0}{q_3}$	0

This gives the leading digit of the quotient 9999 and starts the procedure for the next one.

6.2.2 Integral Mechanization of the Division Algorithm

In an integral accumulator we can do the equivalent of adding or subtracting D from the content of the accumulator by transmitting digits from the X-register to A, and by shifting digits in A we can do the equivalent of multiplying or dividing by r. From these operations we need to form terms $R_{i+1} - qr^i D$ for $q = 1, 2, \ldots, q_i + 1$ and test their signs. Assume that the partial remainder R_{i+1} has been determined and is in a $2n$-digit closed accumulator A. We first shift left in A by $2n - i$ positions, so that the digits in A represent an integer which is congruent to $R_{i+1} r^{2n-i}$ modulo $r^{2n} - 1$. We then subtract D and left-shift an additional i positions. Let R^* be the resultant digits in A. Then

$$R^* \equiv (R_{i+1} r^{2n-i} - D) r^i \pmod{r^{2n} - 1}.$$

Since $R_{i+1} r^{2n} \equiv R_{i+1} \pmod{r^{2n} - 1}$, we may write

$$R^* \equiv R_{i+1} - D r^i \pmod{r^{2n} - 1}.$$

We note that in an open accumulator this procedure will not work since, in this case, $R_{i+1} r^{2n} \equiv 0 \pmod{r^{2n}}$. The resultant R^* represents the partial remainder corresponding to an initial trial value for q_i of $q = 1$. We must be able to determine the sign from R^*. If we are to do this correctly from the leading digit, the following condition must be satisfied

$$\left| R_{i+1} - D r^i \right| < r^{2n-1}.$$

Furthermore, since we repeat the above process for successive values of $q = 1, 2, \ldots, q_i + 1$, we need to investigate the magnitude of each $R_{i+1} - q D r^i$. For $q = 1, 2, \ldots, q_i$,

$$0 \le R_{i+1} - q D r^i < r^{i+1} D - D r^i = (r-1) r^i D,$$

and for $q = q_i + 1$,

$$-D r^i \le R_{i+1} - q D r^i < 0.$$

Since $0 < D < r^n$, we can state that

$$\left| R_{i+1} - q D r^i \right| < r^{n+i+1}, \quad r > 2, \tag{6.9}$$

or

$$\left| R_{i+1} - q D r^i \right| < r^{n+i}, \quad r = 2. \tag{6.10}$$

The maximum value of i is $n - 1$, and it follows that for any base r, a $(2n + 1)$-digit accumulator will be adequate for distinguishing signs of partial remainders. For binary arithmetic, a $2n$-digit accumulator will be adequate. To ensure the correctness of the basic algorithm, we also check the sign of $N - r^n D$. We know only that $|N - r^n D| < r^{2n}$, and $2n$ digital positions are not adequate to determine the sign even in the binary case. We need a $(2n + 1)$-digit accumulator, one digital position more than we had previously assumed.

Starting with the integer R_{i+1} in a $(2n + 1)$-position closed integral accumulator, the next step of the algorithm calls for a closed left shift of $(2n + 1 - i)$ positions which is the same as shifting right i positions. The complement of the content of the X-register is then transmitted into A and the accumulator shifted left i positions. This left shift balances the equivalent right shift, so that $R_{i+1} - D r^i$ can be obtained and its sign tested. This sequence is repeated for successive values of q until the sign of $R_{i+1} - q D r^i$

6.2 BASIC DIVISION ALGORITHM APPLIED TO MACHINE DIVISION

is negative. The digit q_i is obtained as one less than the last count of the number of subtractions.

Example 6.10 We illustrate the algorithm for $n = 2$, $r = 10$, and a closed, $(2n + 1) = 5$-position accumulator with extension facilities. The dividend is 2254 and the divisor is 98. The subtraction is by means of the nine's complement of 98. The initial left shift is $2n + 1 - (n - 1) = 4$, or a circular right shift of one.

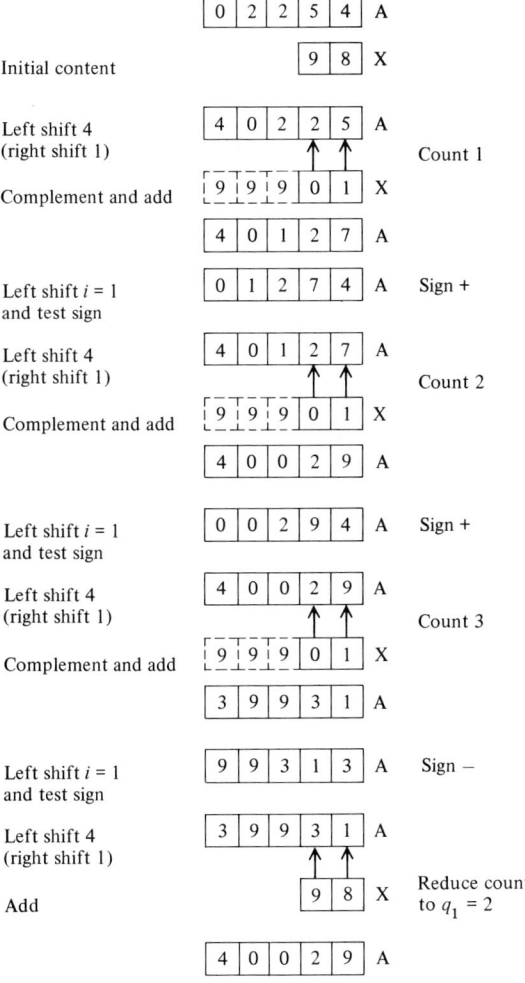

This sequence reflects the division of 2254 by 98 in the form

$$
\begin{array}{r}
2 \\
98\overline{)2254} \\
196 \\
\hline
294
\end{array}
$$

where we have obtained the first quotient digit two and (after a left shift of one) the next partial remainder 294.

At each step the right shift of i positions before the transmission from the X register to A is followed by a left shift of i positions. The last left shift places the sign digit in the highest order digital position. If the sign digit can be examined in a position other than the highest order, this left shift can be omitted, and the initial right shift before the *next* subtraction can also be omitted. Omission of the previous left shift, in effect, carries it out.

Once we have obtained a negative partial remainder from the process described above, we add Dr^i to correct for the superfluous subtraction. The last left shift prior to testing the sign must be balanced by an equal right shift before the corrective transmission. In order to obtain R_i we must follow this correction by another left shift of the same amount i. Symbolically:

Sign negative: $\quad$ A $\quad \boxed{R_{i+1} - (q_i + 1)Dr^i}$

Right shift of i: $\quad$ A $\quad \boxed{R_{i+1}r^{-i} - (q_i + 1)D}$

Add D from the X-register: $\quad$ A $\quad \boxed{R_{i+1}r^{-i} - q_iD}$

Left shift of i: $\quad$ A $\quad \boxed{R_{i+1} - q_iDr^i = R_i}$

When we have R_i, we can again proceed as above with a left shift of $2n + i - (i - 1)$ positions and so on. If we utilize detection of sign in other positions than the highest order, we can avoid much of the shifting back and forth, but a larger shift is necessary before the first step since there is no previous shift to use. This initial shift is left $(n + 2)$ positions, which is equivalent to a right shift of n positions, followed by a left shift. Employing this point of view permits a statement of the algorithm in a uniform way, that is, by prefacing each step, including the first, with a left shift of one position. No special shift will be required at the conclusion after R_0 is computed, since the value of i is zero at this point.

If we are to use this simplified version with less shifting, we will have to decide which digital position we should observe in order to know the sign of any result. We recall from the bound given in (6.9) that for $i = n - 1, n - 2, \ldots, 0$, we have, for a base $r > 2$,

$$|R_{i+1} - qDr^i| < r^{n+i+1}.$$

This means that in the digital expansion of these intermediate partial remainders, coefficients of powers of r with exponents greater than or equal to $n + i + 1$ can contribute nothing to the absolute value. Thus, they must all be equal to each other and to zero if the integer is nonnegative and to $r - 1$ if it is negative. Consequently, the coefficient of r^{n+i+1} carries all of the information about the sign of the content of A. Ordinarily, this digit would be $n + i + 2$ digits from the right in the full accumulator. Without the final left shift of i positions, it is i positions farther to the right, that is, the coefficient of r^{n+1}. For $r = 2$, the right side of (6.10) is r^{n+i}, and the sign in the binary case can be determined from the coefficient of r^n.

We summarize this last integral implementation of the basic division procedure, including steps necessary for the incorporation of an overflow check.

1. Start with N in the accumulator and D in the X-register.
2. Shift right n positions and subtract D. If the digit corresponding to r^{n+1} in A is zero, halt and indicate overflow. If it is $r - 1$, add D and proceed with the determination of the q_i.
3. For $i = n - 1, n - 2, \ldots, 0$ determination of q_i consists of the following:
   a) Shift left one position.
   b) Successively subtract D, checking the sign by observing the coefficient of r^{n+1} in A.
   c) When the coefficient of r^{n+1} becomes $r - 1$, add D and proceed to the next value of i.
   d) Store $q_i =$ one less than the total number of subtractions in the proper position in the quotient register.

At the conclusion of this algorithm R will be in the accumulator. For the binary case the sign determination can be made in the $2n$-digital position.

Example 6.10 To illustrate the procedure, we will take $r = 2$, $n = 4$, $N = 10000110$, and $D = 1111$ (see table below).

The results are $Q = 1000$ and $R = 1110$. This reflects in decimal $8(15) + 14 = 134 = N$. We note that as in the fractional case, for $r = 2$, the counting and correcting procedures can be simplified.

6.3 DIVISION WITH NEGATIVE OPERANDS

In multiplication, when we introduced negative factors, we noted that it was possible to apply the basic algorithm developed for positive operands to the absolute values of the factors to obtain the absolute value of the product. The sign can then be appended after an examination of the signs of the factors. In this section we will employ a similar procedure in division, when the dividend and divisor are not restricted to positive values. That is, we will replace negative dividends and divisors by their absolute values, and then employ the algorithm developed above to obtain a quotient and a remainder. We will then examine the signs of N and D to convert the quotient and remainder obtained from the absolute values to the actual quotient and remainder.

					D	1	1	1	1

Sign ↓ ↑ ↑ ↑ ↑

Purpose	Operation	Sign	q	Numerical content of A								
				OF	A_7	A_6	A_5	A_4	A_3	A_2	A_1	A_0
	Initial state			0	1	0	0	0	0	1	1	0
Overflow check	Shift right 4 positions			0	1	1	0	0	1	0	0	0
	Subtract	Neg	$q_i = 0$ $i \geq 4$	0	1	0	1	1	1	0	0	1
	Add			0	1	1	0	0	1	0	0	0
Determine q_3	Shift left			1	1	0	0	1	0	0	0	0
	Subtract	Pos	$q_3 = 1$	1	1	0	0	0	0	0	0	1
Determine q_2	Shift left			1	0	0	0	0	0	0	1	1
	Subtract	Neg	$q_2 = 0$	0	1	1	1	1	0	1	0	0
	Add			1	0	0	0	0	0	0	1	1
Determine q_1	Shift left			0	0	0	0	0	0	1	1	1
	Subtract	Neg	$q_1 = 0$	1	1	1	1	1	0	1	1	1
	Add			0	0	0	0	0	0	1	1	1
Determine q_0 and R	Shift left			0	0	0	0	0	1	1	1	0
	Subtract	Neg	$q_0 = 0$	1	1	1	1	1	1	1	1	0
	Add			0	0	0	0	0	1	1	1	0

In the previous discussion we have studied the following problem: Given integers $N \geq 0$ (dividend) and $D > 0$ (divisor), find integers Q (quotient) and R (remainder), such that

$$N = QD + R, \quad 0 \leq R < D.$$

We know from the division algorithm of Chapter 1 that there exist unique nonnegative Q and R solving this problem. When the signs of N and D are not restricted, the problem is reformulated in the following way: Given integers N and $D \neq 0$, find integers Q and R such that

$$N = QD + R, \quad |R| < |D|.$$

From Chapter 1 we know that this last problem does not have a unique solution unless the sign of R is specified in advance. The usual specification is that the sign of R is to be the same as the sign of N. The specification $R \geq 0$ has also been used. We will give results for both cases below.

Given general integers N and $D \neq 0$, we will replace them by $|N|$ and $|D|$. We may then apply the results of the division algorithm which assures the existence of unique integers Q^* and R^*, such that

$$|N| = Q^*|D| + R^*, \quad 0 \leq R^* < |D|.$$

In the following table we show how the quotient Q and remainder R for the original N and D can be obtained from Q^* and R^*.

Case		Remainder nonnegative		Sign of remainder = Sign of dividend	
Sign of N	Sign of D	R	Q	R	Q
+	+	R^*	Q^*	R^*	Q^*
−	+	$-R^* + \|D\|$	$-(Q^* + 1)$	$-R^*$	$-Q^*$
−	−	$-R^* + \|D\|$	$Q^* + 1$	$-R^*$	Q^*
+	−	R^*	$-Q^*$	R^*	$-Q^*$

It is clear that the results given in the table for the case of $R \geq 0$ do not apply for a negative dividend when $R^* = 0$. In this case the alternate results should be used.

Choosing the sign of the remainder the same as that of the dividend leads to a simpler algorithm, since all that need be done is to append signs to the results of the absolute-value division, and $R^* = 0$ is not special. This choice also simplifies matters if we wish to generate additional quotient digits by using the digits of the remainder and dividing again, since successive quotients will always have the *same* sign. A further disadvantage of the nonnegative remainder is that, in the case $N < 0$, the correction to Q^* may invalidate the overflow check.

If operands brought from storage to the arithmetic section are represented as absolute-value and sign numbers, we need only segregate the signs and apply the algorithms developed in Section 6.2 directly to the absolute values. If operands are represented in complement form, those which represent negative integers will have to be complemented. In all cases appropriate sign indicators will be saved. When complement numbers which retain a leading sign digit are converted to absolute value and sign, the leading digit of each operand will be zero. If we think of the dividend as the product of two such operands, two of its leading digits will be zero. In this case the earlier analysis concerning overflow positions can be revised. The overflow positions are not needed, since existing positions serve the purpose because of the leading zeros.

6.3.1 Algorithms for Complement Division

In this section we will assume that the sequences of digits used to represent N, D, Q, and R are given a complement interpretation; that is, the operands will represent either absolute values or complements of absolute values, depending on sign. The technique for developing a division algorithm valid for operands and results represented in this manner will be similar to that used in the absolute-value case. We will replace Q by the digits of its complement representation in the defining equation for division. We will then define a sequence of partial remainders which will satisfy inequalities which yield the desired remainder condition, provided that the digits of the representation of Q are properly specified.

In using complement representations, we shall assume that all basic operands are so restricted that the leading digit can serve as a sign digit. With this assumption, we can unambiguously treat each operand as representing the true value. The assumption that the quotient Q is representable as an n-digit complement with the leading digit serving as a sign digit is equivalent to the assumption that $|Q| < r^{n-1}$, which in turn implies that $|N| < r^{n-1}|D|$. To see this we need only note that

$$|N| = |QD + R| \le |Q||D| + |R| \le (r^{n-1} - 1)|D| + |R|,$$

and that

$$|R| < |D|.$$

On the other hand, if $|N| < r^{n-1}|D|$, and we define division so that the remainder has the same sign as the dividend, or is zero, then it must also be true that $|Q| < r^{n-1}$. If $N < 0$, then it follows from the assumption on $|N|$ that

$$-r^{n-1}|D| < QD + R < 0.$$

The condition on the remainder requires that $R = -|R|$ and we may rewrite the last inequality as

$$-r^{n-1}|D| + |R| < QD.$$

If we divide through by $|D|$ and keep in mind that $|R| < |D|$, we obtain

$$-r^{n-1} + |R|/|D| < \pm Q = -|Q| \quad \text{or} \quad |Q| < r^{n-1}.$$

The argument for the case $N \ge 0$, $R = |R|$, follows in a similar way. Thus, we may state the following theorem.

Theorem 6.1 *If the remainder is chosen to have the same sign as the dividend, or is zero, a necessary and sufficient condition that the quotient Q be representable as an n-digit complement is that*

$$|N| < r^{n-1}|D|. \tag{6.11a}$$

The test against quotient overflow provided by Theorem 6.1, namely, that $|N| - r^{n-1}|D| < 0$, is a simple one. It guarantees, as it should, that $|Q| < r^{n-1}$. Although (6.11a) remains a necessary condition when R is chosen from $0 \le R < |D|$, a stronger condition on $|N|$ is needed for sufficiency in this case.

Theorem 6.2 *If R is chosen so that $0 \le R < |D|$, $|Q| < r^{n-1}$ implies $|N| < r^{n-1}|D|$ and*

$$|N| \le (r^{n-1} - 1)|D| \tag{6.11b}$$

implies that $|Q| < r^{n-1}$.

The proof of necessity given for Theorem 6.1 holds here also. If $N \geq 0$ and R is chosen as above, then the hypotheses of Theorem 6.1 are satisfied and the conclusion of that theorem applies. If $N < 0$, then

$$N = -|N| = -|Q||D| + R \geq -(r^{n-1} - 1)|D|.$$

Therefore,

$$|Q||D| \leq (r^{n-1} - 1)|D| + R = r^{n-1}|D| + R - |D| < r^{n-1}|D|,$$

so that

$$|Q| < r^{n-1}.$$

The inability of condition (6.11a) to guarantee that there will be no quotient overflow when $N < 0$ and we choose $0 \leq R < |D|$, is shown by the following example. Suppose $n = 3$, $r = 10$, and we have $N = -199$ and $D = -2$. The dividend certainly satisfies (6.11a) since $199 < 10^2(2)$. On the other hand, for $0 \leq R < D$ we have $-199 = -2(100) + 1$, which gives a quotient $Q = 100 \not< 10^2$, and we have overflow. Condition (6.11b) is, of course, violated, since $(r^{n-1} - 1)|D| = 198 < 199 = |N|$.

The algorithms which we develop are for the digit-by-digit determination of Q, and we will state and prove two forms of the defining equation for division, in which the digits of the complement representation of Q appear explicitly. The forms of these equalities and the reason for the use of the forms are by no means obvious in advance. Their utility, however, will be apparent when we consider how to find the digits of Q. Let the quotient be represented by the sequence of digits $q_{n-1}q_{n-2}\ldots q_0$. In the case of the radix-complement, this form proves to be

$$N = \left[-\left(\frac{r}{r-1}q_{n-1} - 1\right)r^{n-1} + \sum_{i=0}^{n-2}(rq_i - (r-1))r^i - 1\right]Dr^{-1} + R. \tag{6.12}$$

The following equation is valid when the radix-less-one complement is to be used:

$$N = \left[-\left(\frac{r}{r-1}q_{n-1} - 1\right)r^{n-1} + \sum_{i=0}^{n-2}(rq_i - (r-1))r^i \right.$$

$$\left. + \left(\frac{r}{r-1}q_{n-1} - 1\right)\right]Dr^{-1} + R. \tag{6.13}$$

Equations (6.12) and (6.13) can be verified readily. Let $q_{n-1} = 0$. Then

$$|Q| = \sum_{i=0}^{n-1} q_i r_i = Q,$$

and both of the expressions in the square brackets in Equations (6.12) and (6.13) reduce to

$$r^{n-1} + \sum_{i=0}^{n-2} q_i r^{i+1} - \sum_{i=0}^{n-2} (r-1)r^i - 1$$

$$= r^{n-1} + r\sum_{i=0}^{n-1} q_i r^i - \sum_{i=0}^{n-2} (r-1)r^i - 1.$$

This last expression is seen to be equal to rQ, since

$$r^{n-1} - 1 = \sum_{i=0}^{n-2} (r-1)r^i.$$

Therefore, in the case $q_{n-1} = 0$, both (6.12) and (6.13) become $N = QD + R$, as required. For the case $q_{n-1} = r - 1$, $Q = -|Q|$, and for the r's complement,

$$|Q| = r^n - \sum_{i=0}^{n-1} q_i r^i,$$

while for the $(r-1)$'s complement,

$$|Q| = r^n - \sum_{i=0}^{n-1} q_i r^i - 1.$$

The expression in square brackets in Equation (6.12) becomes

$$-r^n + r\sum_{i=0}^{n-2} q_i r^i,$$

and the corresponding part of (6.13) becomes

$$-r^n + r\sum_{i=0}^{n-2} q_i r^i + r.$$

If we extend the summation to the term $(n-1)$ in each of the two expressions above, and compensate by subtracting the term $(r-1)r^n$ from each, we get, for the radix complement case,

$$r\sum_{i=0}^{n-1} q_i r^i - r^{n+1} = -r\left(r^n - \sum_{i=0}^{n-1} q_i r^i\right) = -r|Q| = rQ,$$

and for the radix-less-one complement case,

$$-r\left(r^n - \sum_{i=0}^{n-1} q_i r^i - 1\right) = -r|Q| = rQ.$$

Therefore, both (6.12) and (6.13) reduce to the required form $N = QD + R$.

We now consider the determination of the digits q_i, using either (6.12) or (6.13) appropriate to the modulus employed, together with the results of Theorems 6.1 and 6.2. The technique will be to devise a sequence of partial remainders defining the q_i in a manner similar to that used in the absolute value and sign case. We define the sequence of partial remainders obtained by successively transposing terms to the left in either of the expressions (6.12) or (6.13).

$$\begin{aligned} R_n &\equiv N, \\ R_{n-1} &= R_n + \left(\frac{r}{r-1} q_{n-1} - 1\right) r^{n-2} D, \\ R_{n-2} &= R_{n-1} - [rq_{n-2} - (r-1)] r^{n-3} D, \\ &\vdots \\ R_i &= R_{i+1} - [rq_i - (r-1)] r^{i-1} D, \\ &\vdots \\ R_0 &= R_1 - [rq_0 - (r-1)] r^{-1} D. \end{aligned} \quad (6.14)$$

The partial remainder R_0 is not in general equal to the true remainder R. For radix complementation we have, from (6.12).

$$R = R_0 + Dr^{-1} \qquad (6.15)$$

and for radix-less-one complementation, from (6.13),

$$R = R_0 - \left(\frac{r}{r-1} q_{n-1} - 1\right) Dr^{-1}. \qquad (6.16)$$

Before investigating the properties of the remainders given by formulas (6.15) and (6.16), we first consider the problem of choosing the digits q_i which appear in the partial-remainder formulas (6.14). Since q_{n-1} is to determine the sign of Q, we have an obvious rule for its specification.

1. If R_n and D have the same sign, set $q_{n-1} = 0$.
2. If R_n and D have opposite signs set $q_{n-1} = r - 1$.

With q_{n-1} chosen in this manner, we have

$$-(r-1)r^{n-2}|D| \le R_{n-1} < (r-1)r^{n-2}|D|.$$

However, if we take all possible cases of the signs of R_n and D separately, we can get somewhat sharper inequalities. We summarize them below.

Case	Inequality on R_{n-1}
$R_n \geq 0, D > 0$	$-r^{n-2}\|D\| \leq R_{n-1} < (r-1)r^{n-2}\|D\|$
$R_n < 0, D < 0$	$-(r-1)r^{n-2}\|D\| < R_{n-1} < r^{n-2}\|D\|$
$R_n \geq 0, D < 0$	$-(r-1)r^{n-2}\|D\| \leq R_{n-1} < r^{n-2}\|D\|$
$R_n < 0, D > 0$	$-r^{n-2}\|D\| < R_{n-1} < (r-1)r^{n-2}\|D\|$

In the case in which R_n is nonnegative we have

$$0 \leq R_n < r^{n-1}|D|, \tag{6.17}$$

which we assumed in order to ensure that the quotient Q can be correctly represented in complement form. If $D > 0$, then $q_{n-1} = 0$ and $D = |D|$. Therefore,

$$\left(\frac{r}{r-1} q_{n-1} - 1\right) r^{n-2} D = -r^{n-2} D = -r^{n-2}|D|.$$

Since, in this case, $R_{n-1} = R_n - r^{n-2}|D|$, addition of (6.17) gives

$$-r^{n-2}|D| \leq R_{n-1} < (r-1)r^{n-2}|D|.$$

This establishes the first entry in the table above. If R_n and D are of opposite sign and R_n is nonnegative, we have $q_{n-1} = r - 1$, $D = -|D|$, and $0 \leq R_n < r^{n-1}|D|$, so that

$$\left(\frac{r}{r-1} q_{n-1} - 1\right) r^{n-2} D = -(r-1)r^{n-2}|D|$$

and

$$-(r-1)r^{n-2}|D| \leq R_{n-1} < r^{n-2}|D|.$$

This establishes the result for the case $R_n \geq 0, D < 0$. The remaining cases are readily deduced from similar computations, which we leave to the reader.

The selection rules for the remaining q_i are defined so that the corresponding partial remainders R_i satisfy not only the inequality

$$-(r-1)r^{i-1}|D| \leq R_i < (r-1)r^{i-1}|D|,$$

but also the analogs of the sharper forms tabulated for $i = n - 1$. For each q_i these inequalities take the following form.

Case	Inequality on R_i
$R_{i+1} \geq 0, D > 0$	$-r^{i-1}\|D\| \leq R_i < (r-1)r^{i-1}\|D\|$
$R_{i+1} < 0, D < 0$	$-(r-1)r^{i-1}\|D\| < R_i < r^{i-1}\|D\|$
$R_{i+1} \geq 0, D < 0$	$-(r-1)r^{i-1}\|D\| \leq R_i < r^{i-1}\|D\|$
$R_{i+1} < 0, D > 0$	$-r^{i-1}\|D\| < R_i < (r-1)r^{i-1}\|D\|$

We turn now to the implications that selection of the q_i in this manner has for the remainder R. If the q_i are selected in turn, so that the inequalities of the table hold, we finally arrive at R_0 satisfying corresponding conditions with $i = 0$. We then use R_0 to compute the remainder R from (6.15) or (6.16). It is of utmost importance to check that the condition $|R| < |D|$ is satisfied. We first consider the case of radix complementation, in which R is given by (6.15). Assume that R_1 and D are both negative. Then $D = -|D|$, and from the table, with $i = 0$, we have

$$-(r-1)r^{-1}|D| < R_0 < r^{-1}|D|.$$

Adding $Dr^{-1} = -|D|r^{-1}$ to this inequality to complete formula (6.15), we get $-|D| < R < 0$. Proceeding in a similar manner for the other three cases, we construct the following table.

Case	Remainder condition
$R_1 \geq 0, D > 0$	$0 \leq R < |D|$
$R_1 < 0, D < 0$	$-|D| < R < 0$
$R_1 \geq 0, D < 0$	$-|D| \leq R < 0$
$R_1 < 0, D > 0$	$0 < R < |D|$

We note that, for radix complementation, R turns out to have the same sign as D. Thus, the sign of the remainder can be determined in advance. If we impose uniqueness by the remainder condition $0 \leq R < |D|$, we need make no correction when D is positive, but must correct when D is negative by adding $|D|$ (subtracting D) to form R and adjusting by one to obtain Q. This will take care of all cases in which the nonnegative remainder condition is not at first satisfied. The sign of N need not be remembered. However, if we impose uniqueness by selecting the solution in which R has the same sign as N, we must compare the sign of N with the sign of D. If they are opposite, we adjust R by adding $|D|$ when D is negative and subtracting $|D|$ when D is positive. Of course, Q must also be changed to compensate for the adjustment of the remainder. Unfortunately, even after we make the indicated adjustment, the remainder which results may equal $\pm |D|$ and so not satisfy the remainder condition. Therefore, it seems preferable in the case of radix complementation to obtain uniqueness by the remainder condition $0 \leq R < |D|$.

Using formula (6.16) for R, we can make a similar analysis for the remainder in the radix-less-one complement case. For example, let $R_1 \geq 0$, $D > 0$. Again we use the appropriate inequality for R_i from the table with $i = 0$. There are two subcases, $N \geq 0$ and $N < 0$. If $N \geq 0$, we must have $q_{n-1} = 0$. Then

$$-\left(\frac{r}{r-1} q_{n-1} - 1\right) Dr^{-1} = Dr^{-1} = |D|r^{-1}.$$

Addition of this to the inequality on R_0, gives $0 \le R < |D|$. If $N < 0$, then $q_{n-1} = r - 1$ and so

$$-\left(\frac{r}{r-1} q_{n-1} - 1\right) Dr^{-1} = -(r-1)|D|r^{-1}.$$

Addition of this term to the inequality on R_0 yields $-|D| \le R < 0$. Proceeding in a similar manner for each of the three remaining cases, we derive six more inequalities. The results are summarized in the table below.

R_1	N, D	Remainder condition
$R_1 \ge 0$	$N \ge 0, D > 0$	$0 \le R < |D|$
$R_1 < 0$	$N \ge 0, D > 0$	$0 < R < |D|$
$R_1 < 0$	$N \ge 0, D < 0$	$0 < R < |D|$
$R_1 \ge 0$	$N \ge 0, D < 0$	$0 \le R < |D|$
$R_1 \ge 0$	$N < 0, D > 0$	$-|D| \le R < 0$
$R_1 < 0$	$N < 0, D > 0$	$-|D| < R < 0$
$R_1 < 0$	$N < 0, D < 0$	$-|D| < R < 0$
$R_1 \ge 0$	$N < 0, D < 0$	$-|D| \le R < 0$

We note that for radix-less-one complementation the remainder will have the same sign as the dividend. Thus, the sign of the remainder can be determined in advance. Unfortunately, there are cases in which $R = -|D|$ is possible, which violates the general remainder condition. If we impose uniqueness by the condition $R \ge 0$, we must add $|D|$ to form R whenever $N < 0$ and adjust by one to obtain Q. This will eliminate the case $R = -|D|$.

For either type of complementation we can determine the sign of the remainder *in advance*. Hence, we know which of the two solutions to the division problem we will get directly, before any remainder correction and corresponding quotient adjustment is made. If we choose the solution corresponding to $R \ge 0$, the general remainder condition is always satisfied. However, if we choose the solution in which the remainder is to have the same sign as the dividend, cases can arise in which $|R| = |D|$ and the general remainder condition will not be satisfied. Thus, in contrast to the absolute-value and sign case, in which it is preferable to choose the sign of the remainder to be the same as that of the dividend, in the complement case the condition $R \ge 0$ appears to be the preferred one.

It is easy to see that the condition $|R| = |D|$ arises when N is equal to an integral multiple of D. In this case there is a unique solution corresponding to $R = 0$. If we insist that the sign of R be the same as that of N when N is negative, then we must represent R as negative zero. This situation can be avoided by comparison of R and D at the conclusion of the division process, and correction and adjustment when they are found to be equal in absolute value.

We should bear in mind that if we choose $R \geq 0$, the condition $|R_n| < r^{n-1}|D|$ may not imply $|Q| < r^{n-1}$; that is, the quotient adjustment necessitated by a remainder correction may lead to a division overflow. However, Theorem 6.2 provides an alternative condition sufficient to ensure that this will not occur. Since Q is formed by storing its digits one at a time, as they are determined by the selection rules we will set down below, it can be retained in an ordinary storage register. The process of adjusting Q may require the use of a register with accumulative properties. This could prove to be inconvenient, since the accumulator itself will be occupied by partial remainders and the remainder. To avoid this inconvenience we can carry out the adjustment in advance by appropriately modifying the dividend N before starting the division procedure. We will base this on the fact that we know in advance when adjustment of Q is required.

The remainder and the quotient are to be corrected when R is negative. For r's complementation we know beforehand that R will be negative if the divisor D is negative and for $(r-1)$'s complementation if the dividend N is negative. Assume that we have either of these cases and that $N = QD + R$. If we subtract $|D|$ from both sides, we have

$$N - |D| = (QD - |D|) + R$$
$$= (Q \pm 1)D + R.$$

According to this last equation, if in place of N we start with the adjusted dividend $N - |D|$, we will, on division by D, obtain directly the quotient Q adjusted by one and the original remainder R. This remainder is incorrect and must be corrected by the addition of $|D|$. We note that the correction of the dividend by $-|D|$ does not affect the sign of the divisor D and that if $N < 0$, $N - |D| < 0$. Therefore, making this change before dividing will not change the conditions which produce a negative remainder. Thus, the division of $N - |D|$ by D will produce the same remainder as the division of N by D. This justifies the grouping of $-|D|$ with QD, instead of with R, in the equation above. The fact that the process produces the correct quotient adjustment in all cases follows from

$$N = (QD - |D|) + (R + |D|)$$
$$= (Q \pm 1)D + (R + |D|).$$

Here we have added and subtracted $|D|$ on the right in the original division equation. This amounts to first making the correction to the remainder and then adjusting the quotient.

We note that for $N < 0$ the choice of $0 \leq R < |D|$, the more stringent overflow test (6.11b) should be used. In this case, the corrected dividend N' is $N - |D| = -|N| - |D|$ and the condition states that

$$|N'| = |D| + |N| \leq r^{n-1}|D|.$$

Thus, except for possible equality, the simpler test (6.11a) can be used.

We note that in formulas (6.14) the partial remainder R_0 is expressed in terms of r^{-1}, as is the remainder R in formulas (6.15) and (6.16). In all of the other expressions for partial remainders only nonnegative powers of the base appear. Since we are doing integral arithmetic, it is not convenient to deal with both negative and nonnegative powers of r in the evaluation of R. We can eliminate the nonnegative powers of the base by combining the last two steps in which R_0 and R are evaluated, that is, by proceeding directly from R_1 to R. Combination of first (6.14) and (6.15), and second (6.14) and (6.16), yields, respectively,

$$R = R_1 - (q_0 - 1)D$$

and

$$R = R_1 - (q_0 - 1)D - (q_{n-1}/(r - 1))D.$$

We will use these last two expressions to evaluate R. This eliminates negative powers of the base, but makes the last step in the evaluation of partial remainders different in form from intermediate steps.

We have defined a sequence of partial remainders involving each q_i, and we have imposed certain conditions on these remainders.

We will now specify rules for the selection of q_i so that these conditions hold for $i = n - 2, \ldots, 0$. We have already given the rule for selecting q_{n-1} and shown that the inequalities of the conditions hold for $i = n - 1$. The selection rule for q_i, $i = n - 2, n - 1, \ldots, 0$, is the following:

1. If R_{i+1} and D have the same sign, subtract Dr^i successively from R_{i+1} until a change of sign occurs. The number of subtractions gives the value of q_i.
2. If R_{i+1} and D have opposite signs, set $q_i = 0$.
3. In either case, to obtain R_i for the next step, add Dr^i and then subtract Dr^{i-1}.

In establishing the validity of these rules, we assume that digits q_i have been determined so that the partial remainders R_i satisfy the appropriate inequalities for $i = n - 2, n - 3, \ldots, j + 1$. We then show that this implies that R_j will also satisfy the inequalities of the table if q_j is chosen according to the rule just set down. There are four cases to consider. We will outline the details for one of them. The remaining cases can be handled in a similar manner.

Let R_{j+1} and D have the same sign and both be nonnegative. Then we have $D = |D|$ and $R_{j+1} \geq 0$. From the assumptions about R_{j+1} we have

$$0 \leq R_{j+1} < (r - 1)r^j|D|. \tag{6.18}$$

6.3 DIVISION WITH NEGATIVE OPERANDS

Since we have subtracted $r^j D$ from R_{j+1} exactly q_j times to cause a change of sign, and since $R_{j+1} - (q_j - 1)r^j D \geq 0$ we have the following:

$$-r^j |D| \leq R_{j+1} - q_j r^j D < 0. \tag{6.19}$$

From inequality (6.18), $1 \leq q_j \leq r - 1$ and is a digit base r. If we add $r^j D$ to and subtract $r^{j-1} D$ from all terms in (6.19), we have

$$-r^{j-1}|D| \leq R_j < (r-1)r^{j-1}|D|,$$

which is the appropriate inequality.

We now present some examples for the case $r = 10$, $n = 3$. In each example, ten's complements will be used, as will the remainder condition $0 \leq R < |D|$. The quotient overflow check will be assumed but will not be shown.

Example 6.12 Let $N = 001501$ and $D = 050$.

i	Operation	Quantity calculated	Numerical calculation	Sign	q^i	Count
3	Initial state	R_3	001501	+		
2	$-10D$		999500		0	
		R_2	001001	+		
1	$-10D$		999500			
			000501	+		1
	$-10D$		999500			
			000001	+		2
	$-10D$		999500			
			999501	−		3
	$+10D$		000500		3	
			000001			
	$-D$		999950			
		R_1	999951	−		0
0	$+D$		000050		0	
		R	000001			

We find $Q = 030$ and $R = 1$, corresponding to $(030)(050) + 1 = 001501 = N$.

Example 6.13 We will redo Example 6.12 for the case $N < 0$, $D > 0$, that is, for $N = -1501$, $D = 050$.

178 DIVISION 6.3

i	Operation	Quantity calculated	Numerical calculation	Sign	q_i	Count
3	Initial state	R_3	998499	—		
2	$+10^2 D$		005000		9	
			003499			
	$-10D$		999500			
		R_2	002999	+		
1	$-10D$		999500			
			002499	+		1
	$-10D$		999500			
			001999	+		2
	$-10D$		999500			
			001499	+		3
	$-10D$		999500			
			000999	+		4
	$-10D$		999500			
			000499	+		5
	$-10D$		999500			
			999999	—		6
	$+10D$		000500		6	
			000499			
	$-D$		999950			
		R_1	000449	+		
0	$-D$		999950			
			000399	+		1
	$-D$		999950			
			000349	+		2
	$-D$		999950			
			000299	+		3
	$-D$		999950			
			000249	+		4
	$-D$		999950			
			000199	+		5
	$-D$		999950			
			000149	+		6
	$-D$		999950			
			000099	+		7
	$-D$		999950			
			000049	+		8

6.3 DIVISION WITH NEGATIVE OPERANDS

i	Operation	Quantity calculated	Numerical calculation	Sign	q_i	Count
	$-D$		999950			
			999999	$-$		9
	$+D$		000050		9	
		R	000049			

The result is 969 corresponding to the quotient $Q = -031$ and $R = 049$, so that $(-031)(050) + 049 = -1501 = N$.

Example 6.14 We now apply the algorithm to the case $N = +1501$, $D = -050$, illustrating the *a priori* quotient adjustment.

i	Operation	Quantity calculated	Numerical calculation	Sign	q_i	Count
3	Initial state	R_3	001501	$+$		
	Adjust quotient		999950			
		$R_3 - \|D\|$	001451	$+$		
2	$+10^2 D$		995000		9	
			996451			
	$-10D$		000500			
		R_2	996951	$-$		
1	$-10D$		000500			
			997451	$-$		1
	$-10D$		000500			
			997951	$-$		2
	$-10D$		000500			
			998451	$-$		3
	$-10D$		000500			
			998951	$-$		4
	$-10D$		000500			
			999451	$-$		5
	$-10D$		000500			
			999951	$-$		6
	$-10D$		000500			
			000451	$+$		7
	$+10D$		999500		7	
			999951			
0	$+D$		000050		0	

(continued)

(Example 6.14 *continued*)

i	Operation	Quantity calculated	Numerical calculation	Sign	q_i	Count
		R_1	000001	+		
0	$+D$		999950		0	
			999951	−		
	Adjust remainder		000050			
			000001	+		

The result is 970 corresponding to $Q = -030$ and $R = 1$ so that

$$(-030)(-050) + 1 = 001501 = N.$$

Example 6.15 As a final decimal example, we take $N = -001501$, and $D = -050$ using nine's complements.

i	Operation	Quantity calculated	Numerical calculation	Sign	q_i	Count
3	Initial state	R_3	998498	−		
	Adjust quotient		999949			
2		$R_3 - |D|$	998448	−	0	
	$-10D$		000500			
		R_2	998948	−		
1	$-10D$		000500			
			999448	−		1
	$-10D$		000500			
			999948	−		2
	$-10D$		000500			
			000449	+		3
	$+10D$		999499		3	
			999948			
	$-D$		000050			
		R_1	999998	−		
0	$-D$		000050			
			000049	+		1
	$+D$		999949		1	
		$R_1 = R$	999998	−		
	Adjust remainder		000050			
		$R_1 + |D| = R$	000049	+		

The result is $Q = 031, R = 49$, so that $(031)(-050) + 49 = -001501 = N$.

When $r = 2$, the selection rules for q_i and the rules for computing partial remainders can be stated in a simpler way. Since $q_i = 0$ or $q_i = 1$, we know immediately that when R_{i+1} and D have the same sign, $q_i = 1$ and so a sign change would be achieved by exactly one subtraction of $r^i D$. The next step is an addition of $r^i D$, and the subtraction and addition annul each other and need not be carried out, so that the only step in determining R_i is the subtraction of $r^{i-1} D$. When R_{i+1} and D are of opposite sign and $r = 2$, the correction needed to determine R_i is of the form

$$r^i D - r^{i-1} D = (r - 1) r^{i-1} D = r^{i-1} D.$$

Thus, we add r^{i-1} to R_{i+1} in order to get R_i.

We now summarize the complement division algorithm for the binary case. We select the solution for which $R \geq 0$, and assume that the quotient overflow test has been made and no overflow is indicated.

Binary Complement Division Algorithm

1. QUOTIENT ADJUSTMENT
    a) For two's-complement arithmetic, if $D > 0$, no adjustment is needed. If $D < 0$, subtract $|D|$ from N.
    b) For one's-complement arithmetic, if $N \geq 0$, no adjustment is needed. If $N < 0$, subtract $|D|$ from N.

2. DETERMINATION OF q_{n-1}, R_{n-1}
    a) If N and D have the same sign, set $q_{n-1} = 0$. Form R_{n-1} by subtracting $2^{n-2} D$ from $R_n = N$.
    b) If N and D have opposite signs, set $q_{n-1} = 1$. Form R_{n-1} by adding $2^{n-2} D$ to $R_n = N$.

3. DETERMINATION OF q_i $(i = n - 2, \ldots, 0)$
    a) If R_{i+1} and D have the same sign, set $q_i = 1$.
    b) If R_{i+1} and D have opposite signs, set $q_i = 0$.

4. DETERMINATION OF R_i $(i = n - 2, \ldots, 1)$
    a) If $q_i = 1$, form R_i by subtracting $2^{i-1} D$ from R_{i+1}.
    b) If $q_i = 0$, form R_i by adding 2^{i-1} to R_{i+1}.

5. DETERMINATION OF R FROM R_1
    a) For two's-complement arithmetic, do nothing when $q_0 = 1$ and add D to R_1 when $q_0 = 0$. If $D > 0$, this yields $R \geq 0$. If $D < 0$, adjust by adding $|D|$ to obtain the remainder.
    b) For one's-complement arithmetic, obtain R by adding D to R_1 when $q_0 = q_{n-1} = 0$, and subtracting D from R_1 when $q_0 = q_{n-1} = 1$. If $q_0 \neq q_{n-1}$, $R = R_1$. If $N \geq 0$, no remainder adjustment is necessary. If $N < 0$, add $|D|$ to obtain a nonnegative remainder.

182 DIVISION 6.3

Part (2) of the algorithm can be incorporated into (3) and (4), that is, we can eliminate (2) and extend the index i to $n - 1$ in (3) and (4) by adding the following statement at the conclusion:

Complement the digit obtained for $i = n - 1$ to get q_{n-1}.

We present a number of binary examples using two's complements. We take $n = 5$, $|N| = 0001111001_2 = 121_{10}$, $|D| = 01010_2 = 10_{10}$.

Example 6.16 We first take the case $N > 0$, $D > 0$.

i	Operation	Quantity calculated	Numerical calculation	Sign	q_i
4	Initial state	R_5	0001111001	+	0
	$-2^3 D$		1110110000		
3		R_4	0000101001	+	1
	$-2^2 D$		1111011000		
2		R_3	0000000001	+	1
	$-2D$		1111101100		
1		R_2	1111101101	−	0
	$+D$		0000001010		
0		R_1	1111110111	−	0
	$+D$		0000001010		
		R	0000000001		

The result is $Q = 01100_2 = 12_{10}$, $R = 1$, corresponding in decimal to $(12)(10) + 1 = 121 = N$.

Example 6.17 Let $N < 0$ and $D > 0$.

i	Operation	Quantity calculated	Numerical calculation	Sign	q_i
4	Initial state	R_5	1110000111	−	1
	$+2^3 D$		0001010000		
3		R_4	1111010111	−	0
	$+2^2 D$		0000101000		
2		R_3	1111111111	−	0
	$+2D$		0000010100		
1		R_2	0000010011	+	1
	$-D$		1111110110		
0		$R_1 = R$	0000001001	+	1

The result is seen to be 10011_2 corresponding to $-01101_2 = -13_{10}$, and

6.3 DIVISION WITH NEGATIVE OPERANDS

$R = 01001_2 = 9_{10}$, corresponding in decimal to

$$(-13)(10) + 9 = -121 = N.$$

Example 6.18 We now consider the case for $D < 0$, and $N < 0$, in which both a quotient and remainder adjustment will be necessary.

i	Operation	Quantity calculated	Numerical calculation	Sign	q_i
4	Initial state	R_5	1110000111		
	Adjust quotient		1111110110		
		$R_5 - \|D\|$	1101111101	$-$	0
	$-2^3 D$		0001010000		
3		R_4	1111001101	$-$	1
	$-2^2 D$		0000101000		
2		R_3	1111110101	$-$	1
	$-2D$		0000010100		
1		R_2	0000001001	$+$	0
	$+D$		1111110110		
0		R_1	1111111111	$-$	1
	Adjust remainder		0000001010		
		$R = R_1 + \|D\|$	0000001001	$+$	

The result is $Q = 01101_2 = 13_{10}$ and $R = 01001_2 = 9_{10}$, corresponding in decimal to $(13)(-10) + 9 = -121 = N$.

Example 6.19 To complete the examples in two's-complement arithmetic, we consider the case $N \geq 0$, $D < 0$.

i	Operation	Quantity calculated	Numerical calculation	Sign	q_i
4	Initial state	R_5	0001111001		
	Adjust quotient		1111110110		
		$R_5 - \|D\|$	0001101111	$+$	1
	$+2^3 D$		1110110000		
3		R_4	0000011111	$+$	0
	$+2^2 D$		1111011000		
2		R_3	1111110111	$-$	1
	$-2D$		0000010100		

(continued)

(Example 6.19 *continued*)

i	Operation	Quantity calculated	Numerical calculation	Sign	q_i
1		R_2	0000001011	+	0
	$+D$		1111110110		
0		R_1	0000000001	+	0
	$+D$		1111110110		
			1111110111	−	
	Adjust remainder		0000001010		
		R	0000000001	+	

The result is 10100_2 giving $Q = -01100_2 = -12_{10}$ and $R = 00001$, corresponding in decimal to $(-12)(-10) + 1 = 121 = N$.

Example 6.20 We consider one binary example for one's-complement arithmetic with $N < 0, D > 0$.

i	Operation	Quantity calculated	Numerical calculation	Sign	q_i
4	Initial state	R_5	1110000110	−	
	Adjust quotient		1111110101		
		$R_5 - \|D\|$	1101111100	−	1
	$+2^3 D$		0001010000		
3		R_4	1111001100	−	0
	$+2^2 D$		0000101000		
2		R_3	1111110100	−	0
	$+2D$		0000010100		
1		R_2	0000001001	+	1
	$-D$		1111110101		
0		R_1	1111111110	−	0
	Adjust remainder		0000001010		
		$R = R_1 + \|D\|$	0000001001	+	

The result is 10010_2 to give $Q = -01101_2 = -13_{10}$ and $R = 01001_2 = 9_{10}$ corresponding in decimal to $(-13)(10) + 9 = -121 = N$.

The algorithms which we have developed in this section can be implemented for either the general base r or for the special binary case by means of techniques we have already considered in Sections 6.2.1 and 6.2.2. We could use either a fractional or an integral accumulator. Since the operations involved consist of shifting remainders right or left, additions and subtractions of Dr^k, and testing of signs, nothing new is required. The only differences involved concern the order of taking specific steps. For this reason we will not consider the mechanization of complement division in detail. Some of the particular aspects we will leave to the reader to devise in the exercises.

EXERCISES

SECTION 6.1

6.1 Show that the equation $x = x(2 - Nx)$ has a solution $x = 1/N$. Thus, show that the iteration $x_{n+1} = x_n(2 - Nx_n)$ may produce successive approximations to the reciprocal of N with only additions (subtractions) and multiplications.

6.2 Represent the iteration in Exercise 6.1 in binary and apply the iteration to find successive binary approximations to $\frac{1}{3}$. Use $x_0 = (0.1)_2$ and carry the iteration to x_4.

6.3 Compare the results of Exercise 6.2 with the result of dividing to form $(\frac{1}{11})_2$.

6.4 Use the approximate reciprocals computed in Exercise 6.2 to approximate $(\frac{5}{3})_{10}$ in binary. Compare the results with the binary expansion of the fraction.

6.5* Find a useful range of values for x_0 such that the iteration of Exercise 6.1 will converge to $1/N$.

6.6 Find the binary expansion of $(\frac{5}{3})_{10}$ by successive subtractions using complements. The answer should involve an integral quotient and remainder. What steps are necessary to continue this process?

SECTION 6.2

6.7 Relate the process of checking values of q as potential values of q_{n-1} to ordinary long division. What do the partial remainders in (6.6) correspond to in division?

6.8 Apply the results of Exercise 6.7 to find the quotient and remainder in dividing 12345 by 267 in decimal. Identify each trial quotient digit and each partial remainder.

6.9 If the digit-by-digit determination of the quotient Q is carried out by testing each value of q until the correct one is reached (instead of making a shrewd guess as in long division) what advantages are there to using binary?

6.10 If the condition (6.8) is violated because $D = 0$, what result would the algorithm produce? Is there any way one can detect division overflow or violation of (6.8) without testing (6.8) in advance?

SECTION 6.2.1

6.11 Redo Example 6.5 under the assumption that the accumulator is subtractive.

6.12 The fractional accumulator acquires its name from the fact that the sequences of digits involved are often interpreted as fractions. Are the algorithms described correct if the base point is assumed to be immediately to the right of the first digit? Consider in particular this interpretation for Example 6.8 and Example 6.9.

SECTION 6.2.2

6.13 Do Example 6.10 for a fractional accumulator and compare the step-by-step process involved with that in the example.

6.14 Carry out and show all steps of the integral division algorithm described for the decimal case with $n = 4$, $N = 13254691$, and $D = 8642$. Is the overflow position actually required in this case?

6.15* What bounds, if any, would guarantee correct results in the integral mechanization if no overflow position is provided? Can you find a range of values for N and D which of necessity require the overflow position?

SECTION 6.3

6.16 Justify the results given in the first table of Section 6.3.

6.17 In the general case where $N < 0$ and $D > 0$, how does the basic algorithm for $N \geq 0$ apply? Why does the case of a zero remainder need to be considered separately?

6.18 Whether or not operands are represented as absolute value and sign or as complements, the possible combinations of sign for N and D are $(\pm)$ and $(\pm)$. If binary digits are used in each case to represent plus or minus, what arithmetic operation determines the sign of the quotient? Is this sign altered by the choice of the range for the remainder?

SECTION 6.3.1

6.19 Devise a counterexample, base two, for the sufficiency of condition (6.11a) in case $0 \leq R < |D|$.

6.20* Since condition (6.11b), that

$$N \leq (r^{n-1} - 1)|D|,$$

guarantees (6.11a) that

$$N < r^{n-1}|D|,$$

why is the stronger condition in Theorem 6.2 needed?

6.21* What are the digital implications of Example 6.19 for $r = 2$? That is, if

$$N < (100\underbrace{\ldots 000}_{n})|D|$$

why is it not necessarily true that

$$N \leq (\underbrace{111\ldots 111}_{n-1})|D|,$$

or is it?

6.22 Complete the justification of the remaining entries in the table preceding condition (6.17).

6.23* In the radix-complement case, is there any way of determining whether the remainder is $\pm|D|$, and if so, is there any way of determining the necessary corrections?

6.24* In the $(r-1)$'s-complement case, when $|R| = |D|$ and we must choose $R = 0$, is there any difficulty with the choice of negative zero if the dividend is negative?

6.25* What modifications, if any, are required (complement division) in the mechanization of division according to the binary case given in Section 6.2.1 and 6.2.2?

6.26* Generalize the results of Exercise 6.25 to base r.

CHAPTER 7

FIXED AND FLOATING-POINT ARITHMETIC; SCALING

7.1 INTRODUCTION

In Chapter 1 we pointed out that a complete arithmetic computation on two operands represented by a finite number of digits consists of three parts. These are:

1. The digital algorithm for obtaining the digits of the result;
2. The application of the appropriate rule of signs; and
3. The determination of the correct location of the base point of the result.

For each of the four arithmetic operations we have already considered ways to obtain the appropriate result. We have also shown how, by means of complement arithmetic, the second part of the computation can be included in the first. In this chapter we discuss techniques used for the third stage of a calculation, that is, locating the base point.

Before initiating a formal approach to this problem, we will examine the way in which we usually treat it in long-hand decimal arithmetic. Suppose that we have two operands X and Y, each consisting of a finite sequence of decimal digits with a decimal point located somewhere relative to the sequence. Since algebraic sign can be treated by complement arithmetic, we concern ourselves here only with nonnegative operands, and will assume for the moment that the result of the operation is also nonnegative. The rules which we usually apply to locate the decimal point are probably something like the following.

1. For addition and subtraction, align the decimal points in X and Y and use the same position to locate the point in $S = X + Y$ or $D = X - Y$.
2. For multiplication, count the number of digits to the right of the decimal point in each of X and Y. Let these counts be s and t, respectively. To locate the decimal point in $P = XY$ place it so that there are $s + t$ digits to its right.
3. For division, count the number of digits to the right of the decimal point in the divisor Y. Count the same number of places to the right of the decimal point in the dividend X. This locates the decimal point in $Q = X/Y$, if each quotient digit is positioned in the proper place.

Example 7.1 We wish to form the sum and difference of the two operands (decimal) $X = 1.25$ and $Y = 0.625$. For the sum $X + Y$ and the difference $X - Y$, we align the points, and proceed in the following way:

$$
\begin{array}{cccc}
X & 1.25 & & 1.25 \\
Y & +0.625 & & -0.625 \\
S & \overline{1.875} & D & \overline{0.625}
\end{array}
$$

Since, in the actual process of determining the digits of S and D, we ignore the decimal point once the operands are aligned, that is, we treat X and Y as though they were integers, this is equivalent to doing the integral arithmetic

$$
\begin{array}{cc}
1250 & 1250 \\
+\ 625 & -\ 625 \\
\overline{1875} & \overline{625}
\end{array}
$$

In each case we needed to supply the final zero, which, although not written, was implicit in the mixed number 1.25. This appears explicitly in the integer 1250. To locate the base points in S and D we align them with the common position of the points in X and Y.

On closer inspection we see that to ignore the base point in the digit-by-digit operation and still keep the proper alignment we have multiplied X and Y by appropriate powers of ten. To reduce Y to an integer we must use 10^3 as a smallest factor to obtain 625. To reduce $X = 1.25$ to an integer we could multiply it by 10^2 to get 125. To align the decimal points, however, we must use the *same power of ten* in each case, 10^3, so that we get

$$
\begin{aligned}
X \cdot 10^3 &= 1250 \\
Y \cdot 10^3 &= \ \ 625 \\
(X + Y) \cdot 10^3 = S \cdot 10^3 &= 1875
\end{aligned}
$$

Therefore,
$$S = (1875) \cdot 10^{-3} = 1.875.$$

Multiplying by 10^{-3} aligns the point in the result with the common position of the point in X and Y. Thus, to consider only the strictly digital operation, we multiply each operand by a suitable power of ten to reduce it to an integer, but to align the decimal points we must use the same power of ten for both X and Y. A similar discussion applies to subtraction.

Example 7.2 The actual digital arithmetic to find the product XY for the operands of Example 7.1 is

$$
\begin{array}{r}
125 \\
625 \\
\hline
625 \\
250 \\
750 \\
\hline
78125
\end{array}
$$

We then locate the decimal point by counting two digits to the right of the point in X, three to the right in Y, and putting $2 + 3 = 5$ places to the right in $P = XY$. There was no need to align the points.

In multiplication, as in addition and subtraction, we reduce the operands to integers by multiplying by powers of ten. Thus, $X \cdot 10^2 = 125$ and $Y \cdot 10^3 = 625$. We form

$$(X \cdot 10^2)(Y \cdot 10^3) = (XY) \cdot 10^5 = P \cdot 10^5 = 78125,$$

so that

$$P = (78125) \cdot 10^{-5} = 0.78125.$$

We see that counting digits to the right of the decimal point solves the problem of determining the minimum powers of ten necessary as factors to reduce X and Y to integers. Counting $2 + 3 = 5$ digits to the right of the point in $P = XY$ is simply an application of the additive law of exponents.

Example 7.3 In division, applying Rule 3, we would proceed as follows:

$$Y = 0.625 \overline{\smash{)}\begin{array}{l} 2. = Q \\ 1.250 = X \\ \underline{1\,250} \\ 0 \end{array}}$$

The strictly digital operation, independent of the decimal point, is

$$625 \overline{\smash{)}\begin{array}{l} 2 \\ 1250 \\ \underline{1250} \\ 0 \end{array}}$$

For division we again introduce appropriate powers of the base ten. We form the integers $X \cdot 10^3 = 1250$ and $Y \cdot 10^3 = 625$, to give

$$(Y \cdot 10^3)/(X \cdot 10^3) = (Y/X) \cdot 10^{3-3} = (Y/X) \cdot 10^0 = Q.$$

In this case, as in addition and subtraction, we reduce X to the integer $X \cdot 10^3 = 1250$ instead of $X \cdot 10^2 = 125$. This is in accordance with Rule 3, and gives an exponent of zero on application of the subtractive rule for exponents. This guarantees an integral quotient.

Example 7.4 In Example 7.3 the division was exact, so that the operation produced a zero remainder. In practice, if the remainder is not zero, we apply the same rule for locating the decimal point but continue the digit-by-digit division process to as many places as we wish to retain in the fractional part of the quotient. To describe this technique in terms of integral operations we need a slight alteration in the procedure of Example 7.3. We recall that integral division provides an integral quotient and remainder. If we interpret the digits retained in the answer as the quotient in this process, then for $X = 1.3$ and $Y = 72.51042$, we might proceed as follows, to obtain a five-digit answer:

$$
\begin{array}{r}
5\,5.777 \\
1.3\,\overline{)\,72.5\,1042} \\
65 \\
\hline
7\,5 \\
6\,5 \\
\hline
1\,0\,1 \\
9\,1 \\
\hline
1\,00 \\
91 \\
\hline
94 \\
91 \\
\hline
3
\end{array}
$$

In strictly integral form we have

$$
\begin{array}{r}
55777 \\
13\,\overline{)\,725104} \\
65 \\
\hline
75 \\
65 \\
\hline
101 \\
91 \\
\hline
100 \\
91 \\
\hline
94 \\
91 \\
\hline
3
\end{array}
$$

and we retain the integral quotient 55777, ignoring the integral remainder 3 and the final digit 2 in Y. We then place the decimal point between the 5 and 7.

We can describe what we have done in Example 7.4 in terms of reducing X and Y to integers by introducing appropriate powers of 10. We form

$$X \cdot 10^1 = (1.3) \cdot 10^1 = 13$$

and

$$Y \cdot 10^4 = (72.51042) \cdot 10^4 = 725104.2,$$

of which we use only the integral part

$$\bar{Y} \cdot 10^4 = (72.5104) \cdot 10^4 = 725104.$$

Thus, the reduction of the dividend to an integral operand requires both multiplication by a power of ten and *truncation* of the value of Y to an approximate value $\bar{Y}$. If we assume that $\bar{Y}$ is, for practical purposes, to replace Y, we have

$$(\bar{Y} \cdot 10^4)/(X \cdot 10^1) = (\bar{Y}/X) \cdot 10^3 = Q \cdot 10^3 = 55777,$$

so that $Q = Q \cdot 10^0 = (55777) \cdot 10^{-3} = 55.777$. In this case the final reduction of the integral digits to the correct form $Q \cdot 10^0$ involves multiplying by 10^{-3} (which is done automatically by the application of Rule 3). We see, then, that this rule, combined with continued division, is equivalent to reducing the dividend Y to an integer by multiplying by a power of ten greater than or equal to that used to reduce the divisor X to an integer, and placing the final decimal point so that the net exponent of ten on $Q = Y/X$ is zero. In practice, the excess of the power of ten used for Y over that used for X determines the number of digits to be retained to the right of the decimal point in Q. We note that both X and Y may require truncation, and ignoring a nonzero remainder in effect truncates Q. There is the possibility of using a rounding procedure in this operation. We reserve the discussion of this for Chapter 8.

Although automatic procedure, from long habit, has probably made many people forget why they use certain rules for locating the decimal point, we see that it is really a question of finding an appropriate power of ten to reduce a mixed operand to an integer, and then applying rules of exponents to place the decimal point in the result. For addition and subtraction we form $X \cdot 10^s \pm Y \cdot 10^s = (X \pm Y) \cdot 10^s$ so that a *common* power of ten is required. For multiplication Rule 2 is equivalent to $(X \cdot 10^s)(Y \cdot 10^t) = (XY) \cdot 10^{s+t}$, where s and t may be different. In division we use

$$(Y \cdot 10^s)/(X \cdot 10^t) = (Y/X) \cdot 10^{s-t},$$

with $s - t \geq 0$ to give an integral result. In any case, the digit-by-digit operations treat the sequences of digits as though they represented integers with the base points fixed at the right-hand end of the sequences.

The technique of locating the base point by using appropriate powers of the base, which we have applied in Example 7.1 through 7.4, is called *scaling*. The powers of the base are referred to as *scale factors*. This treatment can be extended to any base and is the general method used in dealing with base points in computers. The proper manipulation of scale factors can be accomplished by the programmer, externally to the machine, or by the computer itself, internally. In either case the required digit-by-digit operations proceed as though they were on operands with the base point in some fixed *reference position*. The simplest reference position is that for integers, the right-hand end of the sequence of digits. This particular choice is not necessary, but in all cases a fixed location is assumed.

If the base point is located by external scaling, the internal operands are sequences of digits with an understood base point in the same fixed reference position relative to all digital sequences. The programmer can then adjust the base point from its fixed location by means of his external notes about the scale factor associated with each operand. This mode of operation is referred to as *fixed point* operation and the operands as *fixed point* operands. By means of programming or of circuitry it is possible for a computer itself to do the necessary steps to locate the base point. In this case, a record of the scale factor is stored internally as is the fixed point operand, and adjusted by the program or by circuitry. This mode of operation in which the machine locates the base point is known as *floating point*, and the operands as *floating-point* operands. In spite of the terminology it should be remembered that both fixed and floating-point operations represent the same treatment (with necessary modifications) of the same problem. A more accurate description would be to call fixed-point operation programmer-scaled, and floating-point operation machine-scaled. In both cases the *digital* operations are always on fixed-point operands. In the following sections we will discuss both modes of scaling.

7.2 SCALE FACTORS

The digit-by-digit aspects of the arithmetic operations by machine methods are carried out as though all operands had the same fixed location for the base point. An easy location to envisage is at the right-hand end of the sequence, the digital algorithms then being applied to integers; however, *any* fixed reference position can be used provided the base point in the final result is placed by the appropriate technique. In practice two locations are normally preferred, the right-hand end for *integral operation* or the left-hand end (exclusive of sign digit) for *fractional operation*. In either case, an individual operand X consists of an infinite sequence of digits, base r, of which we may use only a finite number, say n. Thus, we must truncate the infinite sequence X to a finite sequence of n digits. This may or may not involve a rounding procedure. The infinite sequence X will have a base point whose position is determined by the value of X. The permissible machine operands, however, all have the same fixed location for the base point and the same number of digits n, which are independent of the value of X. The actual operand for digital operation, then, is a number X^* consisting of exactly n digits base r with an *a priori* specified position for the base point. When necessary, insignificant zeros are provided in X^* so that it has precisely n digits. This number X^* is to represent X and thus we will call it the machine representation of X. Thus, X^* and X will contain the same sequence of digits except for insignificant zeros and/or digits omitted in a truncation of X. Otherwise, they differ only in the position of the base point. After truncation, to obtain either sequence of digits from the other requires only a shift of the point relative to the digits, or alternatively, a shift of the digits relative to the point.

Shifting the point, base r, involves multiplication or division by r. This can be accomplished by using factors which are integral powers of the base, say r^s, and we call such powers *scale factors*. The exponent s is called a *scale-factor exponent*. To obtain X^* we first multiply X by an appropriate scale factor and truncate the result to the correct number of digits. Thus,

$$X^* = \text{The truncated version of } [X \cdot r^s].$$

Example 7.5 Suppose we have eight-digit decimal machine arithmetic with integral operation and we wish to use the highest-order digit as a sign digit. Then we might obtain the machine version of $\pi = 3.1415926\ldots$ by means of a scale factor 10^6 in the following manner.

$$X^* = \text{The truncated version of } [\pi \cdot 10^6 = 3141592.6\ldots] = 03141592.$$

In general the reduction of X to a finite sequence of digits involves the introduction of an error. The process of *rounding* may reduce this error and some writers use the terms truncation and rounding to distinguish between the case where all but a finite number of digits are simply deleted and the case where the final digit retained may be increased by one. Whichever technique is used, we will call the resultant sequence the truncated version of X, and in Chapter 8 we will examine the errors involved. Mathematically we should distinguish between X and its truncated version. To avoid unnecessary complexity of notation, however, we will introduce Xr^s with no explicit multiplication symbol to mean the scaled and truncated version of X. Thus, we write

$$X^* = Xr^s = \text{The truncated version of } [X \cdot r^s],$$

and

$$X = X^* \cdot r^{-s}.$$

The value $X^* \cdot r^{-s}$ may not give the true value of X. Practically, however, it is the only reconstruction available, and we use no special notation to indicate that it may be an approximation.

Altering the choice of the base-point location in X^* will affect the rule used to locate the base points after arithmetic operations. The effect is a simple one, however. If X_I^* is a particular fixed-point operand whose base point is at the right-hand end of the sequence (an integer), then for any choice of the base point in X^* there is an exponent t such that

$$X_I^* = X^* r^t = (Xr^s)r^t = Xr^{s+t} = Xr^{s'},$$

with $s' = s + t$. Hence, any change of the base point from the integral position involves only an additive constant in the scale-factor exponent. For this reason we will first treat scaling from an integral point of view.

7.2.1 Scale Factors for Integral Operation

In this section we assume that the choice of the base point in the machine version X^* of a number X is at the right-hand end of the sequence of digits, so that X^* is an integer. We further assume that complement arithmetic is used, so that the highest-order digit of X^* is a sign digit. For absolute value and sign arithmetic we need only augment the $n - 1$ digits in X^* by one to make n as the number of digits actually used.

It is apparent that the scale factors introduced in the preceding section are not uniquely determined. For any scale factor r^s applied to X, we can interpret the location of the resultant base point as coincident with that of X^*. Thus, we can convert the decimal number 1.25 to an integer by forming $(1.25)10^2 = 125$ and also by $(1.25)10^3 = 1250$. We can also use $(1.25)10^4 = 12,500$, which is an integer, or $(1.25)10^1 = 12$.

Indeed we can convert 1.25 to an integer by truncation and the scale factor 10^{-1}, yielding zero. The scale factor 10^{-1} is not worth while since it permits no digits to be retained in the integral part except zero. Because 12,500 uses more than four digits exclusive of sign, the scale factor 10^4 would also be of no interest if only four digits were retained in X^*. Thus, a scale factor can be too large, yielding too many digits in X^*, or it can be too small, yielding only insignificant digits in X^*. To obtain a practical scale factor, appropriate conditions must be introduced. We will call a scale factor which is not too large, that is, which does not produce an X^* requiring more digits than a machine operand, a permissible scale factor. Thus, if X^* is to be an integer consisting of exactly n digits (including at least one leading sign digit), then a *permissible scale factor* r^s for which

$$X^* = Xr^s = \text{Integral part of } [X \cdot r^s]$$

is such that X^* satisfies

$$|X^*| < r^{n-1}.$$

This means that

$$|Xr^s| < r^{n-1}.$$

Every X starts with a first nonzero digit, say that corresponding to r^{p-1}. Therefore, with $d_{p-1} \neq 0$ and $d_k = 0$ for $k > p - 1$,

$$|X| = d_{p-1}r^{p-1} + d_{p-2}r^{p-2} + \cdots,$$

and it follows that there is a unique integer p for which

$$r^{p-1} \leq |X| < r^p.$$

The requirement for a permissible scale factor is therefore that

$$|Xr^s| < r^p r^s \leq r^{n-1}.$$

Thus,

$$p + s \leq n - 1 \tag{7.1}$$

gives the range of permissible scale-factor exponents. This does not mean that for a given n and p all $s \leq n - 1 - p$ are *useful* scale-factor exponents.

It means that they are permitted in the sense that the number of digits of the corresponding X^* will not exceed the capacity of the machine. The larger s is, the greater the number of digits in X that can be retained in X^*. In the interests of accuracy, therefore, the largest permissible value of the scale-factor exponent will generally be preferred for *optimal* scaling.

Example 7.6 The number 1.25 in decimal satisfies the condition

$$1 < |1.25| = 1.25 < 10^1,$$

so that $p = 1$. For a five-place complement machine, $n - 1 = 4$ and the restriction on scale-factor exponents is $s + 1 \leq 4$, or $s \leq 3$. Thus, $(1.25)10^2 = 00125$ and $(1.25)10^3 = 01250$ show that 2 and 3 are permissible scale-factor exponents, although $(1.25)10^4 = 12500$ shows that $s = 4 > 3$ is not permissible since the magnitude of the integral part of 12500 cannot be represented with four digits. The scale-factor exponents 0 and -1 are permissible, but lack quality in that

$$\text{Integral part of } [(1.25) \cdot 10^0] = 00001$$

and

$$\text{Integral part of } [(1.25) \cdot 10^{-1}] = 00000$$

are less accurate representations than those corresponding to 2 or 3.

7.2.2 Scale Factors for Nonintegral Operation

The choice of permissible scale factors when the machine representation is an integer proves to be simple. The automatic truncation given by the retention of the integral portion of a number, together with the maximum number of digits to be used and the magnitude of the number to be represented, dictate the controlling mathematical inequality (7.1). If the base point in the machine representation is to be fixed so that X^* is not an integer, the determination of permissible scale-factor exponents in the general case can be based on the results for integers. In particular cases these can be translated to a more convenient form.

Let X^* be a number consisting of exactly n digits including at least one sign digit. Let t be the smallest nonnegative exponent such that X^*r^t is an n-digit integer. Then a permissible scale factor r^s is one for which

$$|X_I^*| = |X^* r^t| = |X r^{s+t}| = |\text{Integral part of } [X \cdot r^{s+t}]| < r^{n-1}.$$

It follows as before that the requirement for permissible scale factor exponents s is

$$s + t + p \leq n - 1, \tag{7.2}$$

where $r^{p-1} \leq |X| < r^p$ and n and t have known specific values.

Example 7.7 Assume five-digit, complement decimal arithmetic, in which all machine numbers are treated as though the decimal point were just to the

7.2 SCALE FACTORS

right of the sign digit. Thus, permissible scale factors will be those which move the decimal point of the original operand X to this location, but the sequence must then be truncated to no more than four digits to the right of the point. Obviously we must impose the condition that $|X \cdot 10^s| < 1$, but this does not indicate the required truncation. Application of condition (7.2), however, yields the following: With four digits to the right of the decimal point we have $t = n - 1 = 4$. Therefore, if $X < 10^p$, permissible scale-factor exponents are those for which

$$p + s + 4 \leq 4.$$

This yields $|X \cdot 10^s| < |10^{p+s}| \leq 10^0 = 1$, as required, and we must also truncate. Thus, for $X = 1.3333\ldots$, $|X| < 10^1$ and $p = 1$, so that permissible values of s are $s \leq -1$. We require of $X^* = X10^s$ that

$$X^* \cdot 10^t = X_1^* = \text{Integral part of } [(1.333\ldots) \cdot 10^{s+t}].$$

Setting $t = 4$ and $s = -1$, we get $X^* = 0.1333$. For permissible scale-factor exponents -2, -3, and -4 we have, similarly,

$$s = -2: \quad X^* = 0.0133$$
$$s = -3: \quad X^* = 0.0013$$
$$s = -4: \quad X^* = 0.0001.$$

In practice we shall apply the inequality (7.2) with t and $n - 1$ fixed and p known for any specific X. We retain the notation Xr^s to represent the properly truncated version of $X \cdot r^s$.

Example 7.8 Assume one's-complement, binary arithmetic with 24-bit integral operands. We wish to find permissible scale-factor exponents for $X = -27.4$. We have $n - 1 = 23$, and since $2^4 < 27.4 < 2^5$, $p = 5$. Thus, inequality (7.1), or inequality (7.2) with $t = 0$, gives

$$s \leq 23 - 5 = 18.$$

We have

$$27.4_{10} = 11011.01100110011\ldots_2,$$

and choosing $s = 18$ we find that

$$X^* = 01101101100110011001,$$

of which the first zero is a sign bit. Thus,

$$X^* = 10010010011001100110.$$

Example 7.9 If the 24-bit operand of Example 7.8 is fractional, that is, with the binary point appearing after the sign bit, $t = 23$ and application of (7.2) to $X = -27.4$ gives $s \leq -5$ or

$$s' = s + t \leq 18.$$

With $s = -5$, $s' = 18$ and X_1^* is the same as the X^* in the preceding example. Multiplication of X_1^* by 2^{-23} yields the value of X^* for the fractional machine

$$X^* = 1.00100100110011001100110.$$

In practice the base point is implicitly understood, and the stored sequence of digits in both examples would be identical: Of course the scale-factor exponent recorded externally would be different in the two cases, with $s = 18$ in the one case and $s = -5$ in the other.

Since p is determined by the first nonzero digit of $|X|$, the value of $p + s$ for integral arithmetic gives the number of digits of $|X|$ retained in the machine representation. This follows from

$$|Xr^s| = |\text{Integral part of } [X \cdot r^s]| = d_{p-1} r^{p+s-1} + \ldots,$$

which shows that X^* has precisely $p + s$ digits. Thus, for given p, the optimal value of s for accuracy in the range $s \leq n - 1 - p$ is the largest value $s = n - 1 - p$. Each decrease of s by one loses one significant digit of X. Similarly, for nonintegral arithmetic, the maximum number of digits in X_1^* is obtained for $s' = n - 1 - p$ where $s' = s + t$. Therefore the maximum accuracy for X^* is achieved for $s = n - 1 - p - t$, so that in Examples 7.7 and 7.8 the representation was optimal. The decrease in accuracy for smaller values of s is obvious in Example 7.7.

In determining scale factors for a given X, p and t can be found and the inequalities (7.1) and (7.2) dictate a permissible range of values for s. Alternatively, these inequalities determine the range of permissible numbers that can be treated with a given scale factor s.

Example 7.10 For the 24-bit arithmetic of Example 7.8, a scale-factor exponent of 15 is to be used. Thus,

$$p \leq 23 - 15 = 8,$$

and we can use $s = 15$ for any $|X| < 2^8$. For the scaling 2^{15}, we get maximum digital representation if $|X| \geq 2^7$ and fewer digits for smaller magnitudes of X.

7.3 SCALING INTEGRAL ARITHMETIC OPERATIONS

If the only problems in scaling were the determination of permissible scale factors for known numbers and finding a suitable range of numbers for a given scale factor, the direct application of the inequalities (7.1) and (7.2) would suffice. Thus, given data $X_1, X_2, \ldots, X_k$ to be stored in the registers of a particular machine as sequences of n digits (including leading sign digit and with implicitly understood base-point location), we can compute the exponents $p_1, p_2, \ldots, p_k$ such that

$$r^{p_i - 1} \leq |X_i| < r^{p_i},$$

and find permissible scale-factor exponents $s_1, \ldots, s_k$. Similarly for a pre-

7.3 SCALING INTEGRAL ARITHMETIC OPERATIONS

assigned scale factor to be applied to $X_1, \ldots, X_k$, we can test each number to see if it is in the appropriate range. Conversely, if we have a machine representation $X^* = Xr^s$ with a known s, we can reconstruct X without worrying about how s was obtained by using

$$X = X^* \cdot r^{-s}$$

as the truncated representation of X.

The problems of finding a suitable format for the storage of data and the reconstruction of correct results from stored data are, however, only the endpoints of the overall problem. Once numbers are stored, we will want to have arithmetic operations performed on them. In turn, the results of these operations may constitute new arithmetic operands for further calculations. If we are to reconstruct the final results and have the operations correctly performed, we must keep a suitable account of the scale-factor exponents at every step. We shall see that this accounting is very much like the technique we used in the long-hand examples in Section 7.1. Because the digital algorithms for the arithmetic operations are essentially equivalent to those for integers, regardless of base point location, we shall first consider the scaling of integral operations.

7.3.1 Scaling Integral Addition and Subtraction

The problem that arises in forming the sum or difference of two numbers X and Y is that the results must be obtained in terms of the machine representations $X^* = Xr^s$ and $Y^* = Yr^t$. Since we here assume that X^* and Y^* are integers, the formation of the sum or difference of these operands is simple in an open or closed accumulator. We see, however, that

$$X^* \pm Y^* = Xr^s \pm Yr^t = (X \pm Yr^{t-s})r^s$$

gives a scaled version of $X \pm Y$ if and only if $r^{t-s} = 1$, that is, if and only if $t = s$. Thus, a basic requirement for addition and subtraction is a common scale-factor exponent. This is exactly the equivalent of aligning the base points for long-hand calculation. In the event that the initial scale-factor exponents are not equal, we must rescale one of the numbers so that they are. Even if we do this, however, making $X^* = Xr^s$ and $Y^* = Yr^s$, we may not produce a correct result. We properly form

$$X^* \pm Y^* = (X \pm Y)r^s$$

with, presumably, r^s a permissible scale factor for X and Y. The common scale factor, however, may not be permissible for the sum or difference, S or D.

Example 7.11 In a five-digit, decimal complement accumulator, we wish to add $X = 99.56$ and $Y = 87.37$. We introduce a common scale factor 10^2 to give

$$X^* = 09956, \quad Y^* = 08737$$

with

$$S10^2 = X^* + Y^* = 18693 = (186.93)10^2,$$

which is correct mathematically. The five digits 18693, however, exceed the four places allotted to nonsign digits, and therefore 10^2 is not a permissible scale factor. This condition produces an error called *overflow*.

In choosing a common scale factor for two operands in addition or subtraction, we must be sure that the common scale-factor exponent is also permissible for the sum or difference. Thus, in Example 7.11 the overflow could be prevented by noting that the value of p for S is $\underline{3}$ and so the permissible scale-factor exponents are $s \leq 1$. Choice of a suitable scale factor exponent for S, X, and Y may require that we rescale both X and Y prior to addition. In all cases we must have a common scale factor before addition or subtraction. The technique of rescaling to form a common, permissible scale-factor exponent is one of multiplying by a power of the base r. It can be accomplished by shifting, since it is readily seen that we decrease a scale-factor exponent for rescaling before addition or subtraction. If we have

$$X^* = Xr^s \quad \text{and} \quad Y^* = Yr^t$$

and s and t are permissible, then

$$s \leq n - 1 - p, \quad t \leq n - 1 - q,$$

where $r^{p-1} \leq |X| < r^p$ and $r^{q-1} \leq |Y| < r^q$. Since s and t may be maximal in the correct range, we cannot safely increase either without further information which we will rarely have. For this reason, we will normally scale down the number with the larger scale-factor exponent, and, as we have seen, in some cases both operands must be scaled down.

Example 7.12 In a six-bit, binary, complement accumulator we assume that we know $X^* = X2^5$ and $Y^* = Y2^7$ are correctly scaled. We also assume that we know $|X + Y| < 1$. Thus, from $s \leq n - 1 - p = 5$, $s = 5$ is permissible for the sum. Since 2^5 may be optimal as a scale factor for X, we may not scale X^* up, so we rescale $Y^* = Y2^7$ to $\bar{Y}^* = Y2^5$. This may produce loss of accuracy, but it is necessary. Specifically, let

$$X^* = 010101 = (0.10101)2^5 \quad \text{and} \quad Y^* = 010101 = (0.0010101)2^7.$$

We rescale to get

$$X^* = 010101 = (0.10101)2^5,$$
$$\bar{Y}^* = 000101 = (0.00101\,01)2^5 = (0.00101)2^5.$$

These operands yield the correct six-digit result

$$S2^5 = X^* + \bar{Y}^* = 011010 = (0.11010)2^5,$$

although we have sacrificed the final two digits in Y^* to form $\bar{Y}^*$.

7.3 SCALING INTEGRAL ARITHMETIC OPERATIONS

In Example 7.12, we assumed that X and Y were correctly scaled. This information, together with the inequalities (7.1) and (7.2), implies that

For X: $\quad p + s = p + 5 \le 5$ and $p \le 0$;

For Y: $\quad p + t = p + 7 \le 5$ and $p \le -2$.

Therefore we know that $|X| < 2^0$ and $|Y| < 2^{-2}$. If we also knew that $|X| < 2^{-2}$, we could have scaled X up, but such information will rarely be available.

The preceding example shows that we may have to sacrifice accuracy in one operand to align base points. If the sum or difference is too large for the scale factors assigned to operands, we will have to sacrifice accuracy in both operands.

Example 7.12 also illustrates a fact of life that arises in external scaling. Since in most cases we will not have *a priori* knowledge of the magnitude of results of arithmetic operations, we must make estimates of them. Thus, in place of an exact value of p for

$$r^{p-1} \le |X \pm Y| < r^p,$$

we will have to estimate by p' in the form

$$? \le |X \pm Y| < r^{p'}.$$

If $p' \ge p$, the results will be correct, although not always optimal. If $p' < p$, the estimate can produce overflow. We note that it will usually be impossible to find a sufficiently good lower bound to guarantee optimal scaling.

Example 7.13 With the arithmetic and operands of Example 7.12 we again assume that $X2^5$ and $Y2^7$ are correctly scaled, but assume no further knowledge about $X + Y$. As shown above, the inequality (7.1) guarantees $|X| < 2^0$ and $|Y| < 2^{-2}$. Thus,

$$|X + Y| \le |X| + |Y| < 2^0 + 2^{-2} = 2^{-2}(2^2 + 1) = 2^{-2}(5) < 2^{-2} \cdot 2^3 = 2^1.$$

In this case, guaranteed only that $|X + Y| < 2$, the best available *estimate* of p is one, and

$$s \le 5 - 1 = 4$$

are the only known permissible scale-factor exponents. Therefore, we scale both operands down. For the values given we replace

$X^* = 010101 = (0.10101)2^5 \quad$ and $\quad Y^* = 010101 = (0.0010101)2^7$

by

$\bar{X}^* = 001010 = (0.1010)2^4 \quad$ and $\quad \bar{Y}^* = 000010 = (0.0010)2^4,$

to give

$$S2^4 = 001100 = (0.1100)2^4.$$

The accuracy of the result 0.1100 is less than that of the result 0.1101 from the preceding example.

The results in Examples 7.12 and 7.13 indicate that we may have to forego optimal scaling in addition and subtraction to obtain a common scale factor or to obtain a correct scale factor for the result, or both. In some cases less than optimal scaling may be imposed by lack of information. In Example 7.12 we assumed that we knew enough about $|X + Y|$ to use optimal scaling, but with the information assumed in Example 7.13, we were not able to do so.

The operation required to rescale operands for addition or subtraction is multiplication by an integral power of r. Since the exponent of the power in almost all cases results in decreasing the net exponent, it will usually be negative. Thus, rescaling will generally involve an end-off right shift with a resultant loss of the digits shifted to the right of the end of the register.

Example 7.14 For the numbers in Example 7.13 the initial form of the two machine operands is

$X^* =$ | 0 | 1 | 0 | 1 | 0 | 1 |,

$Y^* =$ | 0 | 1 | 0 | 1 | 0 | 1 |.

We must rescale X^* to $\bar{X}^*$ by multiplying by 2^{-1}, that is, by a right shift of one. Similarly we rescale Y^* to $\bar{Y}^*$ by a right shift of 3. This gives

$\bar{X}^* =$ | 0 | 0 | 1 | 0 | 1 | 0 | ⓘ → Lost

$\bar{Y}^* =$ | 0 | 0 | 0 | 0 | 1 | 0 | ⑩⓵ → Lost

7.3.2 Scaling Integral Multiplication

In long-hand multiplication it is not necessary to align points, and in scaling the equivalent of this is that we do not require a common scale factor. However, we do need the equivalent of counting all digits to the right of the base point. This is achieved by adding scale-factor exponents. Thus, to form XY from $X^* = Xr^s$ and $Y^* = Yr^t$, we form the product of the integers X^* and Y^* to obtain

$$X^*Y^* = (XY)r^{s+t},$$

with the desired result scaled with exponent $s + t$. The reasons for forming the product X^*Y^* in a $2n$-digit accumulator were pointed out in Chapter 5. With the scaling rule (7.1) adjusted appropriately for the $2n$-digit result, the scale-factor exponent $s + t$ will always be permissible. However, the use of an n-digit accumulator or the subsequent reduction of X^*Y^* to n digits would require that $s + t$ be permissible for an n-digit register. This could necessitate scaling X and Y or $X \cdot Y$ down, with a resultant loss of accuracy.

7.3 SCALING INTEGRAL ARITHMETIC OPERATIONS

Example 7.15 In a four-digit, complement, decimal accumulator we wish to form the product of $X = 0.923$ and $Y = 83.7$. If we use four-digit representations for the factors, we can correctly scale them with exponents 3 and 1 respectively, to give

$$X^* = 0923 = (0.923)10^3 \quad \text{and} \quad Y^* = 0837 = (83.7)10^1.$$

The product, $XY = 77.2551$, has a maximum permissible scale-factor exponent of one for a four-digit register, while direct multiplication yields a scale-factor exponent of four. Thus, we must reduce the scale factors on X or Y or both so that the sum of the exponents is no greater than one. That is, we must reduce the sum by three. If we rescale either X or Y by 10^{-3}, we lose all significant digits and produce a zero product. If we rescale X to 10^1 and Y to 10^0, we would have

$$\bar{X}^* = 0009 = (0.9|23)10^1 \quad \text{and} \quad \bar{Y}^* = 0083 = (83.|7)10^0,$$

which would give a product $\bar{X}^* \bar{Y}^* = 0747 = (74.7)10^1$. This is not a satisfactory representation for the product which, if rounded for representation in four digits, would be $0773 = (77.3)10^1$.

In what follows we will assume that products are formed in a $2n$-digit accumulator from n-digit factors. If the correctly scaled factors are X^* and Y^*, we have

$$|Xr^s| = |X^*| < r^{n-1} \quad \text{and} \quad |Yr^t| = |Y^*| < r^{n-1},$$

so that

$$|XY|r^{s+t} = |X^*Y^*| < r^{2n-2} < r^{2n-1}$$

and no overflow can occur. Thus, $s + t$ is a permissible exponent for the double-length accumulator. The only scaling problem that arises, then, is that of rescaling and truncating a $2n$-digit product for storage in an n-digit register. We will here assume simple truncation of excess digits without rounding. The problem is then the following: Having formed X^*Y^* in the double-length accumulator, we must scale it by a factor $r^{-\rho}$ so that

$$\text{Integral part of } [|X^*Y^*|r^{-\rho}] < r^{n-1}. \tag{7.3}$$

It is apparent that reduction of X^*Y^* requires moving the base point to the left and then truncating the digits occurring to the right of the base point. This implies multiplication by a negative power of r so that $\rho > 0$ or is zero if no reduction is necessary. Since the implied base point is at the right of the register, we can move its relative position to the left by shifting the digits to the right. Thus, ρ represents the amount of right shift required. The shift is end-off to achieve truncation. We note that since

$$|X^*Y^*| < r^{2n-2},$$

and
$$|X^*Y^*|r^{-\rho} < r^{2n-2-\rho} \leq r^{n-1},$$
any choice $\rho \geq n - 1$ will guarantee that (7.3) is satisfied. However, a smaller value of ρ may suffice, and we will usually choose the smallest value possible.

Let p be the integral exponent such that
$$r^{p-1} \leq |XY| < r^p.$$
Then
$$|X^*Y^*|r^{-\rho} = |XY|r^{s+t-\rho} < r^{s+t-\rho+p},$$
and it follows that
$$\rho \geq s + t - n + 1 + p \qquad (7.4)$$
defines the range of values of ρ such that (7.3) is satisfied. Choosing equality in (7.4) provides the minimum permissible right shift and so allows maximum retention of digits. Each additional right shift causes deletion of one more digit. Therefore, equality in (7.4) is the optimal choice for ρ.

Example 7.16 For the factors of Example 7.15 we have $s = 3, t = 1, n = 4$, and $p = 2$. The product $X^*Y^* = (0923)(0837)$ formed in an eight-digit register is 00772551. Inequality (7.4) gives $\rho \geq 3$. If we use the optimal $\rho = 3$, we obtain
$$X^*Y^*10^{-3} = 00772|.551 = 00772,$$
of which we ignore the leading insignificant zero, to obtain as the final four-digit result 0772. Noting that the scale factor exponent associated with the result is $s + t - \rho = 1$, we find the product to be
$$XY = (0772) \cdot 10^{-1} = 77.2.$$
This is exactly the truncated (not rounded) version of $XY = 77.2551$. Use of permissible, but not optimal values $\rho = 4$ and 5 would give
$$XY = (0077) \cdot 10^0 = 77 \quad \text{and} \quad XY = (0007) \cdot 10^1 = 70,$$
with obvious loss of accuracy.

If we know the exact value of p in the inequality $r^{p-1} \leq |XY| < r^p$, the above procedure suffices. In most practical cases, however, we have only an estimate p' for p. If $p' \geq p$, we can still get correct, but not always optimal results.

Example 7.17 Suppose that for the operands in Example 7.16 we had not known the exact values but only their scale factors. If the scale factors are permissible,
$$\begin{array}{ll} |X|10^3 < 10^3 & \text{or} \quad |X| < 1, \\ |Y|10^1 < 10^3 & \text{or} \quad |Y| < 10^2. \end{array}$$

7.3 SCALING INTEGRAL ARITHMETIC OPERATIONS

Thus
$$? \leq |XY| < 10^2.$$

The question mark in this last inequality is to show that we have only an estimate $p' = 2$ of p. It happens that $p' = p$ in this case, and so we can use the optimal right shift $\rho = 3$.

In the preceding example the estimate p' of p was exact. Frequently this will not be the case.

Example 7.18 In a decimal complement machine which forms an eight-digit product from four-digit operands, we have factors $X = 12.1$ and $Y = 10.2$. These are optimally scaled to give

$$X^* = 0121 = X10 \quad \text{and} \quad Y^* = 0102 = Y10.$$

The product satisfies $10^2 \leq XY < 10^3$ and so $p = 3$ and $\rho \geq 2$. If we use the optimal right shift $\rho = 2$, the eight-digit machine product $X^*Y^* = 00012342$ reduces to 0123 with a scale factor 10^0. If the only information we have is that $s = t = 1$ are correct scale-factor exponents, however, we can deduce only that

$$|X|10^1 < 10^3 \quad \text{or} \quad |X| < 10^2,$$
$$|Y|10^1 < 10^3 \quad \text{or} \quad |Y| < 10^2,$$

so that
$$? \leq |XY| < 10^4.$$

This gives an estimate $p' = 4$ which is correct but not optimal. Using this value of p' in place of p in (7.4), however, we find that $\rho \geq 3$, so at best we lose one more digit and obtain a four-digit machine product 0012 scaled 10^{-1}. From this we find $XY = 120$.

We have seen that although $p' \geq p$ always gives correct results from the point of view of product overflow, optimal results are obtained only if $p' = p$. If $p' < p$, that is, if we underestimate the magnitude of the product, the smallest right shift defined by (7.4) will drop too few digits and overflow can result. It is important, therefore, to get the best bounds on $|XY|$ obtainable.

7.3.3 Scaling Integral Division

Just as we approached division in Chapter 5 as the reverse of multiplication so we can approach scaling division as the reverse of scaling multiplication. We may write

$$\text{Dividend} = (\text{Divisor})(\text{Quotient}) + \text{Remainder},$$

where the dividend and divisor are known and we seek to find the digits of the quotient and remainder. To find these by reversing the operation of multiplication, we could find the second factor required in multiplication for a given first factor and a known product. This implies that we know the $2n$-digit product and one n-digit factor and are looking for the second n-digit factor. This formulation of the problem requires a remainder of zero which, in general, does not occur. However, to phrase the more general problem in these terms, we can subtract the remainder from the dividend in the Division Algorithm to obtain

$$[\text{Dividend} - \text{Remainder}] = [\text{Divisor}][\text{Quotient}] \quad (7.5)$$

$$\downarrow \qquad\qquad\qquad \downarrow \qquad\qquad \downarrow$$

$$2n\text{-Digit Product} \qquad n\text{-Digit Factor} \quad n\text{-Digit Factor}$$

We must still obtain the remainder (numerically less than the divisor), but we may consider it as the residue left over from the division operation. The problem of scaling for division can thus be thought of as the problem of providing the appropriately scaled $2n$-digit product to yield the correctly scaled n-digit factors, and we can attempt to reverse the steps of the preceding section.

Although the $2n$-digit product was introduced in Chapter 4 so that residue-class multiplication could be interpreted as ordinary multiplication, we still need to retain $2n$ product digits in the present division context. This point is illustrated in the following example.

Example 7.19 In a four-digit, complement, decimal machine we wish to find X/Y with $X = 350$, $Y = 0.5$. By ordinary longhand methods we find $X/Y = Q = 700$. The four-digit machine versions of X and Y are

$$X^* = 0350 = (350)10^0 \quad \text{and} \quad Y^* = 0500 = (0.5)10^3,$$

and we wish to obtain

$$Q^* = 0700 = (700)10^0.$$

If we leave X^* as a four-digit integer, however, we have, after substitution in the Division Algorithm,

$$0350 = (0500)(Q^*) + R^*.$$

Since we are dealing with nonnegative integers $Q^* = 0000$ and $R^* = 0350$, this means a scaled version of the quotient equal to zero, with no significance. This results because the net scale-factor exponent (-3) is too small. On the other hand, if we convert X^* to a $2n$-digit number by scaling up by 10^3, we get a new machine version of X,

$$X^* = 00350000 = (350)10^3.$$

Substitution in relation (7.5) gives, correctly,

$$00350000 = (0500)(0700),$$

with

$$Q^* = \frac{X10^3}{Y10^3} = Q10^0 = (700)10^0.$$

To see the steps required for scaling division, assume that initially $X^* = Xr^s$ and $Y^* = Yr^t$ are both n-digit integers. We seek a scaled n-digit machine version of Q, say $Q^* = Qr^u$. To convert X^* to $2n$-digits we place it, properly extended, into a $2n$-digit accumulator and then scale up by r^λ by moving digits to the left λ positions. Substitution into (7.5) in terms of machine numbers gives

$$X^* r^\lambda - R^* = Y^* Q^*.$$

Recalling that the machine numbers are merely scaled versions of the original operands, we may rewrite this equation

$$Xr^{s+\lambda} - Rr^v = YQr^{t+u},$$

where $R^* = Rr^v$. This is the scaled version of $X - R = YQ$ and will be correct if and only if

$$s + \lambda = v = t + u.$$

It follows that the machine form of the remainder is scaled the same as the $2n$-digit dividend, and that the net scale factor exponent for the machine quotient is $u = s + \lambda - t$. This reflects subtraction of scale-factor exponents on division, after conversion of the dividend to $2n$-digit form.

To be a permissible scale-factor exponent for Q, u must satisfy the fundamental condition (7.1), that is, $p + u \leq n - 1$, with

$$r^{p-1} \leq |Q| < r^p.$$

Thus, (7.1) takes the form

$$p + s + \lambda - t \leq n - 1$$

or

$$\lambda \leq t - p - s + n - 1. \quad (7.6)$$

The inequality (7.6) defines the restriction on the left shift λ used to scale up the dividend before division. The choice of equality in (7.6) gives maximum retention of quotient digits, since for fixed s and t, $u = s + \lambda - t$ is maximized and thus is optimal for the maximum value of λ. Choice of λ less than the maximum permissible value produces less than optimal scaling of the quotient and loss of accuracy. Violation of (7.6) leads to too large a scale factor and to possible quotient overflow.

Example 7.20 We wish to form 2.33/34.2 in a four-digit, complement, decimal machine. The mathematical result is $Q = 0.06812\ldots$, of which we can retain at most three digits. Since $10^{-2} \leq |Q| < 10^{-1}$, we can have

$p = -1$. In machine form

$$X^* = 0233 = (2.33)10^2 \quad \text{and} \quad Y^* = 0342 = (34.2)10^1.$$

Therefore in inequality (7.6) we may use $s = 2$, $t = 1$, and $n = 4$ to get $\lambda \leq 3$. To divide, we first extend X^* to double length and then shift left $\lambda = 3$ positions to produce

$$X^*10^3 = 00233000 = (2.33)10^5.$$

Application of the Division Algorithm results in

$$00233000 = (0342)(0681) + 0098.$$

Thus, $Q^* = 0681 = Qr^u$ and $R^* = 0098 = Rr^v$. Since $u = s + \lambda - t = 4$ and $v = s + \lambda = 5$, $Q = (0681)10^{-4} = 0.0681$ and $R = (0098)10^{-5} = 0.00098$. A smaller value of λ, that is a left shift of less than three, although permissible, would give fewer quotient digits. The remainder represents the residue of the usual longhand division process stopped after three quotient digits.

In division, proper scaling requires that we have a bound on the result in the form $r^{p-1} \leq |X/Y| < r^p$. However, we frequently do not know p and must introduce an approximation to p, say p', which to be correct must satisfy $p' \geq p$. We can see from (7.6) that as p increases, the bound on λ decreases. Therefore $p' > p$ will result in a smaller left shift than optimal. On the other hand $p' < p$ will give too large a shift before division, leading to quotient overflow. In general, determining p' will require that we have an *upper* bound on X and a *lower* bound on Y. To see this we note that if we know only that $|X| < A$ and $|Y| < B$, we can say nothing about X/Y, but if we know that $|X| < A$ and $|Y| > B$, we have

$$|Y^{-1}| < B^{-1} \quad \text{and} \quad |X/Y| = |XY^{-1}| < A/B.$$

Example 7.21 We know that $X^* = X10^4$ and $Y^* = Y10^3$ are correctly scaled in a four-digit, complement decimal machine. This guarantees $|X| < 10^{-1}$ and $|Y| < 1$ but tells nothing about $|X/Y|$. On the other hand, if we know that X and Y are *optimally* scaled, it follows that

$$10^{-2} \leq |X| < 10^{-1} \quad \text{and} \quad 10^{-1} \leq |Y| < 1,$$

so that $|Y^{-1}| \geq 10$ and $|X/Y| < 1$ and we can apply the techniques developed above to choose a correct scale factor.

The consequences of a weak or an incorrect estimate p' of p are illustrated in the following example.

Example 7.22 Consider the division problem 2.33/34.2 of Example 7.20. There we utilized the known value $Q = 0.06812\ldots$ to obtain an exact value of p. Suppose, however, we know only that $X^* = X10^2$ and $Y^* = Y10$. We might be tempted to estimate $p' = 1$. Since $p = -1$, $p < p'$ and this is correct. The result of the integral division 0233 by 0342 would yield a quotient of zero. This is due to the fact that with $\lambda = 0$ the net scale factor $u = s - t = 1$ and $Q10 = 0$, so all significance is lost.

On the other hand, suppose we use an estimate $p' = -2$. This would yield from (7.6), $\lambda \leq 4$. If we use the equality sign, the extension and shifted value of X^* becomes 02330000 and the result of dividing 2330000 by 342 is an integral quotient of 6812, too large for a four-digit register.

7.3.4 Combined Integral Operations

In practical applications of scaling it will be necessary to take account of many combinations of the four arithmetic operations. No new conditions are involved over and above those we have discussed, and the steps for each operation are those already outlined. The important thing is to have bounds for the absolute values of the results of the operations at each stage. This permits the choice of a correct scale-factor exponent at each point of the computation. In general, we will attempt to use optimal scaling, where possible, to minimize the inherent error due to the use of a fixed number of digits. In some cases optimal scaling may be impossible. For example, obtaining common scale factors in addition and subtraction or lack of a sufficiently restrictive bound on an operand may preclude the use of optimal scale-factor exponents. In other cases we may be willing to sacrifice some accuracy for the sake of expediency in scaling. We will not pursue the topic in depth, but we will illustrate the general technique with an example.

Example 7.23 We wish to tabulate values of the function

$$f(x) = \frac{4.6x^2 - 0.9x + 4.9}{0.1 + x^2}$$

for values of x in $0 \leq x \leq 1$, in increments $\Delta x = 0.01$. We need to obtain upper bounds for all numbers which arise. To be useful for the techniques developed above, these upper bounds need to be expressed as powers of the base r. For purposes of illustration we will assume a machine in which $r = 2$, the highest-order digit is a sign digit, a standard register is 48 digits, and the accumulator is 96 digits for multiplication and division. To use optimal scaling on x would require that we subdivide its range of values into distinct subsets. For example, $x = 0.01$ and $x = 0.02$ would require distinct optimal scale-factor exponents s_1 and s_2. We have

$$2^{-7} < 0.01 < 2^{-6} \quad \text{and} \quad 2^{-6} < 0.02 < 2^{-5}.$$

Therefore

$$s_1 \leq 47 - (-6) = 53,$$
$$s_2 \leq 47 - (-5) = 52.$$

If we use this technique, it will complicate the scaling. Instead we choose expediency and note that since $|x| \leq 1 < 2$, a permissible choice of scale-factor exponents for all values of x in $0 \le x \le 1$ is given by $s \le 47 - 1 = 46$. For the constants of the numerator

$$2^2 < 4.6 < 2^3$$
$$2^{-1} < 0.9 < 2^0$$
$$2^2 < 4.9 < 2^3$$

and we could use optimal exponents 44, 47, and 44 respectively. Since $x \le 1$, $4.9x^2$ and $0.9x$ can have the same upper bounds as 4.9 and 0.9. For addition, however, we need a common scale factor. Since the maximum value of the numerator represents a sum equal to 8.6 at $x = 1$, we can scale all terms in the numerator at 2^{43}. The constants 4.6 and 0.9 enter as factors, so no common scale factor is required and we retain the optimal scaling 2^{44} and 2^{47} for them. To this point, then, the scaling is

$$(x)2^{46},$$
$$(4.6)2^{44},$$
$$(0.9)2^{47},$$
$$(4.6x^2)2^{43},$$
$$(0.9x)2^{43},$$
$$(4.9)2^{43},$$
$$(4.6x^2 - 0.9x + 4.9)2^{43}.$$

In the denominator, with the same scaling on x, we have

$$0.1 \le 0.1 + x^2 < 2,$$

and we can scale 0.1, x^2, and $0.1 + x^2$ at 2^{46}. Since

$$|4.6x^2 - 0.9x + 4.9| \le 8.6$$

and

$$|0.1 + x^2| \ge 0.1,$$

the function $f(x)$ satisfies

$$|f(x)| \le 8.6/0.1 = 86 < 2^7,$$

and we scale $f(x)$ at 2^{40}. Thus, we have

$$(0.1)2^{46},$$
$$(x^2)2^{46},$$
$$(0.1 + x^2)2^{46},$$
$$f(x)2^{40}.$$

We now compute, in order, x^2, $4.6x^2$, $0.9x$ and $4.6x^2 - 0.9x + 4.9$. In the double-length accumulator, the machine product has the form

$$(x2^{46})(x2^{46}) = (x^2)2^{92}.$$

To rescale this to $(x^2)2^{46}$ requires a right shift of 46 positions. We form the product $4.6x^2$ in the double-length accumulator as

$$(4.6)2^{44}(x^2)2^{46} = (4.6x^2)2^{90}.$$

This requires a right shift of 47 positions to give $(4.6x^2)2^{43}$. Similarly, we form $(0.9)2^{47}(x)2^{46} = (0.9x)2^{93}$ and right shift 50 positions to obtain $(0.9x)2^{43}$. With common scale factors for all terms we can now compute

$$(4.6x^2)2^{43} - (0.9x)2^{43} + (4.9)2^{43} = (4.6x^2 - 0.9x + 4.9)2^{43},$$

and save the result. For the denominator we use the already computed (and saved) $(x^2)2^{46}$ and $(0.1)2^{46}$ to form

$$(x^2)2^{46} + (0.1)2^{46} = (x^2 + 0.1)2^{46}.$$

Using $p' = 7$ as an estimate for p we can determine $\lambda \leq 43$ from (7.6). Since we have already fixed the scaling of $f(x)$ at 2^{40}, however, we need a factor 2^{λ} such that

$$\frac{(4.6x^2 - 0.9x + 4.9)2^{43+\lambda}}{(0.1 + x^2)2^{46}} = f(x)2^{40}.$$

Therefore, prior to division we put the numerator in the accumulator and left shifts λ positions so that $43 + \lambda - 46 = 40$, that is, $\lambda = 43$.

In the preceding example we carried out the computation in the most obvious fashion to illustrate the scaling procedures. We did not attempt to use optimal scaling throughout, since this would have considerably complicated the steps involved. We also note that the scaling steps are dependent on the method of computing. If, for example, we had computed the numerator as

$$(4.6x - 0.9)x + 4.9,$$

which requires fewer machine operations, the scaling at each stage would have varied from that outlined.

7.4 SCALING FOR NON INTEGRAL OPERATIONS

The implied location of the base point relative to machine numbers need not be fixed at the rightmost position as it is with integers. Any fixed location can be assumed. Moreover, in accordance with Section 7.2.2 the only change in the integral scaling procedures developed above is to alter all integral scale-factor exponents by the constant exponent value required to move the base point from its integral position to the assumed location. Thus, let X^* be the machine representation for a nonintegral base point location and t be the smallest exponent such that X^*r^t is an integer. Then, if s' is a suitable scale-factor exponent for the integral machine representation of X, $s = s' - t$ is a suitable scale-factor exponent for X^*. This follows from the equations

$$X^* = Xr^s$$

and
$$X^* r^t = X r^{s+t} = X r^{s'}.$$

The main deviation from the integral case in general use is the fractional. This implies a base-point location immediately to the left of all digits that can contribute to the magnitude of the operand. Thus, in an n-digit complement machine, the base point is understood to be immediately after the leading sign digit and preceding all of the remaining $n - 1$ digits. In this case $t = n - 1$ and, if s' is permissible for integral operation, $s = s' - n + 1$ is permissible for fractional operation.

Example 7.24 In a 48-digit, binary, complement, fractional machine we wish to represent the number $X = 5.7$. Since $2^2 \leq |X| < 2^3$, $p = 3$ and the permissible integral scale-factor exponents are $s' \leq 44$. In this case, $t = 47$, and the conversion rule to fractional scale-factor exponents gives $s \leq -3$ as permissible.

Although fractional scaling can be based on integral scaling by the above technique, in practice the actual conversion is somewhat tedious. Moreover, since floating-point operation is often based on direct fractional scaling, it is worth while to consider the basic inequalities which are the direct fractional formulations of those we used for integral scaling. We again assume an n-digit machine number of which the leading digit is a sign digit, with the base point between this sign digit and the next digit to the right. Then $t = n - 1$, and permissible fractional scale-factor exponents s satisfy

$$s = s' - n + 1,$$

where s' is a permissible integral scale-factor exponent. If X is an operand for which $r^{p-1} \leq |X| < r^p$, then values of s' are given by

$$s' + p \leq n - 1,$$

so that

$$s + n - 1 + p \leq n - 1,$$

and

$$s + p \leq 0 \tag{7.7}$$

is the basic inequality for fractional scaling. In using (7.7) to form the machine operand X^*, we will truncate or fill out digits in $X \cdot r^s$ to obtain precisely $n - 1$ digits after the point, and then append the sign digit.

Example 7.25 For the value $X = 5.7$ of Example 7.24 we have $p = 3$. Thus, a direct application of the inequality (7.7) yields $s \leq -3$. The result is the same as in Example 7.24 but without the necessity of starting with integral exponents and converting through the subtraction of 47.

To re-express (7.7) in an appropriate form for each of the fundamental arithmetic operations, suppose that $X^* = Xr^s$ and $Y^* = Yr^t$ are fractional machine representations. We wish to form the sum (or difference)

7.4 SCALING FOR NON INTEGRAL OPERATIONS

$S = X + Y$, the product $P = XY$, and the quotient $Q = X/Y$. Let S^*, P^*, and Q^* be the ultimate n-digit machine results with

$$S^* = Sr^u, \qquad P^* = Pr^u, \qquad Q^* = Qr^u.$$

For addition or subtraction, as in the integral case, we must line up the base point, to obtain $s = t = u$. If $r^{p-1} \leq |S| < r^p$, equation (7.7) takes the form $u + p \leq 0$. Obtaining the scale-factor exponent u may require rescaling one or both of the operands X and Y before the addition or subtraction operation.

Formation of a fractional $2n$-digit product of two n-digit operands offers no more of a scaling problem than does the integral counterpart. The net exponent $s + t$ is always permissible, and, as with integral operation, the rescaling required is that involved in reduction to n-digit form. Contrary to the integral procedure which requires placement of the $(n - 1)$ most significant digits to the left of the base point, however, the $(n - 1)$ digits retained must be positioned to the right of the point. This means that the digits retained will be in the left-hand end of the register rather than the right-hand end. Thus, in fractional operation the right shift of integral operations will usually be replaced by a left shift to remove unnecessary sign digits in order to keep maximum accuracy. Let this left shift be λ positions to multiply by r^λ. Then, with $r^{p-1} \leq P < r^p$, we have

$$u = s + t + \lambda,$$

and the condition

$$u + p \leq 0$$

requires that

$$\lambda \leq -(p + s + t). \tag{7.8}$$

Optimal scaling, as in preceding cases, corresponds to the equality. This guarantees the removal of a maximum number of leading insignificant sign digits and retention of $(n - 1)$ product digits. The minus sign on the right of (7.8) need not concern us. If $|X| < r^{p_1}$ and $|Y| < r^{p_2}$, then $|XY| < r^{p_1 + p_2}$, so that $p \leq p_1 + p_2$. On the other hand, $s + p_1 \leq 0$ and $t + p_2 \leq 0$. Thus, $s + t + p \leq 0$ and $-(s + t + p) \geq 0$, so that the upper bound on λ is nonnegative. In the event that a permissible negative value is used, it may be interpreted as a right shift which will reduce the number of product digits retained.

Fractional division requires the placement of an n-digit dividend in the high-order end of the accumulator and extension to the lower half. Prior to division the dividend may have to be scaled up with a left shift, or down with a right shift, depending on the relative magnitudes of dividend and divisor. If we define this rescaling operation as a left shift of λ, with negative values interpreted as a right shift of $|\lambda|$, then $u = s - t + \lambda$. Since $u + p \leq 0$, we have for $r^{p-1} \leq |Q| < r^p$, $s - t + \lambda + p \leq 0$, and

$$\lambda \leq t - s - p. \tag{7.9}$$

In inequalities (7.7), (7.8), and (7.9) we will, as in the integral case, usually have to estimate the value of p. If p' is the estimate, then $p' \geq p$ gives correct results, but too large a value of p' may result in loss of accuracy. The consequences of using $p' < p$ will usually be incorrect results.

We illustrate fractional scaling by applying inequalities (7.7), (7.8), and (7.9) directly to the data of some previous examples.

Example 7.26 For the six binary-digit arithmetic and operands of Example 7.13, we know that $|X| < 1$, $|Y| < 2^{-2}$, and $|X + Y| < 2$. Using (7.7) we determine the permissible range of scale-factor exponents s, t, and u for X, Y, and $X + Y$, respectively, as $s \leq 0$, $t \leq 2$, and $u \leq -1$. For addition and subtraction, however, we must have $u = s = t$. Therefore we choose $s = t = u = -1$. Originally $X = 0.10101$ and $Y = 0.0010101$ and, scaled optimally, these take the machine form

$$X^* = 0.10101 = (0.10101)2^0,$$
$$Y^* = 0.10101 = (0.0010101)2^2.$$

Rescaling for addition with an end-off right shift we get the new machine version of X and Y

$$X^* = 0.01010 = (0.1010\ 1)2^{-1},$$
$$Y^* = 0.00010 = (0.0010\ 101)2^{-1}.$$

We then form the sum $X^* + Y^* = 0.01100 = (0.1100)2^{-1}$.

In Chapter 5 we introduced the concept of the fractional accumulator. The basic difference between the fractional and integral accumulator in multiplication is that the digits of the multiplicand are transmitted into the high-order end of the $2n$-digit accumulator rather than into the low-order. Thus, considering X and Y as integers, we form first in the cleared accumulator

$$y_0 X r^n + 0,$$

and follow this by a right shift. This yields $y_0 X r^{n-1}$, to which we then add $y_1 X r^n$, followed by a right shift to give $y_1 X r^{n-1} + y_0 X r^{n-2}$. Repetition of this procedure for $y_2, y_3, \ldots, y_{n-1}$ finally gives, after the final right shift following transmission of $y_{n-1} X r^n$ to A,

$$y_{n-1} X r^{n-1} + y_{n-2} X r^{n-2} + \cdots + y_0 X r^{n-n} = X(y_{n-1} r^{n-1} + \cdots + y_0) = XY.$$

Thus, in Chapter 5 we developed algorithms for an integral multiplication operation with integral operands for both the integral and the fractional accumulators. If all operands are to be interpreted as fractions, however, we must think of the base point as being immediately to the right of the first (sign) digit. Thus, if X_f, Y_f, are such fractional interpretations for the factors, we have

$$X = X_f \cdot r^{n-1} \quad \text{and} \quad Y = Y_f \cdot r^{n-1}.$$

7.4 SCALING FOR NON INTEGRAL OPERATIONS

The integral multiplication algorithm gives us

$$XY = X_f Y_f r^{2n-2}.$$

We set $P_f = X_f Y_f$, with the integral version of P_f given by

$$P = P_f \cdot r^{2n-1}.$$

Thus, $P = X_f Y_f r$. In effect this says that the digit-by-digit algorithm is the same as that given for integers, but operands and results interpreted as fractions require that the final digits be positioned differently. A fractional multiplication operation must do the equivalent of multiplying the integral result by r. This can be achieved by means of a final left shift or by omission of the final right shift. Because of this we may find two different multiplication operations in a computing machine. In any case, we assume here that the correct positioning is implicit in the operation itself and so has no effect on the scaling steps described above.

Example 7.27 In forming the eight-binary-digit product of the integers $X = 7 = 0111$ and $Y = 5 = 0101$ in a fractional accumulator, we obtain from the integral multiplication operation 00100011. This result can correctly be interpreted as $7 \times 5 = 35 = 100011$. If we interpret both the factors and the product as having the binary point right of the first digit, however,

$$(0.111)(0.101) = (0.100011);$$

that is,

$$\tfrac{7}{8} \cdot \tfrac{5}{8} = \tfrac{35}{64}.$$

To be correctly positioned the final result in A should be 01000110. This can be obtained in a fractional multiplication operation by the application of a left shift of one position to the result of the integral operation.

Example 7.28 In an eight-place, fractional, decimal accumulator we form the product of the two numbers $X = 9.2$ and $Y = 80$. With $p = 1$ and 2, respectively, we can correctly, but not optimally, choose to scale $X^* = X10^{-2} = 0092$ and $Y^* = Y10^{-3} = 0080$, with the points assumed to be just to the right of the leading sign digit. If we form the product of the digits 0092 and 0080 employing an integral multiplication operation in an eight-digit accumulator, we get 00007360. The fractional multiplication operation, however, requires a left shift of these digits one position, giving 00073600. Application of (7.8) with $p' = 3$ gives a left shift, $\lambda \leq 2$, and for optimal scaling, $\lambda = 2$. The left shift of two yields 07360000 with the scale factor 10^{-3}. Unscaling this result, we obtain

$$XY = (0.7360000)10^3 = 736.$$

In Chapter 6 we developed algorithms for both the integral and fractional accumulators, presupposing integral results. Let X, Y, Q, R, respectively,

represent the integral dividend, divisor, quotient, and remainder. If these operands and results are to be interpreted as fractions, we must again think of the base point as being immediately to the right of the first (sign) digit. Thus, if Y_f and Q_f are fractional interpretations of the divisor and quotient, we have
$$Y = Y_f r^{n-1}, \qquad Q = Q_f r^{n-1},$$
and
$$X - R = Y \cdot Q = Y_f \cdot Q_f \cdot r^{2n-2}.$$

If X_f and R_f are the fractional versions of the dividend and remainder,
$$X_f - R_f = Y_f \cdot Q_f,$$
and the integral form of this is
$$(X_f - R_f) \cdot r^{2n-1} = (Y_f \cdot Q_f) r^{2n-1} = (X - R)r.$$
Therefore, we may write
$$X - R = \left((X_f - R_f) \cdot r^{2n-1}\right) \cdot r^{-1}.$$

In effect this says that if the integral division operation is to apply correctly to fractional operands, the equivalent of a right shift of one position must be applied to the fractional dividend before initiation of the operation. The resulting fractional remainder would then appear in the double-length accumulator shifted right one position from its normal fractional location. In this position its associated scale-factor exponent will be $s + \lambda - 1$. The effect of removing the remainder from the low-order end of the accumulator into an n-digit register, however, is to multiply it by r^n if the fractional interpretation is retained. In this case the scale-factor exponent of the remainder will become $n + s + \lambda - 1$.

In fractional scaling of multiplication we assume a fractional multiplication operation, that is, automatic positioning of the product in the accumulator to place the base point just to the right of the first digit. Similarly, in fractional scaling of division we assume a fractional division operation, that is, an operation which includes the equivalent of a right shift of the scaled dividend prior to initiation of the procedure of Chapter 5.

Example 7.29 We wish to form the quotient of $X = 3.23$ and $Y = 1.51$ with $r = 10$ and $n = 4$ in fractional form. We will first place the extended version of X in an eight-digit register. Optimal scaling gives the four-digit machine representations as $X^* = (3.23)10^{-1} = 0323$ and $Y^* = (1.51)10^{-1} = 0151$. Since $1 \leq |X/Y| < 10$, we can take $p = 1$ in (7.9) to determine $\lambda \leq -1$. The optimal choice of $\lambda = -1$ specifies a right shift of one position before division. Thus, we put X^* into the double-length accumulator and extend it as 03230000 and right shift to get 00323000. The fractional division operation then produces $Q^* = 0213$ scaled at 10^{-1} and $R^* = 00000137$ scaled at 10^{-3}

in the eight-digit accumulator or $R^* = 0137$ scaled at 10 in a four-digit register. Applying the fractional interpretation to the machine operands and unscaling, we obtain

$$Q = (0.213) \cdot 10 = 2.13,$$
$$R = (0.137) \cdot 10^{-1} = 0.0137.$$

Thus,
$$3.23 = (2.13)(1.51) + 0.0137.$$

To get the integral equivalent of this division, we recall that the fractional division operation introduced one more right shift of the dividend. The initial dividend is therefore 00032300, and for the integers involved.

$$00032300 = (0151)(0213) + 00000137.$$

7.5 FLOATING-POINT OPERATION

In the previous sections we have considered how machine operands, all with the same *a priori* fixed location for the base point, can be made to represent numbers whose base point location may vary. This is done by means of a scale factor of the form r^s so that $X^* = Xr^s$. To reconstruct an operand from its machine representation, it is necessary to unscale in the form $X = X^*r^{-s}$. For external scaling the person using the computer does the accounting necessary to associate an appropriate scale-factor exponent with each machine number. In this sense each number X consists of a pair of numbers (X^*,s) of which the machine, internally, is responsible for the first and the programmer, externally, the second. We see from (7.1), (7.2), and (7.3) that the details of arithmetic on scale-factor exponents are very simple and can easily be done by the computer (only the exponent is needed since the base r is implicit). Since the specific machine operands are stored internally, the computer can also perform for each one the equivalent of determining p. Thus, if each internal fixed-point number is always associated with an internally stored integral exponent needed to relocate the base point, the machine can treat the combined pair as a single number with a variable base-point location. When the steps of scaling are carried out internally by the machine using these numbers, the mode of operation is referred to as *floating-point* and the numbers as *floating-point* numbers. A more accurate description would be machine-scaled operation, but the term floating-point is almost universally used.

For external scaling, permissible scale-factor exponents are chosen by means of the basic inequality $p + s \leq a$, where $r^{p-1} \leq |X| < r^p$. $X^* = Xr^s$ is the machine representation of X and a is determined by the number of digits used and the fixed base-point location in X^*. All of the preceding discussion has merely been an elaboration of this in which a given estimate of p gives a theoretically infinite set of permissible scale factor exponents (not all of which are useful). From this set an appropriate one can be chosen by the programmer on the basis of a set of criteria (such as aligning base points, or expediency, or some other which he himself supplies). If the scaling is to be done internally, however, the machine must be given a consistent way of choosing a single exponent at each step. This is done by means of normalizing the machine operand X^*, which, in effect, determines p precisely and selects the equality in all of the relevant inequalities.

By a *normalized* operand X^* represented by n digits, with the leading digit a sign digit, we mean one for which the digit in the $(n-1)$st position is not also a sign digit. That is, the two high-order digits are not equal.

This definition is applicable no matter where the implicit base point is located relative to the digits. The specific form of mathematical requirement imposed by normalization will be dependent on this location. We shall here consider only two cases, the integral and the fractional. For n-digit integral complement arithmetic, an n-digit integer X^* is normalized if and only if

$$r^{n-2} \le |X^*| < r^{n-1}. \tag{7.10}$$

Example 7.30 For a four-digit decimal machine the number $X = 2.3$ for integral representation permits any scale-factor exponent s for which $s \le 2$. The choices $s = 2, 1, 0$ give, respectively,

$$X^* = 0230 = (2.3)10^2,$$
$$X^* = 0023 = (2.3)10^1,$$
$$X^* = 0002 = (2.3)10^0,$$

and all negative values of s give

$$X^* = 0000.$$

Only the representation for the optimal choice of $s = 2$ satisfies

$$10^2 \le |X^*| = 0230 < 10^3,$$

and so it is the normalized operand.

If t is the scale-factor exponent required to convert a normalized n-digit (not necessarily integral) machine number X^* to a normalized n-digit integer $X^* r^t$, then

$$r^{n-2} \le |X^* r^t| < r^{n-1},$$

and X^* satisfies

$$r^{n-2-t} \le |X^*| < r^{n-1-t}.$$

For fractional representation, $t = n - 1$, and

$$r^{-1} \le |X^*| < 1, \tag{7.11}$$

called a normalized fraction. Conversely, an n-digit fractional representation with a leading sign digit not equal to the adjacent digit must satisfy (7.11). Therefore (7.11) is the mathematical form of the normalization condition in the fractional case.

A floating-point operand is a pair of numbers consisting of an n-digit normalized machine representation X^* and an integral exponent x such that, for $X \neq 0$, $X = X^* r^x$. We note that the notation $X^* r^x$ implies, as before, the truncation of X to exactly n digits including at least one sign digit. We introduce the notation

$$X = (X^*, x).$$

For integral normalization

$$X^* = \text{Integral part of } [Xr^{-x}],$$

and

$$r^{n-2} \leq |X^*| < r^{n-1}.$$

Therefore, if we find p by the usual inequality

$$r^{p-1} \leq |X| < r^p,$$

we have

$$r^{n-2} \leq |X^*| = |Xr^{-x}| < r^{p-x},$$

and it follows that $x \leq p - n + 1$. Similarly, $x \geq p - n + 1$ and we must use the equality $x = p - n + 1$, that is, $-x = n - 1 - p$. This means that normalizing always picks the unique optimal scale-factor exponent for X. The *negative* of this exponent is then recorded along with the normalized X^* as the floating-point machine representation.

Example 7.31 In decimal floating-point arithmetic, the normalized integer is allocated six digits, and the exponent part of the pair is allocated three. The number $\frac{16}{3} = 5.333\ldots$ is to be represented as a floating-point operand. To normalize we must have $10^4 \leq X^* < 10^5$, so we set

$$X^* = 053333 = \text{Integral part of } X10^4,$$

and

$$X = (053333)10^{-4} = (053333, 995).$$

The exponent -4 is expressed in 9's complement form.

For the fractional floating-point representation of X,

$$X = X^* r^x = (X^*, x),$$

where X^* is normalized fractionally, so that $r^{-1} \leq |X^*| < 1$. If p is defined in the usual way $x = p$, giving optimal fractional scaling of X.

Example 7.32 If the representations of Example 7.31 were fractional we would need $|X^*| \geq 0.1$ for normalization. Thus,

$$X^* = 0.53333 = X10^{-1} \quad \text{and} \quad X = (0.53333, 001).$$

Since the base point does not appear explicitly, the normalized part is 053333 in both cases, but the exponents differ.

The number of digits allocated to exponents imposes restrictions on the range of operands that can be represented in floating-point form. If m digits, including a sign digit, are allocated, then exponents x must satisfy

$$0 \le |x| < r^{m-1};$$

that is,

$$-r^{(m-1)} < x < r^{m-1}.$$

If exponents $x \ge r^{m-1}$ are generated, the resultant condition is called exponent overflow. If $x \le -r^{(m-1)}$, the condition is called exponent underflow. For integral operation, we combine $r^{n-2} \le |X^*| < r^{n-1}$ and $r^{-(m-1)} < x < r^{m-1}$ to get

$$r^{n-2-r^{(m-1)}} < |X| = |X^* r^x| < r^{n-1+r^{m-1}}$$

as the permissible range on X for floating point representation.

Even the smallest permissible number has a positive absolute value since we have so far excluded zero from floating-point representation. Zero will need to be included, but requires special consideration. This stems from two causes. There is no normalized version of zero, since it does not have any nonsign digits. Furthermore, even if we define the nonnormalized version $X^* = 0$, we still have

$$X = 0 = X^* r^x = 0 r^x,$$

valid for *any* exponent x. Thus, the fact that zero cannot be normalized prevents a unique choice of the exponent x. We shall assume here that the most obvious values are assigned to X^* and x; that is, each is given the value zero, so that

$$0 = (0, 0).$$

This assumption may not correspond to the treatment of zero in some computing machines. Since X^* is not normalized, it is apparent that whatever normalization procedures are used must be modified to accommodate this fact.

Floating-point operands are stored as ordered pairs. Thus, the four arithmetic operations on these operands become a form of arithmetic on ordered pairs. We will present here a brief discussion of the arithmetic steps required in this method of internal machine scaling. In each case we will assume n digital locations for the normalized X^* and m digital locations for the integral exponent x.

The treatment of floating-point operation for arithmetic with absolute values and separately stored signs does not differ materially from that of the above discussion. In the case of fractional normalization we can think of all

n digits of X^* as being to the right of the base point with no sign digit needed to the left. The normalization condition (7.11), however, remains the same. For integral normalization, all n digits of X^* are potentially significant, and in this case the condition (7.10) is replaced by

$$r^{n-1} \leq |X^*| < r^n. \tag{7.12}$$

In effect, then, the absolute-value case is essentially the same as the complement case with $n - 1$ replaced by n where it occurs explicitly in integral normalization and implicitly in the inclusion of the additional digit in a normalized fraction. In either case the number is normalized if and only if the leading digit is different from zero.

7.5.1 Floating-Point Addition and Subtraction

We wish to form the floating-point representation of $X \pm Y$ where $X = (X^*, x)$ and $Y = (Y^*, y)$. We face the same scaling problems as for external scaling. We must first align the base points by providing equal exponents. Each increase of one in an exponent (multiplying X by r) requires a compensating multiplication by r^{-1} to preserve equality. That is with $X = X^* r^x r r^{-1}$ we form

$$X = (X^* r^{-1}) \cdot r^{x+1}.$$

Multiplication of X^* by r^{-1} requires a right shift of one position. We do not decrease the exponent, which is optimal, since this would require a compensating left shift of X^*, producing overflow. Thus, if $x \neq y$, say $x > y$, we form for Y the operand $Y = Y^* r^y r^{-x} r^x = (Y^* r^{y-x}) r^x$ so that $Y = (\bar{Y}^*, x)$ where $\bar{Y}^* = Y^* r^{y-x}$, requiring a right shift of $y - x$ in Y^*. In the event that $y - x > n$, all significance in Y is lost and it is equivalent to zero. If addition of X^* and $\bar{Y}^*$ occurs in an exactly n-digit accumulator we will need to reduce each exponent to $x - 1$ with an additional right shift of one for each normalized number. If we assume, however, the equivalent of at least one overflow accumulator position, we can form directly

$$X \pm Y = (X^*, x) \pm (\bar{Y}^*, x) = (X^* \pm \bar{Y}^*, x),$$

since this reflects $X \pm Y = X^* r^x \pm \bar{Y}^* r^x = (X^* \pm \bar{Y}^*) r^x$ with base point aligned. The result, however, may not be a legitimate floating-point form, since $X^* + \bar{Y}^*$ may not be normalized. The lack of normalization can result from a carry into the sign position or because a result of small absolute value produces more than one leading sign digit. The first condition requires a single right shift of $X^* + \bar{Y}^*$ to normalize and a compensating increase in the final exponent part of the result. If the second condition produces s leading sign digits, normalization requires a left shift of $s - 1$ positions and a decrease of $s - 1$ in the exponent of the result. The shift and compensation in an exponent are required in the first case because the upper-bound condition of (7.10) or (7.11) is violated, and we have

$$X \pm Y = (X^* \pm \bar{Y}^*)r^x = ((X^* \pm \bar{Y}^*)r^{-1})r^{x+1}$$
$$= ((X^* \pm \bar{Y}^*)r^{-1}, x + 1).$$

Too many leading sign digits or $s > 1$ implies violation of the lower-bound condition in (7.10) or (7.11) and

$$X \pm Y = (X^* \pm \bar{Y}^*)r^x = ((X^* \pm \bar{Y}^*)r^{s-1})r^{x-s+1}$$
$$= ((X^* \pm \bar{Y}^*)r^{s-1}, (x - s + 1)).$$

If $s = n$, normalization is not possible and the result is zero. In the case of a zero operand, the choice of the form (0, 0) may result in a compensating right shift of the other operand to increase its exponent to zero. This would lead to unnecessary loss of significance, and, practically, the use of the smallest permissible exponent instead of zero might have been a better choice.

Example 7.33 The numbers $X = 1.213$ and $Y = 35.4$ are to be added in floating-point, decimal nine's-complement arithmetic using five digits for the normalized value and three digits for exponent. For integral normalization

$$X = (01213)10^{-3} = (01213, 996),$$
$$Y = (03540)10^{-2} = (03540, 997).$$

We right shift X^* by one and correspondingly increase the exponent from -3 to -2, so that

$$X = (00121, 997) \quad \text{and} \quad Y = (03540, 997),$$

and then form $X + Y = (03661, 997) = 36.61$. No normalizing is necessary. For fractional normalization we would have

$$X = (0.1213, 001), \quad Y = (0.3540, 002),$$

with

$$X = (0.0121, 002),$$
$$Y = (0.3540, 002),$$
$$X + Y = (0.3661, 002) = 36.61.$$

Example 7.34 The numbers 23.45 and -23.44 are to be added in the floating-point arithmetic and format of Example 7.33. For integral normalization,

$$X = (02345, 997), \quad Y = (97655, 997).$$

No change in either exponent is required to align the point for addition, which yields

$$X + Y = (02345 + 97655, 997)$$
$$= (00001, 997).$$

This result is not complete, however, since it is not normalized. A left shift of three and a compensating decrease in the exponent is required to give

$$X + Y = (01000, 994) = (01000)10^{-5} = 0.01.$$

Note that accuracy has deteriorated from four digits to one. For the fractional version of this computation

$$X = (0.2345, 002), Y = (9.7655, 002),$$
$$X + Y = (0.0001, 002) = (0.1000, 998) = (0.1000)10^{-1} = 0.01.$$

As illustrated in Example 7.34 the normalized part and the exponent of a floating-point representation each has its own algebraic sign, and these are independent of each other.

The problem of normalizing in floating-point is solved internally. One way of doing this is to sense when the two highest-order digits are distinct, that is, when the leading nonzero digit in the absolute value is different from zero. Since the highest-order digit is exclusively a sign digit, the process of normalizing can be accomplished in a straightforward way by sensing when the two highest-order digits are distinct. A count of the number of shifts required provides the magnitude of the compensation in the exponent. Thus, in Example 7.34, starting with the initial result and compensating for each shift,

$$(00001, 997)$$
$$(00010, 996) \quad \text{one left shift,}$$
$$(00100, 995) \quad \text{two left shifts,}$$
$$(01000, 994) \quad \text{three left shifts,}$$

with the two high-order digits finally satisfying $0 \neq 1$. Had the result been negative we would have had, in order

$$(99998, 997)$$
$$(99989, 996)$$
$$(99899, 995)$$
$$(98999, 994),$$

where $9 \neq 8$ is the signal that normalization is attained.

The process of normalizing must take special account of zero, since zero cannot be normalized. If the technique outlined above is applied to zero, no combination of shifts would yield $d_{n-1} \neq d_{n-2}$. This condition becomes obvious after $n - 1$ left shifts do not yield the desired result. In this case the operand should be converted to the standard form defined for zero.

Example 7.35 Using the decimal floating-point arithmetic of Examples 7.33, 7.34 and integral normalization, the number 234.51 is to be subtracted

from the number 234.52. We have
$$X = (02345, 998),$$
$$Y = (02345, 998),$$
$$X - Y = (00000, 998).$$

Since four left shifts still yield $d_3 = d_4$, the number is recognized as zero and the result is converted to the standard form (00000, 000).

The results of Example 7.35 point out a further complication in floating-point computation with zero. Besides the fact that there may be two versions (positive and negative zero) in $(r - 1)$'s-complement arithmetic, there is no unique exponent. This is further complicated, as above, by nonzero numbers which appear to be zero because of limited accuracy. The true result for $X - Y$ in Example 7.35 is 0.01, which could have been represented in the format utilized but was not obtained because there were too few digits to represent the operands to full accuracy.

7.5.2 Floating-Point Multiplication

For addition and subtraction the distinction between integral and fractional normalization in floating-point operation is not significant. The steps for internal scaling are the same, and the only real difference lies in the value of the exponents. Since addition and subtraction require only alignment of base points, their exact location is immaterial as long as the exponents align them properly. In multiplication, however, the differences can be more substantial, corresponding to the fact that the product of the normalized factors will be formed in either an integral or fractional accumulator. Furthermore, since either kind of accumulator can be used for either type of normalization, the shifting and exponent compensation required will be influenced by the choices made. In all cases, however, we will assume that the multiplication (and division) operation involved is appropriate to the kind of normalization used. Thus, as with external scaling, the proper positioning of the digits is implicit in the operation itself and does not affect the following discussion.

Example 7.36 The numbers 0.1 and 0.2 are to be multiplied using the same arithmetic and format of Examples 7.33 through 7.35. For integral normalization
$$X = (01000, 995) \quad \text{and} \quad Y = (02000, 995).$$

If the product is formed in a ten-digit accumulator, we have
$$XY = ((01000)(02000), 995 + 995)$$
$$= (0002000000, 991).$$

We obtain the normalized part in the lower five digits of the accumulator with a right shift of three positions. This requires an increase of three in the

7.5 FLOATING-POINT OPERATION

exponent to give

$$XY = (02000, 994) = (02000)10^{-5} = 0.02.$$

On the other hand, for fractional normalization

$$X = (0.1000, 000) \quad \text{and} \quad Y = (0.2000, 000),$$

and the double-length product is

$$XY = (0.0200000000, 000).$$

We obtain the normalized part in the upper five digits of the accumulator by a left shift of one. This requires a decrease of one in the exponent to give

$$XY = (0.2000, 998) = (0.2000)10^{-1} = 0.02.$$

The two normalized parts are the same, but the exponents are different. However, in each case the steps to reduce the product to normalized form are different.

We consider first the case where the normalized part is an integer. Thus,

$$X = X*r^x = (X*, x) \quad \text{and} \quad Y = Y*r^y = (Y*, y),$$

with

$$r^{n-2} \le |X*| < r^{n-1} \quad \text{and} \quad r^{n-2} \le |Y*| < r^{n-1}.$$

Initially, we form the product XY as

$$XY = X*Y*r^{x+y},$$

and $X*Y*$ is double length with $2n$ digits, of which the leading one is a sign digit. We note, however, that

$$r^{2n-4} \le |X*Y*| < r^{2n-2},$$

and is thus not normalized for $2n$ digits. We wish to normalize for n digits and adjust the exponent accordingly. Multiplication by $r^{-(n-1)}$ yields

$$r^{n-3} \le |X*Y*r^{-n+1}| < r^{n-1},$$

which fits an n-digit register but may not be normalized. Thus, we will have to consider two cases:

1. $r^{2n-3} \le |X*Y*| < r^{2n-2}$, and   2. $r^{2n-4} \le |X*Y*| < r^{2n-3}$.

For Case 1, multiplications by $r^{-(n-1)}$ by a right shift of $n-1$ positions gives

$$r^{n-2} \le |X*Y*r^{-n+1}| < r^{n-1}.$$

Therefore $P* =$ Integral part of $[X*Y*r^{-n+1}]$ is normalized. In this case we form

$$XY = X*Y*r^{x+y} = (X*Y*r^{-n+1})r^{n-1}r^{x+y} = P*r^{x+y+n-1},$$

so that $XY = (P^*, x + y + n - 1)$. In Case 2, we multiply by $r^{-(n-2)}$ by a right shift of $n - 2$ positions to get

$$r^{n-2} \leq |X^*Y^*r^{-n+2}| < r^{n-1}.$$

Thus, $P^* =$ Integral part of $[X^*Y^*r^{-n+2}]$ is normalized. Therefore we proceed as above to obtain

$$XY = P^*r^{x+y+n-2} = (P^*, x + y + n - 2).$$

We can distinguish between the two cases by observing the leading digits of X^*Y^*. If there are exactly two sign digits, we have the first case, so we move the normalized part of the product into the n low-order positions by a right shift of $n - 1$. We then add x and y to $n - 1$ to give the final exponent $x + y + n - 1$. If there are exactly three sign digits, only a shift of $n - 2$ positions is required, and we add x and y to $n - 2$ to calculate the final exponent $x + y + n - 2$.

The two cases above do not include the possibility that at least one of the factors is zero. In this case the product X^*Y^* consists of *all* sign digits, and, by definition, $XY = (0, 0)$.

For fractional floating point arithmetic, again we have

$$X = X^*r^x = (X^*, x) \quad \text{and} \quad Y = Y^*r^y = (Y^*, y),$$

but the appropriate normalization inequalities are

$$r^{-1} \leq |X^*| < 1 \quad \text{and} \quad r^{-1} \leq |Y^*| < 1.$$

The initial product is formed as

$$XY = X^*Y^*r^{x+y},$$

with

$$r^{-2} \leq |X^*Y^*| < 1.$$

Again we must consider two cases:

1. $r^{-1} \leq |X^*Y^*| < 1$, and
2. $r^{-2} \leq |X^*Y^*| < r^{-1}$.

In Case 1, $P^* = X^*Y^*$ truncated to the high-order n digits is already normalized. Furthermore, this truncation requires no adjustment of the exponent, so the final result is

$$XY = P^*r^{x+y} = (P^*, x + y).$$

For Case 1 X^*Y^*r satisfies (7.11) and $P^* = X^*Y^*r$ truncated to the high-order n digits is normalized. Thus we left-shift X^*Y^* one position and drop the final n digits. The left shift must be compensated by decreasing the exponent. Thus,

$$XY = X^*Y^*r^{x+y} = (X^*Y^*r)r^{x+y-1} = P^*r^{x+y-1} = (P^*, x + y - 1).$$

7.5 FLOATING-POINT OPERATION

Again we can distinguish between the two cases by the number of sign digits of $X*Y*$. In Case 1 there is exactly one, and in Case 2 there are exactly two sign digits.

Example 7.37 We form the product of 9.8 and 0.86 in five-digit decimal integral floating-point arithmetic. We have

$$X = (09800, 996) \quad \text{and} \quad Y = (08600, 995).$$

In the ten-digit accumulator the initial product is

$$X*Y* = (09800)(08600) = 0084280000.$$

With two leading sign digits we have Case 1. Therefore we right shift by four positions and take $P* = 08428$. The exponent is $x + y + n - 1 = -3$, and

$$XY = (08428, 996) = (08428)10^{-3} = 8.428.$$

For fractional normalization we have

$$X = (0.9800, 001), \quad Y = (0.8600, 000)$$

and

$$X*Y* = 0.842800000.$$

With one sign digit, this is fractional Case 1, and we form $P* = 0.8428$ by deleting the final $n = 5$ digits. The exponent $x + y = 001$ needs no adjustment and we have

$$XY = (0.8428, 001) = 8.428.$$

Example 7.38 The numbers -1.2 and 21.3 are to be multiplied in their integral floating-point form

$$X = (98799, 996), \quad Y = (02130, 997);$$

therefore

$$X*Y* = (98799)(02130) = 9997443999.$$

With three sign digits we have integral Case 2. Therefore we right-shift $n - 2 = 3$ positions to form $P* = $ Integral part of $[X*Y*10^{-3}] = 97443$. In this case exponent the $x + y + n - 2 = -2$ so that

$$XY = (97443, 997) = (-02556)10^{-2} = -25.56.$$

Had the normalization been fractional, we would have had the values

$$X = (9.8799, 001), \quad Y = (0.2130, 002),$$

and obtained

$$X*Y* = (9.8799)(0.2130) = 9.974439999.$$

With two leading zeros this must be normalized by a left shift of one, after which deletion of the last $n = 5$ digits gives $P* = 9.7443$. In this case, the

exponent is $x + y - 1 = 2$, and
$$XY = (9.7443, 002) = -25.56.$$

7.5.3 Floating-Point Division

When we form the quotient of two floating-point operands, the algorithms required will again depend on the assigned base-point location in the normalized part. We consider first the formation of X/Y with

$$X = X^* r^x = (X^*, x) \quad \text{and} \quad Y = Y^* r^y = (Y^*, y),$$

where X^* and Y^* are n-digit normalized integers. As usual for integral division, we place X^* in the lower half of the double-length accumulator and left shift λ positions before dividing. Because of normalization we know

$$r^{-1} = r^{n-2}/r^{n-1} < |X^*/Y^*| < r^{n-1}/r^{n-2} = r.$$

Therefore, we take $p' = 1$ and, with $s = t = 0$, inequality (7.6) yields $\lambda \leq n - 2$. Thus, the best left shift for X^* is $n - 2$ positions. We form the preliminary machine quotient Q_1^* from the Division Algorithm,

$$X^* r^{n-2} = Y^* Q_1^* + R^*,$$

and ignore R^*. This guarantees

$$r^{n-3} < |(X^* r^{n-2})/Y^*| < r^{n-1},$$

so

$$r^{n-3} < |Q_1^*| < r^{n-1}.$$

Again there are two cases to consider.

1. $r^{n-2} < |Q_1^*| < r^{n-1}$ or $1 < |X^*/Y^*| < r$, and
2. $r^{n-3} < |Q_1^*| < r^{n-2}$ or $r^{-1} < |X^*/Y^*| < 1$.

In Case 1, Q_1^* is already normalized and $Q^* = Q_1^*$ is the normalized part of the quotient. Since we have used $X^* r^{n-2}$ as the dividend we have

$$\frac{X}{Y} = \frac{(X^* r^{n-2}) r^{x-(n-2)}}{Y^* r^y} = \frac{X^* r^{n-2}}{Y^*} r^{x-y-(n-2)}.$$

Thus, the final exponent is $x - y - n + 2$ and

$$X/Y = Q^* r^{x-y-n+2} = (Q^*, x - y - n + 2).$$

In Case 2, Q_1^* has an excess sign digit and so requires a left shift of one for normalization, to give $Q^* = Q_1^* r$. This requires a compensating decrease of one in the exponent. That is

$$\frac{X}{Y} = \frac{X^* r^{n-2}}{Y^*} r^{x-y-(n-2)} = \left(\frac{X^* r^{n-2}}{Y^*} r\right) r^{x-y-(n-2)-1},$$

7.5 FLOATING-POINT OPERATION

so with $Q^* = Q_1^* r$ and the exponent $x - y - n + 1$,

$$X/Y = Q^* r^{x-y-n+1} = (Q^*, x - y - n + 1).$$

We can distinguish the two cases by examining the leading digits. In Case 1 there will be exactly one leading sign digit and in Case 2 exactly two.

Example 7.39 We wish to form X/Y_1 and X/Y_2, where

$$X = 23.5 = (02350, 997),$$
$$Y_1 = 1.6 = (01600, 996),$$

and

$$Y_2 = 392.1 = (03921, 998).$$

For X/Y_1 we form, with $n - 2 = 3$, the dividend $X*10^3 = 0002350000$, and the Division Algorithm results in

$$0002350000 = (01600)(01468) + 00012.$$

Thus, $Q_1^* = 01468$ and, with just one sign digit, is already normalized. We set $Q^* = 01468$. The exponent is $x - y_1 - n + 2 = -2$, and so

$$X/Y_1 = (01468, 997) = 14.68.$$

For X/Y_2 we again use $X*10^3 = 0002350000$ as the dividend to get

$$(0002350000) = (03921)(00599) + 01321,$$

from which $Q_1^* = 00599$ with two sign digits is not normalized. By a left shift $Q^* = Q_1^* 10$ and we set $Q^* = 05990$ and then form the exponent as $x - y_2 - n + 1 = -5$. Thus,

$$X/Y = (05990, 994) = 0.0599.$$

For fractional floating point division, the dividend is entered as a normalized fraction. If we denote this extended version by $\bar{X}^*$, then $\bar{X}^* = X^*$ and

$$\frac{X}{Y} = \frac{\bar{X}^* r^x}{Y^* r^y} = \frac{\bar{X}^*}{Y^*} r^{x-y}.$$

Since $\bar{X}^*$ and Y^* satisfy

$$r^{-1} \leq |\bar{X}^*| < 1 \quad \text{and} \quad r^{-1} \leq |Y^*| < 1,$$

the quotient satisfies

$$r^{-1} < |\bar{X}^*/Y^*| < r.$$

Therefore the fractional division operation can produce quotient overflow. Thus $\bar{X}^*$ must be scaled down by a right shift of one position, so that

$$r^{-2} < |(\bar{X}^* r^{-1})/Y^*| < 1.$$

Ignoring the remainder we set the preliminary quotient $Q_1^* = \bar{X}*r^{-1}/Y*$. Again two cases must be considered:

1. $r^{-1} \leq |Q_1^*| < 1,$ and
2. $r^{-2} \leq |Q_1^*| < r^{-1}.$

In Case 1, $Q^* = Q_1^*$ is the normalized part of the quotient and

$$\frac{X}{Y} = \left(\frac{X*r^{-1}}{Y*}\right) r^{x-y+1},$$

so the exponent is $x - y + 1$ and

$$X/Y = (Q^*, x - y + 1).$$

For Case 2, $Q^* = Q_1^* r$ satisfies (7.11) and

$$\frac{X}{Y} = \left(\left(\frac{\bar{X}*r^{-1}}{Y*}\right)r\right) r^{x-y} = Q^* r^{x-y}.$$

Thus, the exponent is $x - y$ so that

$$X/Y = (Q^*, x - y).$$

The two cases are again distinguished by the fact that in Case 1 the preliminary quotient has exactly one sign digit and in Case 2 exactly two.

Example 7.40 With the same operands of Example 7.39, but with fractional normalization, we have

$$X = 23.5 = (0.2350, 002),$$
$$Y_1 = 1.6 = (0.1600, 001),$$

and

$$Y_2 = 392.1 = (0.3921, 003).$$

To ensure against quotient overflow, we form $\bar{X}*10^{-1} = 0.023500000$. This gives, for X/Y_1,

$$Q_1^* = 0.1468,$$

which is already normalized, so

$$Q^* = Q_1^* = 0.1468$$

and the exponent is $x - y_1 + 1 = 2$. Therefore,

$$X/Y_1 = (0.1468, 002) = 14.68.$$

For X/Y_2, $Q_1^* = 0.0599$. To normalize, we shift left one position to get $Q^* = 0.5990$ and the exponent remains $x - y_2 = -1$. Therefore,

$$X/Y = (0.5990, 998) = 0.0599.$$

EXERCISES

SECTION 7.1

7.1 Form the sum $X + Y$, product XY, and quotient X/Y of the decimal numbers $X = 12.3457$ and $Y = 3.2653$. Use the usual method, retaining four digits in the quotient. Describe the integral operations involved and location of the decimal point in terms of powers of ten.

7.2 Interpret the numbers in Exercise 7.1 as octal numbers and repeat the exercise. Notice that the steps for locating the base point are the same. What is the difference in terms of scale factors?

SECTION 7.2

7.3 Suppose we consider a decimal machine with ten digits for each number. We wish to interpret ten-digit sequences as integers with a sign retained. What would be a suitable scale factor to retain as many digits as possible in the representation of $e = 2.7182818284590\ldots$? What sequence of digits would constitute the machine representation X^*?

7.4 For the ten-digit machine of Exercise 7.3 what scale factor would you use if the decimal point is interpreted as being immediately to the right of the first (sign) digit? What would be the sequence of digits in X^*?

SECTION 7.2.1

7.5 Determine the values of p, s, and n in condition (7.1) for the machine and number of Exercise 7.3. Which value of s did you use? What is the effect of using the permissible smaller values of s?

7.6 For an eight-bit binary machine we wish to find suitable scale factors for the decimal number 17.25. What are the permissible values of s? If the optimal value of s is chosen, what is the binary representation X^*? What changes occur if the number is -17.25 and we use one's-complement representation?

SECTION 7.2.2

7.7 For the binary representation of Exercise 7.6, what adjustment to integral scale-factor exponents is required if the binary point is interpreted as following the first (sign) digit? What scale-factor exponents are permissible for the operands of the exercise? What are the resulting digital sequences?

7.8 In terms of condition (7.1) what is the relation between the value chosen for s and the total number of digits (excluding leading zeros) retained from the original operand? How is this relation affected by choosing a different location for the base point?

7.9 For the eight-bit integral machine of Exercise 7.6, a scale factor of $s = 10$ (decimal) is to be used. What is the permissible range of values on operands X?

SECTION 7.3.1

7.10 For seven-place, decimal, complement integral arithmetic we wish to represent

1234.56 and 65.4321 optimally. What scale-factor exponents are appropriate? If the two machine operands are to be added, what adjustments are required in scale-factor exponents?

7.11 Do the scaling steps of Exercise 7.10 for the same operands if the arithmetic is 24-bit binary arithmetic.

7.12 In 48-bit complement arithmetic, X and Y are correctly scaled at 2^{45}. No further information is available. What scaling steps are needed to form $X + Y$?

7.13* Rescaling for addition may result in loss of digits. Suppose an n-digit base-r, complement machine has an operand X correctly scaled at r^s. What range of scale factors for Y would make the operation $X + Y$ appear as though Y is zero (no digits retained)? What is the corresponding relation between X and Y themselves?

SECTION 7.3.2

7.14 The numbers X and Y are optimally scaled at 2^{43} and 2^{51} in a 48-bit complement machine. What scaling steps are required if the product is formed in a 48-bit register? What steps are required if the product is formed in a 96-bit register? Assume integral operation.

7.15* Is there an analog of the situation of Exercise 7.13 for multiplication? That is, for X-scaled r^s is there a range of values of Y which would make XY appear as zero?

SECTION 7.3.3

7.16 For the machine and operands of Exercise 7.14 what scaling steps are required to form X/Y? Assume a 96-bit dividend register.

7.17* Suppose that Y, scaled at r^0, is $Y = 0$. What conditions, if any, for correct division in X/Y are violated? In what sense can the impossibility of division by zero be detected in a computer?

SECTION 7.3.4

7.18 Newton's method applied to $f(x) = x^3 + x - 1 = 0$, with $x_0 = 0.5$, yields the iteration

$$x_{n+1} = x_n - \frac{x_n^3 + x_n - 1}{3x_n^2 + 1}.$$

Describe the scaling steps necessary to carry out this iteration five times on a 60-bit, complement, integral, binary computer. Assume that the computation is carried out as described in the iterative equation.

7.19 Describe the scaling for the problem of Example 7.23 if the numerator is computed as $(4.6x + 0.9)x + 4.9$.

SECTION 7.4

7.20 Repeat Exercise 7.18 for fractional operation.

7.21 Repeat Exercise 7.19 for fractional operation.

EXERCISES

SECTION 7.5

7.22 Suppose we are going to use one's-complement arithmetic in a 36-bit machine in which eight bits are for exponent, including sign, and 28 bits are for the digits of the number represented as a normalized integer. Find the floating-point format for the decimal numbers 2.5, 0.25, -2.5, and -0.25.

7.23 What is the format for the numbers in Exercise 7.22 if the fractional interpretation is used?

7.24* Prove that normalization always yields optimal scaling.

SECTION 7.5.1

7.25 Carry out the analysis of the machine scaling steps for Exercise 7.10 if floating-point operation is used with an integral interpretation.

7.26 Redo Exercise 7.24 using fractional floating-point interpretation.

SECTION 7.5.3

7.27 Describe the steps of computation in Exercise 7.18 for floating-point operation, both integral and fractional.

7.28 Describe the steps of computation in Exercise 7.19 for floating-point operation, both integral and fractional.

CHAPTER 8

ERROR IN COMPUTATION

8.1 INTRODUCTION

In Chapter 1 we considered how a real number can be represented by a sequence of digits, base r. The sequence can always be considered as infinite, although for some numbers the representation may terminate with a last nonzero digit followed by an infinite string of zeros. Except for such numbers, when all nonzero digits are retained, the restriction to computation with a finite number of digits introduces an error. The two most commonly used methods of reducing an infinite string of digits to a terminating set are truncation and rounding. In truncation the first α digits of the number are retained, without regard to the value of those not used, while in rounding, the retained digits may be modified on the basis of this value. Thus, if we reduce the number $\frac{2}{3} = 0.6666\ldots$ to four digits by truncating we would use the approximation 0.6666, but rounding would dictate the choice of 0.6667. We shall consider both of these methods and the magnitude of the error each generates, but for either method we will refer to the error and the concomitant errors which it generates in computation as *round-off error*.

In the application of computation to the numerical solution of mathematical or physical problems, round-off is not the only source of error. Error may be introduced because there is inadequate information about the primary data to be operated on. Thus, for example, in obtaining numbers from physical measurements, the measuring devices used will usually provide results of limited accuracy. Another source of error is that which is implicit in a computational algorithm itself. For computation we will often replace the exact mathematical formulation by an approximation to it; for example, integrals may be replaced by approximating sums and derivatives by difference quotients. This error, while similar in character to round-off error, is distinct, and is often called truncation error.

There may be error introduced by a mistake or blunder in computation, for example, adding five to six and obtaining twelve. This kind of error is much less likely to occur if arithmetic operations are carried out by a computer. Ways of guarding against faulty machine arithmetic include repeating a calculation in a different but equivalent manner and comparing the results. They also include such methods as casting out nines. These techniques are costly, however, and offer no guarantee against new errors resulting from the computational operations associated with the checking procedures. Improvement in machine reliability has, over the years, made undetected machine failure a less serious question than round-off. Therefore we shall give errors due to such causes no further consideration.

Errors in data, round-off errors, and truncation errors are not only independent sources of error, but may very well have interacting effects on the final accuracy. Errors in initial data are properly a topic for the field of application. Truncation errors depend on the numerical algorithms employed, as do the cumulative effects of round-off errors introduced in the calculations. We shall not attempt to discuss this problem. Rather we will restrict the discussion to some basic ideas about round-off errors, their possible consequence in arithmetic computation, and some of the techniques which can be employed to reduce them.

8.2 THE FUNDAMENTAL ROUND-OFF ERROR

We shall consider two forms for the error in approximating one real number by another. Let X be any real number and $\bar{X}$ any approximation to it. By $e(\bar{X})$ we mean the *error* generated in approximating X by $\bar{X}$, given by

$$e(\bar{X}) = X - \bar{X},$$

and by $\rho(\bar{X})$ we mean the *relative error*

$$\rho(\bar{X}) = \frac{e(\bar{X})}{X} = \frac{X - \bar{X}}{X}.$$

To contrast it with the relative error, the error is frequently referred to as the absolute error. The error, however, is a signed number and we are often concerned with its absolute value. Therefore, to avoid confusion we shall continue to refer to it simply as the error.

Let $\bar{X}$ be a finite digital representation which approximates X. We consider two schemes for arriving at $\bar{X}$ when both it and X are represented in the base r. For one scheme we will simply *cut* X to the number of digits in $\bar{X}$, say α. In the other method we will *round* X to α digits. Let

$$r^{p-1} \leq |X| < r^p$$

as determined in the preceding chapter. Then

$$X = +[d_{p-1}r^{p-1} + d_{p-2}r^{p-2} + \cdots + d_{p-\alpha}r^{p-\alpha} + d_{p-\alpha-1}r^{p-\alpha-1} + \cdots]$$

$$= \pm \sum_{j=1}^{\infty} d_{p-j} r^{p-j}, \quad d_{p-1} \neq 0,$$

and we have
a) X *cut* to α digits if we choose $\bar{X} = X_{c\alpha}$, where

$$X_{c\alpha} = \pm [d_{p-1}r^{p-1} + d_{p-2}r^{p-2} + \cdots + d_{p-\alpha}r^{p-\alpha}]$$

$$= \pm \sum_{j=1}^{\alpha} d_{p-j}r^{p-j};$$

b) X *rounded* to α digits if we choose $\bar{X} = X_{r\alpha}$, where

$$X_{r\alpha} = \pm [d_{p-1}r^{p-1} + \cdots + d_{p-\alpha+1}r^{p-\alpha+1} + D_{p-\alpha}r^{p-\alpha}]$$

$$= \pm \left[\sum_{j=1}^{\alpha-1} d_{p-j}r^{p-j} + D_{p-\alpha}r^{p-\alpha} \right],$$

and

$$D_{p-\alpha} = d_{p-\alpha} \text{ if } \sum_{j=\alpha+1}^{\infty} d_{p-j}r^{p-j} < \tfrac{1}{2}r^{p-\alpha} \quad \text{(rounding down)}$$

or

$$D_{p-\alpha} = d_{p-\alpha} + 1 \text{ if } \sum_{j=\alpha+1}^{\infty} d_{p-j}r^{p-j} \geq \tfrac{1}{2}r^{p-\alpha} \quad \text{(rounding up)}.$$

In this latter case a carry may be generated when $d_{p-\alpha} + 1$ is formed.

We note that, in rounding for an even base, the determination of the final value $D_{p-\alpha}$ can be based on inspection of the first digit $d_{p-\alpha-1}$ not retained. If $d_{p-\alpha-1} < r/2$, then

$$D_{p-\alpha} = d_{p-\alpha},$$

while if $d_{p-\alpha-1} \geq r/2$, then

$$D_{p-\alpha} = d_{p-\alpha} + 1.$$

To see this, in the case of rounding down, we note that $r/2$ is a digit. Therefore, if $d_{p-\alpha-1} < r/2$, $d_{p-\alpha-1} \leq (r/2) - 1$. It follows that

$$\sum_{j=\alpha+1}^{\infty} d_{p-j}r^{p-j} < \left(\frac{r}{2} - 1\right)r^{p-\alpha-1} + \sum_{j=\alpha+2}^{\infty} (r-1)r^{p-j}$$

$$= \frac{r^{p-\alpha}}{2} - r^{p-\alpha-1} + r^{p-\alpha-1} = \frac{r^{p-\alpha}}{2}.$$

The strict inequality above is justified by the fact that we have excluded the case of an infinite sequence of maximal digits. This is consistent with the

8.2 THE FUNDAMENTAL ROUND-OFF ERROR

uniqueness defined in Chapter 1. We leave it to the reader to show that the value $d_{p-\alpha-1} \geq r/2$ implies rounding up.

Example 8.1 Consider the decimal number $3.141592\ldots$ and the two cases $\alpha = 6$ and $\alpha = 5$. For $\alpha = 6$ the number is cut to six digits as $X_{c6} = 3.14159$ and rounded to six digits as $X_{r6} = 3.14159$. In this case the two methods of approximation yield the same result. For $\alpha = 5$, however, we may have X cut to $X_{c5} = 3.1415$ and rounded to $X_{r5} = 3.1416$. For the two cases of rounding we rounded down in the first case since $2 < 5$, but in the second case we rounded up, since $9 > 5$. If X were $3.141596\ldots$, rounding to $X_{r6} = 3.14160$ would generate a carry into the fifth digit retained.

In the discussion so far we have considered cutting or rounding as applied to the digits of the absolute value of an operand. If we deal with complements, we can of course convert to absolute value, carry out the appropriate truncation, and recomplement, if necessary. It is cumbersome, and so we introduce below the modifications necessary to deal with complement representations directly. Strictly speaking, a true complement does not require any truncation, because it is already an integer with a finite number of digits. Practically, however, the problem does arise when reducing a double-length product to standard length.

We will assume, as before, that α digits are retained. Superimposed on the complement structure, however, is the modulus as defined by the number of digits in a register, which we take to be n. Since we will need at least one sign digit, we will assume that $\alpha < n - 1$. Let

$$|X| = \sum_{j=1}^{\infty} d_{p-j} r^{p-j}.$$

In the representation we will retain α digits in the form

$$|X|_{c\alpha} = d_{p-1} r^{p-1} + \cdots + d_{p-\alpha} r^{p-\alpha} \quad \text{or}$$

$$|X|_{r\alpha} = d_{p-1} r^{p-1} + \cdots + D_{p-\alpha} r^{p-\alpha},$$

according to whether we cut or round. If $X < 0$, these two numbers must be interpreted as integers in order to form the complement modulo r^n or r^{n-1}. Thus, we first provide a scale factor, $r^{\alpha-p}$ and write

$$r^{\alpha-p}|X| = \sum_{j=1}^{\infty} d_{p-j} r^{\alpha-j}$$

$$= d_{p-1} r^{\alpha-1} + d_{p-2} r^{\alpha-2} + \cdots + d_{p-\alpha} + \sum_{j=\alpha+1}^{\infty} d_{p-j} r^{\alpha-j}.$$

The second sum, starting with $d_{p-\alpha-1}r^{-1}$ is the fractional part of the scaled number and to be deleted. For the scaled version, then, for cutting, we retain the digits $d_{p-1}, \ldots, d_{p-\alpha}$ unmodified. For rounding, we replace $d_{p-\alpha}$ by $D_{p-\alpha}$ where

$$D_{p-\alpha} = d_{p-\alpha} \qquad \text{if} \quad \sum_{j=\alpha+1}^{\infty} d_{p-j} r^{\alpha-j} < \tfrac{1}{2}$$

and

$$D_{p-\alpha} = d_{p-\alpha+1} + 1 \qquad \text{if} \quad \sum_{j=\alpha+1}^{\infty} d_{p-j} r^{\alpha-j} \geq \tfrac{1}{2}.$$

In either case we extend the retained digits to n positions by filling in zeros, so that

$$r^{\alpha-p}|X| = d_{p-\alpha} + d_{p-\alpha+1}r + \cdots + d_{p-1}r^{\alpha-1} + 0 \cdot r^{\alpha} + \cdots$$

$$+ 0 \cdot r^{n-1} + \sum_{j=\alpha+1}^{\infty} d_{p-j} r^{\alpha-j}.$$

We now define the extended $(r - 1)$'s-complement of $|X|$ to be

$$|X|_{ec} = \sum_{j=1}^{\infty} [(r - 1) - d_{p-j}] r^{p-j},$$

that is, all digits individually complemented. The scaled, extended version of this extended complement is

$$r^{\alpha-p}|X|_{ec} = [(r - 1) - d_{p-\alpha}] + [(r - 1) - d_{p-\alpha+1}]r + \cdots$$

$$+ [(r - 1) - d_{p-1}]r^{\alpha-1} + (r - 1)r^{\alpha} + \cdots$$

$$+ (r - 1)r^{n-1} + \sum_{j=\alpha+1}^{\infty} [(r - 1) - d_{p-j}] r^{\alpha-j}.$$

This gives, with $X^* = [r^{\alpha-p}|X_{c\alpha}|]$,

$$r^{\alpha-p}|X|_{ec} = r^{n-1} - X^* + \sum_{j=\alpha+1}^{\infty} (r - 1) r^{\alpha-j} - \sum_{j=\alpha+1}^{\infty} d_{p-j} r^{\alpha-j}$$

$$= r^{n-1} - X^* + 1 - \sum_{j=\alpha+1}^{\infty} d_{p-j} r^{\alpha-j}.$$

8.2 THE FUNDAMENTAL ROUND-OFF ERROR

The term $r^{n-1} - X^*$ represents the $(r-1)$'s-complement of $r^{\alpha-p}|X|_{c\alpha}$ and gives precisely the digits to be retained. For rounding, we use this approximation if we round down $|X|$, with

$$\sum_{j=\alpha+1}^{\infty} d_{p-j} r^{\alpha-j} < \tfrac{1}{2}$$

and thus the error

$$1 - \sum_{j=\alpha+1}^{\infty} d_{p-j} r^{\alpha-j} > \tfrac{1}{2}.$$

If $|X|$ is rounded up so that $D_{p-\alpha} = d_{p-\alpha} + 1$, then the complemented digit is $[(r-1) - (d_{p-\alpha} + 1)]$, and thus we must *subtract* one from the final complement digit retained. This occurs if

$$\sum_{j=\alpha+1}^{\infty} d_{p-j} r^{\alpha-j} \geq \tfrac{1}{2}$$

or the deleted part

$$1 - \sum_{j=\alpha+1}^{\infty} d_{p-j} r^{\alpha-j} \leq \tfrac{1}{2}.$$

If we unscale by multiplication by $r^{p-\alpha}$, we can summarize the rule for complements of absolute values representing negative numbers in the following manner: If we retain α digits in the extended complement of $|X|$, then

a) for cutting we retain the first α extended complement digits and ignore all remaining digits, and fill out with sign digits to n places.

b) for rounding, we define

$$|X|_{ec} = \sum_{j=1}^{\infty} \bar{d}_{p-j} r^{p-j},$$

where $\bar{d}_{p-j} = (r-1) - d_{p-j}$, and we replace $\bar{d}_{p-j}$ by $\bar{D}_{p-j}$, where $\bar{D}_{p-j} = \bar{d}_{p-j}$, if

$$\sum_{j=\alpha+1}^{\infty} \bar{d}_{p-j} r^{p-j} > \frac{r^{p-\alpha}}{2} \quad \text{(round up)}$$

and

$$\bar{D}_{p-j} = \bar{d}_{p-j} - 1 \text{ if } \sum_{j=\alpha+1}^{\infty} \bar{d}_{p-j} r^{p-j} \le \frac{r^{p-\alpha}}{2} \quad \text{(round down)}.$$

In either case we fill out with sign digits to n places.

The rule differs from that for absolute values in that rounding up requires no change in the final digit, while rounding down requires a subtraction of one. For r's complements we need only add one to the final digit of the cut or rounded $(r - 1)$'s complement.

Example 8.2 The numbers $-1.23445\ldots$ and $-1.23455\ldots$ are to be represented, as complement numbers in five-digit nine's-complement arithmetic. Therefore, $\alpha = 4$ digits. The extended complements are

$$8.76554\ldots \quad \text{and} \quad 8.76544\ldots$$

With four digits retained, we round the first extended complement up to

$$8.765,$$

since $.00054\ldots > \frac{10^{-3}}{2}$, and the second one down to

$$8.764,$$

since $.00044\ldots < \frac{10^{-3}}{2}$. For the five-digit machine we fill in a sign digit to give 98765 and 98764, which are really the rounded versions of the original numbers scaled 10^3. If we unscale and complement, we have 01.234 and 01.235 which are the correct rounded absolute values. For ten's complements we would use 98766 and 98765, respectively.

Cutting and rounding both generate an error except in the special case in which $\sum_{j=\alpha+1}^{\infty} d_{p-j} r^{p-j} = 0$, that is, except when all omitted digits are zeros. The error, however, is bounded by a power of r which is dependent on p and on α. The following theorem gives bounds for each case.

Theorem 8.1 *Let X be represented in base r. The error generated is bounded as follows.*

a) *If X is cut to α places, the error satisfies*

$$|e(X_{c\alpha})| = |X - X_{c\alpha}| < r^{p-\alpha}.$$

b) *If X is rounded to α places, the error satisfies*

$$|e(X_{r\alpha})| = |X - X_{r\alpha}| \le \tfrac{1}{2} r^{p-\alpha}.$$

8.2 THE FUNDAMENTAL ROUND-OFF ERROR

Proof. Case (a) is easy, since

$$e(X_{c\alpha}) = \pm \sum_{j=\alpha+1}^{\infty} d_{p-j} r^{p-j},$$

so that

$$|e(X_{c\alpha})| = \sum_{j=\alpha+1}^{\infty} d_{p-j} r^{p-j} < \sum_{j=\alpha+1}^{\infty} (r-1) r^{p-j} = r^{p-\alpha}.$$

For case (b) if $D_{p-\alpha} = d_{p-\alpha}$ because

$$\sum_{j=\alpha+1}^{\infty} d_{p-j} r^{p-j} < \tfrac{1}{2} r^{p-\alpha}$$

then

$$|e(X_{r\alpha})| = \sum_{j=\alpha+1}^{\infty} d_{p-j} r^{p-j} < \tfrac{1}{2} r^{p-\alpha}.$$

If $D_{p-\alpha} = d_{p-\alpha} + 1$, because

$$\sum_{j=\alpha+1}^{\infty} d_{p-j} r^{p-j} \geq \tfrac{1}{2} r^{p-\alpha},$$

we have

$$e(X_{r\alpha}) = X - X_{r\alpha} = \pm \sum_{j=\alpha+1}^{\infty} d_{p-j} r^{p-j} \pm r^{p-\alpha},$$

with the sign of the summation determined by the sign of X. Thus,

$$|e(X_{r\alpha})| = \left| r^{p-\alpha} - \sum_{j=\alpha+1}^{\infty} d_{p-j} r^{p-j} \right|.$$

But

$$-\sum_{j=\alpha+1}^{\infty} d_{p-j} r^{p-j} \leq -\tfrac{1}{2} r^{p-\alpha},$$

so that

$$|e(X_{r\alpha})| \leq r^{p-\alpha} - \tfrac{1}{2} r^{p-\alpha} \leq \tfrac{1}{2} r^{p-\alpha}.$$

We note that for the error bound in rounding, the equality sign can occur if and only if the base r is even, say $r = 2b$, and

$$\sum_{j=\alpha+1}^{\infty} d_{p-j} r^{p-j} = b r^{p-\alpha-1};$$

that is, if the first digit not used in x is $b = r/2$, followed by all zeros. In this case rounding up *or* down would yield an error of exactly $\frac{1}{2}r^{p-\alpha}$ in magnitude. The definition of rounding given here consistently treats this case by always rounding up. This avoids ambiguity, but one should consider that this may introduce a noticeable error bias in a sequence of many computations. If we consider signed numbers and rounding up, however, we can argue that we would probably have about half and half of positive and negative numbers and a bias would tend to even out. In any case, this very special situation of equality is not likely to occur normally with a high enough frequency to be of serious consequence.

It is apparent that rounding is, in general, preferable to cutting a number to α digits, since we are guaranteed an error in the former case no greater in magnitude than one-half the guaranteed error bound in the latter. Furthermore, if we consider absolute values only, the practice of cutting always yields a positive error in the range $0 < e(X_{c\alpha}) < r^{p-\alpha}$, which may introduce a bias. In rounding, the magnitude of the error is in the range

$$0 < |e(X_{r\alpha})| \leq \tfrac{1}{2} r^{p-\alpha},$$

while the errors satisfy

$$-\tfrac{1}{2} r^{p-\alpha} \leq e(X_{r\alpha}) \leq \tfrac{1}{2} r^{p-\alpha}$$

with, presumably, about half of them negative and half positive. This fact indicates a decreased bias for round-off only in dealing with numbers which are all of one sign. In choosing between round-off or cutting, however, bias is likely to be a consideration only in special circumstances, and other more prosaic factors such as the increased cost of circuitry or programming required for rounding may determine the choice.

8.2.1 Relative Round-Off Errors

If all quantities involved in a computation are in a reasonably restricted range, the error itself is useful in determining accuracy. On the other hand, a comparison of an error of one foot in measuring the diameter of the moon with one foot in measuring the diameter of an electron is hardly useful. A more practical guide is given in many cases by the relative error (often quoted as a percentage when multiplied by one hundred). The bounds on relative error for the two schemes of reducing X to α digits are given in the following theorem.

8.2 THE FUNDAMENTAL ROUND-OFF ERROR

Theorem 8.2 *If X is represented base r, then the relative error is bounded as follows.*

a) If X is cut to α digits, the relative error satisfies

$$|\rho(X_{c\alpha})| = \left|\frac{e(X_{c\alpha})}{X}\right| < r^{-\alpha+1}.$$

b) If X is rounded to α digits, the relative error satisfies

$$|\rho(X_{r\alpha})| = \left|\frac{e(X_{r\alpha})}{X}\right| \leq \tfrac{1}{2}r^{-\alpha+1}.$$

Since $r^{p-1} \leq |X| < r^p$, we have

$$\frac{1}{r^p} < \frac{1}{|X|} \leq \frac{1}{r^{p-1}}.$$

Hence from Theorem 8.1:

a) $|\rho(X_{c\alpha})| = \left|\frac{e(X_{c\alpha})}{X}\right| \leq \left|\frac{e(X_{c\alpha})}{r^{p-1}}\right| < \frac{r^{p-\alpha}}{r^{p-1}} = r^{-\alpha+1}$;

b) $|\rho(X_{r\alpha})| = \left|\frac{e(X_{r\alpha})}{X}\right| \leq \frac{|e(X_{r\alpha})|}{r^{p-1}} \leq \frac{1}{2}\frac{r^{p-\alpha}}{r^{p-1}} = \tfrac{1}{2}r^{-\alpha+1}.$

The relative error is more closely tied to the total number of digits actually retained than is the error itself. This fact gives rise to the commonly used terminology in which accuracy is described in terms of "the number of *places* to the right of the base point." We shall use places in this sense, that is, the number of digits to the right of the point, which should not be confused with the *total* number of digits retained.

Example 8.3 In the so-called scientific notation, numbers may be represented by a sequence of digits with the decimal point immediately to the right of the first nonzero digit, multiplied by a power of ten. Thus, the three-digit approximation to the speed of light of 186,000 miles/second might be represented as 1.86×10^5. If this is a cut or rounded version of the true value with $\alpha = 3$, the magnitude of the relative error is either less than 10^{-2} or $(\tfrac{1}{2})10^{-2}$, corresponding to the fact that two places are retained following the point.

More generally the idea of places to the right of the base point can be related to the relative error in the following way.

Example 8.4 If $r^{p-1} \leq |X| < r^p$, then

$$|X| = d_{p-1}r^{p-1} + \cdots = (d_{p-1} \cdot d_{p-2} d_{p-3} \ldots)r^{p-1}.$$

If exactly α digits are retained, then $\alpha - 1$ are to the right of the base point and Theorem 8.2 states that the relative errors are bounded by $r^{-(\alpha-1)}$ or

$\frac{1}{2}r^{-(\alpha-1)}$. We note that this scheme of notation corresponds to a floating-point form for an operand.

It is unfortunate that no really standard terminology has evolved in reference to round-off error. Thus, some authors would refer to the approximation in Example 8.3 as "two-place" accuracy with a round-off error of no more than $(\frac{1}{2})10^{-2}$. Others would call it "three-place" accuracy because three digits are retained. In any event the true round-off error as defined here is bounded in magnitude by $(\frac{1}{2})10^3$. In interpreting the intended meaning of any statement with regard to round-off error we should be careful to determine whether it is the error or the relative error and precisely what "places" means. We can only warn, *caveat lector*.

8.2.2 Significant Digits

A common designation of accuracy in computation is in terms of "significant" digits. Intuitively it is plausible to think of a digit as being significant if it is the correct digit in the given location for the method of approximation we are using. Thus, if $\pi = 3.141592\ldots$ and we are interpreting it as the true value with infinitely many digits, then, as long as each digit in the string is the correct digit, it can be considered as significant. Similarly, in the approximations 3.1415 and 3.1416 we can consider that all five digits are significant if our criterion for correctness is mere cutting in the one case and rounding in the other.

Unfortunately, using a completely intuitive approach to the concept of significant digits can lead to ambiguity, and it can also lead to error in estimating the accuracy of an approximation. Consider the number $\frac{4}{3} = 1.333\ldots$ as a decimal representation. For either cutting or rounding, we would probably think of the representation 1.33 as containing three significant digits. Merely rewriting this number as 01.33 would not tempt many to say that there are four significant digits. The leading zero is regarded as "insignificant," and in this case certainly has no effect on the accuracy of the approximation. If we use 0.133 as an approximation to $\frac{4}{30}$, a similar argument applies, since again the leading zero can safely be dispensed with. Suppose, however, we consider $\frac{4}{300}$ and represent it by .0133. Is it still true that the leading zero is insignificant? This depends on the point of view. It is certainly significant in determining the location of the decimal point and is correct in the given location. Also it does affect the true error in terms of the fact that it determines the value of p for which $10^{p-1} \leq |X| < 10^p$, and this is not true in either of the preceding cases. On the other hand, the relative error in each case is the same, namely, 0.0025. In this sense, it is the nonzero digits 1, 3, and 3 which determine the relative accuracy.

Example 8.5 If we interpret leading zeros as significant in determining accuracy we can arrive at unsatisfactory results. For example, if we interpret 0.0133 as four significant digits in representing $\frac{4}{300}$, we would certainly

8.2 THE FUNDAMENTAL ROUND-OFF ERROR

interpret 1333.0 as four significant digits in the representation of $\frac{4000}{3}$. If we divide one four-digit representation by the other, we have

$$(0.0133)/(1333.0) = 0.0000099\ldots$$

If we cut or round this last number to four "significant" digits we obtain 0.0000, not a very satisfactory representation of the true quotient.

Since leading zeros, whether required or not, determine at most only the location of the base point, it seems reasonable to exclude them as significant and to rely on the relative error as defining significance. In effect, this says that we should consider as significant only that string of digits starting with the first nonzero digit as determined by p in $r^{p-1} \leq |X| < r^p$. Although we will not consider leading zeros as significant, the same is not necessarily true of final zeros.

Example 8.6 Consider the number 52.196 rounded to four digits. The result is 52.20 where the final zero must certainly be considered as significant as the final nine if we round 52.193 to 52.19.

To make the ideas discussed above precise we will use the following definition of significant digits.

Definition 8.1 Let $r^{p-1} \leq |X| < r^p$ so that

$$X = \pm[d_{p-1}r^{p-1} + d_{p-2}r^{p-2} + \cdots],$$

with $d_{p-1} \neq 0$. If $\bar{X}$ is any α-digit representation,

$$\bar{X} = \pm[\bar{d}_{p-1}r^{p-1} + \bar{d}_{p-2}r^{p-2} + \cdots + \bar{d}_{p-\alpha}r^{p-\alpha}],$$

we define each digit $\bar{d}_{p-j}, j = 1, \ldots, \alpha$ as significant, if and only if

a) $|\rho(\bar{X})| = \dfrac{|X - \bar{X}|}{|X|} < r^{-\alpha+1}$

when cutting is the criterion, and

b) $|\rho(\bar{X})| = \dfrac{|X - \bar{X}|}{|X|} \leq \tfrac{1}{2}r^{-\alpha+1}$

when rounding is the criterion.

Although rounding is apparently the more accurate technique, and many people would consider part (a) of Definition 8.1 as invalid, it seems reasonable to take a practical approach. Many modern computer systems provide automatic rounding (usually at some cost in computing time) while others merely cut. Thus, selecting the concept of significance on the basis of the operation available appears to be reasonable.

8.2.3 Register Size and Round-Off Error

The dependence of the accuracy of individual operands on register capacity is apparent. The maximum number of digits which can be retained is dictated by the number of digital positions allocated to an operand. This, in turn, will be a function of the mode of operation, either fixed point with external scaling or floating point by means of internal scaling. If a single register is used for a single operand, a higher degree of accuracy is obtainable by external scaling since none of the digits needs to be used to represent the exponent.

For purely integral operation where all operands are interpreted as integers, the question of accuracy as such does not arise. In all cases the representation is exact, or at worst is congruent to the true operand, so that if the conditions considered in the previous chapters are satisfied, the congruence becomes an equality. In this case the representation is exact. In discussing accuracy, then, of individually stored operands, we will consider them as representing approximations to real numbers. The position of the base point is determined by an exponent of the base r, either an external scale-factor exponent or an internal floating-point exponent.

In either mode of operation, the exponent has no bearing on the relative error of the approximation. Its function is merely to record the number of positions the base point must be moved from some fixed location to its proper place. Theoretically, for external scaling, this produces no limitations, since scale-factor exponents can be of any size. For internal scaling, the limitation imposed is not one of accuracy but one of magnitude. This is based on the fact that the integral exponent is limited by the number of digits allocated to it, as considered in Chapter 7.

Although the exponent does not affect the relative error, it certainly does have an effect on the absolute error. Suppose that

$$X = X^* r^x,$$

where both X and X^* are exact and the integer x is the floating-point exponent. If X^* is approximated by $\bar{X}^*$ with a relative error $\rho(\bar{X}^*)$, then X is approximated by $\bar{X}^* r^x$. We have the exact relationship

$$X = [\bar{X}^* + X^* \rho(\bar{X}^*)] r^x,$$

approximated by

$$\bar{X} = \bar{X}^* r^x.$$

The error in approximating X is

$$e(\bar{X}) = X^* \rho(\bar{X}^*) r^x = e(\bar{X}^*) r^x,$$

so that the true error is the true machine error multiplied by r^x. On the other hand, since

$$\rho(\bar{X}) = \rho(\bar{X}^*),$$

the relative error in approximating X is the same as the relative error in its machine representation.

8.2 THE FUNDAMENTAL ROUND-OFF ERROR

Since the accuracy is dictated by the number of digits allocated to an operand, it will depend on scaling. If n digits, exclusive of scale-factor or floating-point exponent, are used, and if one of these digits is a sign digit, then the maximum number of significant digits is $n - 1$.

For external scaling the optimal scale-factor exponent need not be used. For $r^{p-1} \leq |X| < r^p$ the permissible scale factors as given in Chapter 7 are

$$s \leq n - p \quad \text{or} \quad s \leq (n-1) - p$$

if a sign digit is retained. The optimal choice of the equality yields error bounds as given for floating-point operation. If we assume that a sign digit is retained, then the maximum number of digits is $\alpha = n - 1$, corresponding to $s = (n-1) - p$. Each decrease of one in scale-factor exponent reduces the number of digits retained by one, with $\alpha = s + p$ for integral scaling. The relative error is thus

$$r^{-(s+p)+1} \quad \text{or} \quad r^{-(s+p)+1}/2,$$

depending on cutting or rounding. Since s is the negative of the floating-point exponent, the true error is the machine true error multiplied by r^{-s}.

Example 8.7 In a 48-bit machine, with a sign digit retained for fixed-point integral arithmetic we have a maximum of $\alpha = 47$ binary digits, with relative errors bounded by 2^{-46} or 2^{-47}, depending on cutting or rounding. This assumes a scale-factor exponent of $s = 47 - p$. Each decrease of one reduces the number of digits by one. For the number $X = 5.2$, we have $p = 3$ and $s \leq 44$. For $s = 44$ the above bounds apply, but for $s = 40$, for example, the relative error bounds are 2^{-42} and 2^{-43}.

For internal scaling the optimal choice is always made through normalization. Thus, theoretically, if X^* is the normalized part of the machine representation, the relative error can satisfy

$$|\rho(\bar{X}^*)| < r^{-(n-1)+1} = r^{-n+2}$$

if cutting is used, and

$$|\rho(\bar{X}^*)| \leq \tfrac{1}{2} r^{-n+2}$$

if rounding is used. These bounds are valid only if all $n - 1$ digits are significant. If the operand is obtained as a result of a computation (for example, if alignment of the base point in addition or subtraction is required), these bounds may no longer be applicable.

Example 8.8 In a 48-bit machine, 12 bits are used for the exponent and 36 bits for the coefficient in floating-point operation. With one's-complement operation one digit each in exponent and coefficient is reserved for sign. Thus, maximum significance is 35 binary digits, and ideally the relative error satisfies, with rounding,

$$\rho(\bar{X}^*) \leq \tfrac{1}{2} 2^{-34} = 2^{-35} = (2.9\ldots)10^{-11}.$$

Therefore, at best, we can expect about ten decimal digit accuracy. The 12-bit exponent size limits the operands to roughly the decimal range 10^{-308} to 10^{308} in absolute value. Although the relative errors do not change very much in this range, the true errors may vary in absolute value from about 10^{-318} to 10^{298}.

8.3 ROUND-OFF ERROR AND THE ARITHMETIC OPERATIONS

To this point we have considered the errors in individual operands introduced by the necessity of reducing them to finite sequences of digits conforming to various formats. We now investigate the consequences of using these machine operands for the individual operations of arithmetic. This is only a point of departure, however, since a machine computation may consist of thousands or even millions of consecutive operations, requiring an estimate of the accumulative effect of the initial round-off error and all of the additional round-off errors introduced as the computation proceeds. This problem is far beyond what can be attempted here. For single operations, however, it is easy to obtain explicit formulas for the resultant errors and even to put bounds on them. In particular if X is approximated by X_α and Y is approximated by Y_β, that is, by α-digit and β-digit sequences respectively, we have:

$$X + Y = X_\alpha + Y_\beta + [e(X_\alpha) + e(Y_\beta)],$$
$$X - Y = X_\alpha - Y_\beta + [e(X_\alpha) - e(Y_\beta)],$$
$$XY = X_\alpha Y_\beta + [X_\alpha e(Y_\beta) + Y_\beta e(X_\alpha) + e(X_\alpha)e(Y_\beta)],$$
$$\frac{X}{Y} = \frac{X_\alpha}{Y_\beta} + \left[\frac{Y_\beta e(X_\alpha) - X_\alpha e(Y_\beta)}{Y_\beta(Y_\beta + e(Y_\beta))}\right].$$

In each case the formula for the resultant error is enclosed in square brackets. These formulas assume that the exact value of the operation on X_α and Y_β is obtained. This assumption ignores the fact that additional round-off error may be introduced by the necessity of cutting or rounding the result of the operation. For example, if $X_\alpha + Y_\beta$ must be cut to γ digits, the true error would be given by the bracketed expression in

$$X + Y = (X_\alpha + Y_\beta)_\gamma + [e((X_\alpha + Y_\beta)_\gamma) + e(X_\alpha) + e(Y_\beta)].$$

In general we will not attempt to evaluate errors in terms of these formulas, since such calculations would themselves be in error. What we will be concerned with is establishing bounds for the errors, based on the formulas.

Example 8.9 Suppose $X = 5.12346$ and $Y = 6.98322$ and we are restricted

8.3 ROUND-OFF ERROR AND THE ARITHMETIC OPERATIONS

to five digits. If we use rounding, we have
$$X_5 = 5.1235 \quad \text{and} \quad Y_5 = 6.9832$$
with
$$|e(X_5)| \le \tfrac{1}{2}10^{-4} \quad \text{and} \quad |e(Y_5)| \le \tfrac{1}{2}10^{-4}.$$

If we form $X_5 + Y_5 = 12.1067$, we can guarantee that the absolute value of the error for this six-digit result is no greater than $\tfrac{1}{2}10^{-4} + \tfrac{1}{2}10^{-4} = 10^{-4}$. The restriction to five digits, however, requires that $X_5 + Y_5$ be rounded to 12.107. This introduces another error bounded in magnitude by $\tfrac{1}{2}10^{-3}$. Thus, without more information, we would have to estimate the magnitude of the overall error as no greater that $\tfrac{1}{2}10^{-3} + 10^{-4} = (0.6)10^{-3}$. Thus, we are not even guaranteed a correctly rounded five-digit result from the two correctly rounded five-digit operands.

If we compute the exact result for the operands in Example 8.8, we see that the final value of 12.107 is correctly rounded. That is, the five-digit sum obtained from the two five-digit operands has five-digit significance. This is due to the fact that two of the errors are of opposite sign and tend to cancel each other. Without *a priori* knowledge or some probabilistic theory of round-off, however, we would not be able to utilize this fact.

In the preceding formulas the number of digits retained in each operand is not necessarily the same. Even when each operand is stored with the maximum number of digits, not all of these digits are necessarily used in an arithmetic operation. The following example illustrates this point.

Example 8.10 In floating-point operation for a base-r machine with n digits (exclusive of exponent), two operands have machine versions
$$X = X_\alpha r^x \quad \text{and} \quad Y = Y_\alpha r^y$$
with $\alpha = n - 1$. If $x \ge y$, we require
$$X_{n-1} r^x + (Y_{n-1} r^{y-x}) r^x = [X_{n-1} + (Y_{n-1} r^{y-x})] r^x,$$
to align base points for addition. If $x = y$, all $n - 1$ digits can be taken for each of the addends. If $y < x$, however, the use of $Y_{n-1} r^{y-x}$ as an addend, instead of Y_{n-1}, introduces a right shift of $x - y$ places and the consequent loss of a like number of digits in Y_{n-1}. If the original $n - 1$ digits are significant the most which can be obtained in the shifted operand is $n - 1 - (x - y)$.

In terms of the actual error, the reduction of the number of significant digits in one of the operands in addition is probably not critical. In Example 8.9, if $n - 1 - (x - y) = 0$, within the accuracy retained, Y is zero for all practical purposes, and no great harm is done. Moreover, in this situation $|Y|$ is small compared to $|X|$ and the relative error will not be severely affected by the loss of accuracy in Y. A more serious increase in relative error can occur with operands which are numerically close together.

Example 8.11 Suppose that in Example 8.10, $r = 10$, $n = 6$ with $x = y = 0$, $X_{n-1} = 0.12346$, and $Y_{n-1} = 0.12345$. Each coefficient is correctly rounded, so that each contains five significant digits. If we form their difference, however, we have 0.00001. Since significance is based on the first nonzero digit, this result has at most one significant digit.

The difficulty encountered in Example 8.11 is due to the fact that although true error may not increase, the relative error does. In the example the normalized version of 0.00001 would be 0.1 with a relative error no greater than $\frac{1}{2}10^{-1}$. On the other hand the original operands have relative errors no greater than $\frac{1}{2}10^{-5}$. The small size of the result of the operation has decreased the number of significant digits by four. Single discrepancies of this sort are not usually important in cases where the true error is the measure of accuracy. The accumulation of such errors can produce serious consequences in situations where the relative error is important. An error of one foot in measuring the distance from earth to a point on the moon would not usually be serious. If two such errors were made in estimating the height of an object on the moon by the length of the shadow cast, however, it is apparent that the loss of relative accuracy could be serious.

8.4 EXACT COMPUTATIONS; INCREASED PRECISION

Although the explicit formulas for errors in individual operands and the formulas for errors in arithmetic operations can be useful on occasion, and are necessary to know, they are not particularly pertinent in the kind of lengthy calculations common to modern computing machines. Exact error analysis in such cases is exceedingly difficult, and where guaranteed results are needed, often leads to estimates which are too crude to be of value. Consider, for example, a computation requiring one million additions. If all errors were in the same direction, and all of the same magnitude, the final error would be one million times the initial error. If each initial error were less than 10^{-5}, the final result could be meaningless. However, common sense says that the errors will usually not be in the same direction. We might assume about half in one and half in the other, or some other reasonable distribution. On this basis a fruitful approach to an estimation of round-off error has been a probabilistic one. We will not consider this or other methods of error analysis in detail. We shall instead recognize that many calculations are carried out without prior error analysis, and so we consider briefly two methods of possibly improving accuracy. One of these methods is to augment precision by using more digits. The other is to exploit the fact that in integral computation, basic operations are always on exact quantities and can be considered to produce exact results.

8.4.1 Exact Computation with Integers

If we ignore the theoretical zeros to the right of the base point, integers can be defined to consist of only a finite sequence of digits. Therefore, if we apply

8.4 EXACT COMPUTATIONS: INCREASED PRECISION

operations to integers, the results produced can always represent exact quantities. Although the exact result may not represent the basic operand we have in mind, we have seen that in many cases it will represent something congruent to it. If we reinforce this fact by placing sufficient restrictions on magnitude to guarantee that congruence is equality, we may tolerate a loss of digits without resultant loss in accuracy. We illustrate this general approach in the following example.

Example 8.12 We have the following set of data for a_i, b_i, c_i, d_i, e_i, and f_i.

i	a_i	b_i	c_i	d_i	e_i	f_i
1	0.11234	0.32156	0.97182	0.21234	0.92156	0.98764
2	0.27653	0.12314	0.91416	0.17652	0.32314	0.47181
3	0.21241	0.13215	0.98765	0.31151	0.12351	0.95112
4	0.32176	0.92354	0.99874	0.31242	0.33215	0.99885
5	0.31151	0.35398	0.45110	0.22177	0.15397	0.91403
Sum	1.23455	1.85437	4.32347	1.23456	1.85433	4.32345

We wish to compute the solution of the system of equations

$$\left(\sum_{i=1}^{5} a_i\right) x + \left(\sum_{i=1}^{5} b_i\right) y = \sum_{i=1}^{5} c_i,$$

$$\left(\sum_{i=1}^{5} d_i\right) x + \left(\sum_{i=1}^{5} e_i\right) y = \sum_{i=1}^{5} f_i,$$

on a five-digit decimal machine with separately stored sign. The equivalent equations, with rounded coefficients, are

$$1.2346x + 1.8544y = 4.3235,$$
$$1.2356x + 1.8543y = 4.3235,$$

which yields, with conventional floating-point operation, the five-decimal digit solution $y = 0.0000$ and $x = 3.5019$ with five-decimal digits. Suppose, however, we consider the equivalent equation with integral coefficients, obtained by multiplying each equation by 10^5. Using consistent residue-class arithmetic, modulo 10^5, we obtain the equations,

$$23455x + 85437y = 32347,$$
$$23456x + 85433y = 32345.$$

Proceeding further with consistent residue-class arithmetic, we can easily determine that $y = 1$ and $x = 2$. We know that these results are at worst congruent to the true results and perhaps exact. In this problem we can easily see that the components of the true solution are bounded by 10^4, so that the results are in fact exact.

The zero error obtained by the residue-class technique of Example 8.11, compared to the approximately 100% error of the floating-point approach, emphasizes the distinct effects of reducing the number of digits by each of the two methods. In rounding, deletion of digits always produces error, whereas in congruence, reduction of the number of digits by a smaller member of its residue class never introduces error.

If equality of the final results in the method of integral operation is required, rather than congruence, the final numbers must satisfy the appropriate restriction on magnitude. Thus, in Example 8.12 the solution, representable by means of integers with no more than five decimal digits, was exact. In practical applications, of course, utilization of this technique may require some *a priori* knowledge of ranges on the quantities involved.

We note, also, that the excessive error introduced in the solution by rounding in Example 8.12 is due to the nature of the matrix of the system of equations. Such matrices are often termed ill-conditioned and they represent one of the serious problems in round-off error analysis.

8.4.2 Multiple Precision

The round-off problem arises from the necessity of using finite strings of digits to approximate infinite strings. Since the accuracy is directly related to the number of digits used, one obvious possibility is to utilize more of them to improve the accuracy. The difficulty in this approach is that the number of digits that can be conveniently used is usually predefined by the machine itself. Thus, we have been concerned in this book with numbers representable by n digits, or possibly in accumulators by $2n$ or more. Although the implication in assumptions of n-digit registers or $2n$-(or more)-digit accumulators is that a given machine will work with digit sequences of fixed length, this is not always the case. Some computers have a fixed operand length, but others are designed to have a variable length in which the number of digits is, to a certain extent, at the disposal of the programmer. Even in such machines, however, there will be a maximum operand length to which a single arithmetic operation, as defined by the machine's instructions, will apply. We will assume, therefore, in the following discussion that the basic building blocks are n-digit sequences which may be combined by the arithmetic operations in $2n$-(or more)-digit accumulators. Arithmetic operations on and with these n-digit operands, and directly obtained from machine instructions, are usually referred to as *single precision*. That is, they involve the precision implicit in the use of a single register to store a single operand.

Single precision implies an accuracy defined by the content of a single register. To increase this accuracy by including more digits requires the use of more than one register to store the digits of an individual operand. Such usage is called *multiple precision*, and specifically double and triple precision and the like, depending on the number of registers used per operand.

8.4 EXACT COMPUTATIONS: INCREASED PRECISION

We formally define a k-fold precision number in an n-digit machine as the sequence of digits in k ordered registers, each of n-digit length in which the ordering of the registers is the same as the normal ordering of the digits themselves. Thus, we can schematically think of the number as represented in blocks of n digits

$$\boxed{n} \quad \boxed{n} \quad \cdots \quad \boxed{n}$$
$$\quad 1 \qquad\quad 2 \qquad\qquad k$$

with an ordering on the blocks and of the digits within the blocks given by

$$\boxed{(n-1)\ldots 0} \quad \boxed{(n-1)\ldots 0} \quad \cdots \quad \boxed{(n-1)\ldots 0}$$
$$\quad k-1 \qquad\qquad k-2 \qquad\qquad\qquad 0$$

This is a generalization of what we ordinarily do with commas to separate groups of one thousand in decimal. The understood location of the base point in such multiple precision sequences is immaterial, since it is handled separately by scaling, either internal or external. Thus, we will, as before, consider such sequences to be integers. The basic problem, then, is how to deal with the arithmetic of integers when the digits are separated into individual blocks of n each. The actual implementation of arithmetic is of course dependent on whether internal or external scaling is employed and on questions relating to sign. We do not describe such implementation but rather consider some basic ideas for nonnegative integers. The reader can extend the discussions of earlier chapters to include scaling and sign. A simple point of view in which to see what is involved in multiple precision arithmetic is afforded by the method of reduction employed in Section 2.7. That is, we consider each block of n digits in a k-fold precision number as a digit in the base $\sigma = r^n$. If we consider the individual digits in each register as the digits base r and the relative digital positions of these determined by the ordering of the registers, we have the number X represented in multiple precision by base r digits d_{ij} in the form

$$X = \sum_{i=0}^{k-1} \sum_{j=0}^{n-1} d_{ij} r^{ni+j},$$

where d_{ij} is the digit in the ith register corresponding to the coefficient of r^j. We can rewrite this as

$$X = \sum_{i=0}^{k-1} \left(\sum_{j=0}^{n-1} d_{ij} r^j \right) r^{ni} = \sum_{i=0}^{k-1} \left(\sum_{j=0}^{n-1} d_{ij} r^j \right) \sigma^i.$$

If we define $D_i = \sum_{j=0}^{n-1} d_{ij} r^j$, it follows, from the inequality $0 \leq d_{ij} < r$, that

$$0 \leq D_i = \sum_{j=0}^{n-1} d_{ij} r^j \leq r^n - 1 = \sigma - 1.$$

Therefore, the D_i are legitimate digits in the base σ. Thus, X is represented as an integer base-σ with at most k nonzero digits. That is,

$$X = \sum_{i=0}^{k-1} D_i \sigma^i,$$

with $0 \leq D_i < \sigma$.

Example 8.13 In a machine which provides only for arithmetic with eight-digit operands, thirteen-digit operands would be dealt with in double precision. Thus, the number 3.141592653898 could be stored in two registers in the form:

| 00031415 | | 92653898 |

The three leading zeros in the left block can be considered as sign digits or as possibly a sign digit and room for two exponent digits. Some might prefer other formats for signs and exponents. From the point of view adopted, we consider the two blocks of the double precision number as its two digits base-10^8. Interpretation of multiple precision numbers as integers base-σ permits the direct application of the material developed and discussed in the preceding chapters by the simple replacement of r by $\sigma = r^n$ and n by k. In particular the relative error, as developed in Section 8.2.3, can be bounded by

$$|\rho(X^*)| < \sigma^{-k+2} = r^{-nk+2n} = r^{-(k-2)n}$$

or

$$|\rho(X^*)| \leq \tfrac{1}{2}\sigma^{-k+2} = \tfrac{1}{2} r^{-(k-2)n}.$$

The precision in terms of relative error is increased not quite k-fold.

In the discussion above we have assumed that the full content of all registers storing a multiple precision operand can be used, that is, each register can store any digit, base-σ. Practically, in single precision floating-point operation some digits will be allotted for the exponent and its sign, and in complement arithmetic a digital position is usually reserved for a sign indicator. These requirements carry over to multiple precision as well. Although we do not intend to deal with them in detail, we have indicated one possible approach in Example 8.13. If we consistently follow the interpretation of multiple precision arithmetic as arithmetic base-$\sigma = r^n$, the usual restriction for sign indication would require that the only meaning-

8.4   EXACT COMPUTATIONS: INCREASED PRECISION   255

ful content of the highest-order register be $\sigma - 1$, consisting of all digits, base-r, equal to $r - 1$ or else be all zeros. Since sign indication requires only one of two combinations, this means that possibly we are not utilizing as many as $r^n - 2$ permissible single precision configurations. We can avoid this inefficiency by allotting as a sign designator only the highest-order base-r digit in the register holding the highest-order digit base-σ. Thus, as usual, a highest-order base-r digit of zero means positive and one of $r - 1$ means negative. This means that all values in the highest-order register which are less than or equal to $r^{n-1} - 1$ would indicate positive numbers, while values r^{n-1} to $r^n - 1$ would correspond to negative numbers. We leave it to the reader to find the improvement in relative error if this technique is used.

Example 8.14 It is customary in binary machines to read the content of registers in groups of three as octal digits. This is the same as considering each number to be a multiple precision number with $n = 3$. Thus, in a 36-bit register, the number can be thought of as a 12-fold multiple precision number with $r = 2$ and $\sigma = 2^3$. For a number ending with ...011101 we think of the final two octal digits as ...35. Using the highest-order octal digit to indicate sign requires assigning three bits for this purpose. If we extend the binary practice of using only the highest-order bit to indicate sign, however, we see that highest-order octal digits 000 through 011 designate positive numbers, while 100 through 111 specify negative numbers.

The idea developed in Example 8.13 can, of course, be carried back to base-r as well. The restriction of sign digit to $r - 1$ or zero is overly stringent. In decimal, for example, we could use high-order digits 0, 1, 2, 3, 4 for non-negative numbers and 5, 6, 7, 8, 9 for negative. The analysis is simpler using only zero and $r - 1$, however, and we have used this fact to make a considerable body of the results less cumbersome. On the other hand, in multiple precision the simplification of the restriction to $\sigma - 1$ or zero may be too costly in storage space to be practical.

The point of view that interprets multiple precision arithmetic as arithmetic on base-σ numbers makes all of the algorithms previously developed valid, although, as we have indicated, a straightforward application may lead to some inefficiency. In constructing algorithms for dealing with such numbers we will be restricted by the fact that the arithmetic ability of a typical computing machine is limited to forming the sum or difference of a pair of σ-digits, the two σ-digit product of a pair of σ-digits, and the single σ-digit quotient of a two σ-digit dividend and one σ-digit divisor. Thus, we will in general have to think of a serial rather than a parallel implementation. In many respects the problem is the same as that which we encountered in Chapter 3 in going from the logic of the binary accumulator to that of a general accumulator base-r. In the present case we wish to extend base-r

arithmetic to base-σ. In Chapter 3 we exploited the fact that we could represent base-r digits in base-two terms, and here we exploit the fact that we can represent base-σ digits in base-r form. Of course, the details of the representations must have an effect on the details of the algorithms employed. If carries are properly noted, we can thus think of the base-r accumulator as a base-σ half-adder. We are therefore faced with the necessity of using the base-r accumulator and existing instructions for base-r operands and devising combinations of these instructions which will in the end produce the correct logic. The challenge is to do this efficiently, but since this is programming, it is not one of the topics in this book.

For addition and subtraction in base-σ, the foregoing ideas can be applied in a relatively straightforward manner. As before, the details of dealing with carries and borrows will cause difficulties. These difficulties may be enhanced if we are using a modulo $r^n - 1$ accumulator to form an r^n half-adder. We illustrate these points with some numerical examples.

Example 8.15 Suppose we wish threefold precision using a three-binary digit additive open accumulator. Consider addition of the two numbers

$$\boxed{0\ 1\ 0}\quad \boxed{1\ 0\ 1}\quad \boxed{1\ 1\ 0}$$

and

$$\boxed{0\ 0\ 0}\quad \boxed{0\ 1\ 0}\quad \boxed{1\ 0\ 0}$$

In octal we can think of these numbers as

$$\boxed{1}\boxed{1}$$
$$2\ 5\ 6$$
$$+\ 0\ 2\ 4$$
$$\overline{3\ 0\ 2}$$

so that register-by-register addition would give

$$\boxed{1}\boxed{1}\boxed{1}\quad\boxed{1}$$

$$\begin{array}{ccc} 0\ 1\ 0 & 1\ 0\ 1 & 1\ 1\ 0 \\ 0\ 0\ 0 & 0\ 1\ 0 & 1\ 0\ 0 \\ \hline 0\ 1\ 1 & 0\ 0\ 0 & 0\ 1\ 0 \end{array}$$

which is properly 302 octal, including a leading octal digit of three, which correctly includes a leading binary-digit zero for a positive number. There are only two octal carries. These go from register to register and therefore come from the highest-order accumulator position, and so must be detected by overflow. The concomitant internal binary carries, however, will be automatic in the accumulator.

We note that if one's-complements are used, the accumulator would normally produce an end-around carry, so that

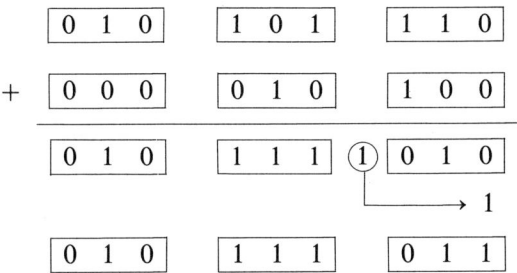

to give the incorrect result in octal 273. The end-around carry at the lowest-order register suppresses the proper carry from the adder and also gives an incorrect residue digit. For $(r-1)$'s-complement arithmetic, any automatic n-digit end-around carry must be suppressed. This carry is not appropriate except from the highest-order digit of the highest-order register to the lowest-order digit of the lowest-order register.

Example 8.16 In the counterpart of Example 8.14, we wish to subtract using one's- and seven's-complement arithmetic. We proceed as follows.

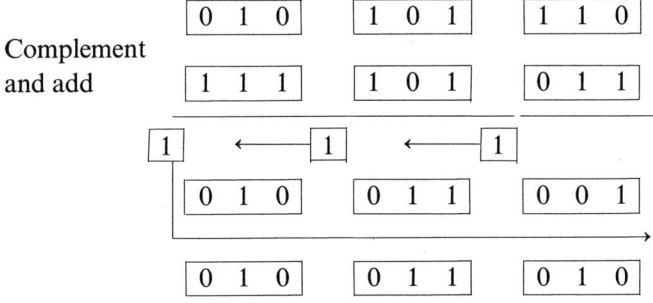

As shown, this reflects the equivalent of octal $256 - 024 = 232$ or, as given in seven's-complement arithmetic, by

$$\begin{array}{r} 2\ 5\ 6 \\ +\ 7\ 5\ 3 \\ \hline \boxed{1}\ 2\ 3\ 1 \\ \hookrightarrow 1 \\ \hline 2\ 3\ 2 \end{array}$$

We assume that individual end-around carries have been diverted to the appropriate next-order octal position, including the overall end-around from highest- to lowest-order octal.

While in $(r - 1)$'s-complement arithmetic, digit-by-digit complementation in each register is equivalent to register-by-register complementation, the same may not be true in r's complements. The most straightforward way of diverting the n-digit end-carry would be simply to use $(r - 1)$'s-complements in a modulo r^n accumulator.

It was true in Example 8.16, as it is in general, that forming the $(r - 1)$'s-complement of each σ-digit is equivalent to forming the $(\sigma - 1)$'s-complement of the k-fold precision number. Unfortunately, the same is not true of r's- and σ's-complements.

Example 8.17 If we complement the subtrahend, $\boxed{000}$ $\boxed{010}$ $\boxed{100}$ of Example 8.16, by means of two's-complements on each octal digit to form the eight's-complement of the octal number, we would have

$$\boxed{0\ 0\ 0} \quad \boxed{1\ 1\ 0} \quad \boxed{1\ 0\ 0}$$

From available operations the multiple precision algorithms for addition, subtraction, and multiplication are straightforward. Division, however, is somewhat more difficult.

We consider the process of division for nonnegative integers. Suppose that N is j-fold and D is k-fold precision, so that

$$N = \sum_{i=0}^{j=1} n_i \sigma^i, \qquad D = \sum_{i=0}^{k-1} d_i \sigma^i,$$

with $n_{j-1} > 0$, $d_{k-1} > 0$, and $k \leq j$. The n_i and d_i are base-σ digits, so that $0 \leq n_i < \sigma$ and $0 \leq d_i < \sigma$. These inequalities guarantee

$$\sigma^{-1} \leq \sigma^{j-k-1} < \frac{N}{D} < \sigma^{j-k+1}. \tag{8.1}$$

We seek Q and R to satisfy the division algorithm in the form

$$N = QD + R,$$

and the inequalities (8.1) on N/D guarantee that we may write the quotient Q in the form

$$Q = \sum_{i=0}^{j-k} q_i \sigma^i,$$

where the q_i, as σ-digits, satisfy $0 \leq q_i < \sigma$ and represent the single precision components of Q.

8.4 EXACT COMPUTATIONS: INCREASED PRECISION

The digits of Q are obtained sequentially, beginning with the most significant σ-digit q_{j-k}. We define a recursive algorithm in the following way: Let $R(j - k) = N$ and assume that digits $q_{m+1}, q_{m+2}, \ldots, q_{j-k}$ have been found. We form the partial remainders

$$R(m) = R(m + 1) - q_{m+1} D\sigma^{m+1} = N - D \sum_{i=m+1}^{j-k} q_i \sigma^i,$$

which satisfies

$$0 \le R(m) < D\sigma^{m+1} < \sigma^{m+k+1}. \tag{8.2}$$

Let the two most significant σ-digits of $R(m)$ be $R_{m+k+1}(m)$ and $R_{m+k}(m)$, and define

$$R^*(m) = R_{m+k+1}(m)\sigma + R_{m+k}(m).$$

We find a trial quotient digit q_m^* by dividing $R^*(m)$ by the leading digit d_k of the divisor. Thus,

$$q_m^* = \text{Integral part of} \left[\frac{R^*(m)}{d_k} \right].$$

The quantity

$$R^* = R(m) - q_m^* D\sigma^m$$

is then computed. This is an initial approximation to $R(m - 1)$ which must satisfy an inequality of the form (8.2). If

$$R^* < 0 \quad \text{or} \quad R^* \ge D\sigma^m,$$

$D\sigma^m$ is added to or subtracted from R^* until $0 \le R^* < D\sigma^m$. For each addition q_m^* must be decreased by one and for each subtraction q_m^* must be increased by one. The final $R^* = R(m - 1)$ and the final $q_m^* = q_m$. Repetition of the procedure generates the digits of Q in order from highest to lowest with $R(-1) = R$.

Example 8.18 Suppose $r = 2$, $n = 3$, so that for multiple precision $\sigma = 2^3$. The fourfold precision product $N = 1412_8 = 001, 100, 001, 010_2$ has been formed from two double precision numbers. We wish to divide it by the double precision divisor $D = 37_8 = 011, 111$. With $j = 4$, $k = 2$, we have

$$R(2) = N = 1412_8 \quad \text{and} \quad R^*(2) = 14_8,$$

so the initial value of $q_2^* = \left[\frac{14}{3} \right]_8 = 4$. It is easily seen, however, that

$$R^* = R(2) - (q_2^* D) 8^2 = 1412_8 - q_2^* \cdot 3700_8$$

satisfies the appropriate inequality only if $q_2^* = 0 = q_2$. We set
$$R^* = R(1) = 1412_8 - 0.8^2 = 1412_8,$$
and form $R^*(1) = 14_8$ and $q_1^* = 4$. In this case
$$R^* = 1412_8 - 4(37_8)8 = -326_8 < 0,$$
so we add $D\sigma = 370_8$ once and must thus decrease q_1^* by one to obtain $q_1 = 3$. This gives
$$R^* = R(0) = 1412_8 - 3(37_8)8 = 42_8 < D\sigma = 370_8,$$
and we form $q_0^* = \begin{bmatrix} 42 \\ 3 \end{bmatrix}_8$. Repetition of the above procedure shows that we must add 37_8 enough times, and hence subtract one from q_0^* enough times, to produce an
$$R^* = 42_8 - 1(37_8) = 3 = R(-1) = R,$$
with $q_2 = 0$, $q_1 = 3$, and $q_0 = 1$ reflecting, in octal,
$$1412 = (031)(37) + 3.$$

The steps of the algorithm are quite analogous to those introduced in Chapter 6. The successive subtractions are facilitated by the built-in division for single precision numbers. The octal digits of Example 8.18 of course reflect combinations of single precision three-bit numbers, and we leave it to the reader to carry out the details of the division using binary arithmetic.

EXERCISES
SECTION 8.2

8.1 If each of the following numbers is approximated by the given terminating decimal representation, determine the error or bounds for the error.
   a) $X = \frac{1}{7}$; $\bar{X} = 0.14285$ b) $X = \frac{1}{7}$; $\bar{X} = 0.14286$
   c) $X = \frac{26}{3}$; $\bar{X} = 8.6667$ d) $X = \sqrt{2}$; $\bar{X} = 1.414$.

8.2 For each of the approximations in Exercise 8.1, determine the relative error or bounds for the relative error.

8.3 Show that for an even base r, if the first digit dropped is at least $r/2$, the rounding should be up.

8.4 In the case of a base r which is odd, can the value of the first digit dropped be used to determine rounding up or down?

8.5 Express $\frac{15}{7}$ as a repeating representation in the bases ten, eight, and two. In each case cut the representation to five digits and round the representation to five digits.

8.6 For each of the six representations of Exercise 8.5 estimate an upper bound on the absolute value of the error in terms of p and α. Compare this with the true result.

EXERCISES

8.7 Suppose $(-\frac{17}{6})$ is to be represented in a six-digit register in decimal, in octal, and in binary. Find the extended complement representation for each base and determine the properly rounded version.

8.8 Redo Exercise 8.7 using absolute values and show that the two forms of rounded results agree.

SECTION 8.2.1

8.9 For the operand of Exercises 8.5 and 8.6 find the estimated bound on magnitude of relative error for each of the bases. Compare it with the true relative error and compare the relative errors in the three bases.

8.10 Generalize the comparison in relative error for a fixed number of digits. That is, given two bases r and r', and fixed α, compare the relative errors in the two cases. In which direction does the error increase, with increasing or decreasing r?

8.11 Find an estimate of the number of digits α and α' required for bases r and r', to provide essentially comparable accuracy. In particular find how many binary digits are required to provide a relative error no greater than that given by ten decimal digits. Include both cutting and rounding procedures.

SECTION 8.2.2

8.12 The approximation $\frac{22}{7}$ is often used to approximate π. If it is represented in decimal determine the number of significant digits it contains, if either cutting or rounding is the criterion. Do the same for its binary representation.

8.13 Represent $\frac{355}{113}$ decimally. If it is used as an approximation to π, how many significant digits can you obtain?

8.14 If the number 0.99991999... (all 9's) is rounded to one, how many of the zeros are significant?

SECTION 8.2.3

8.15 Determine the maximum decimal accuracy in terms of significant digits in a 60-digit, complement, binary register in which a sign digit is retained and external scaling is used.

8.16 For the 60-bit machine of Exercise 8.15 what is the maximum decimal accuracy in terms of significant digits if fractional floating-point operation is used with 12 bits assigned to exponent? Take account of both sign digits.

8.17 For the floating-point arithmetic in Exercise 8.16 what is the range on operands, exclusive of zero? What is the range on relative error? On true error?

SECTION 8.3

8.18 For the operations indicated, suppose that $X = \frac{4}{3}$ is represented decimally by five digits and $Y = \frac{2}{3}$ is represented decimally by five digits and five digits (rounded) are retained in the results. Determine the true and relative errors in the result R.

a) $R = X + Y$ b) $R = X - Y$
c) $R = XY$ d) $R = X/Y$

SECTION 8.4.1

8.19 Show that the errors in the floating-point solution in Example 8.12 are approximately 100%.

8.20 For the system of equations

$$x + \tfrac{1}{2}y + \tfrac{1}{3}z = 1,$$
$$\tfrac{1}{2}x + \tfrac{1}{3}y + \tfrac{1}{4}z = 0,$$
$$\tfrac{1}{3}x + \tfrac{1}{4}y + \tfrac{1}{5}z = 0,$$

use two-digit decimal approximations to the coefficients and find x, y, and z as two-digit solutions using two-digit arithmetic throughout. Compare the results with the exact solutions.

SECTION 8.4.2

8.21 Show that in a binary machine with $n = 3k$ digits and with a leading sign digit zero or one, if we interpret the octal configuration as a k-fold precision number, then leading 0, 1, 2, 3 correspond to nonnegative and 4, 5, 6, 7 to negative operands.

8.22 Extend the idea of Exercise 8.21 to decimal.

8.23 Show generally that for even bases $r > 2$ the use of only 0 and $(r - 1)$ as sign indicators is overly restrictive.

8.24* For an n-digit, base-r, closed accumulator devise the steps required to produce the sum of two k-fold precision operands as a k-fold precision result.

8.25* Establish the validity of inequality (8.1).

8.26* Establish the validity of inequality (8.2).

8.27* Carry out the details of the binary implementation of the division in Example 8.18.

ANSWERS

CHAPTER 1

SECTION 1.2

1.1 *Property 1.* The number zero, written 0, is a nonnegative integer.

   Property 2. Every nonnegative integer has a unique sequel of which it is the antecedent.

   Property 3. The nonnegative integer zero has no antecedent.

   Property 4. If two sequels are equal, then so are their antecedents.

   Property 5. If a set of nonnegative integers A has two properties:
   a) A contains 0;
   b) If A contains any nonnegative integer it also contains its sequel,
   Then A is the entire set of negative integers.

1.2 Let A be the set n of integers satisfying the condition

   a) A contains 1, since

$$\sum_{k=1}^{1} k = 1 = 1(1+1)/2.$$

   b) If A contains n,

$$\sum_{k=1}^{n} k = n(n+1)/2,$$

   and the sequel of n is $n+1$, so

$$\sum_{k=1}^{n} k + (n+1) = \sum_{k=1}^{n+1} k$$

$$= \frac{n(n+1)}{2} + (n+1)$$

$$= \frac{(n+1)(n+2)}{2};$$

   therefore A contains the sequel of n.

1.3 Let A be the set of integers satisfying the condition.

a) A contains 1, since

$$\sum_{k=1}^{1} k^2 = 1 = \frac{1(1+1)(2+1)}{6}.$$

b) If A contains n, then

$$\sum_{k=1}^{n} k^2 = \frac{n(n+1)(2n+1)}{6},$$

and the sequel of n is $n+1$, so

$$\sum_{k=1}^{n} k^2 + (n+1)^2 = \sum_{k=1}^{n+1} k^2$$

$$= \frac{n(n+1)(2n+1)}{6} + (n+1)^2$$

$$= \frac{(n+1)(n+2)(2n+3)}{6};$$

therefore A contains the sequel of n.

1.4 Let A be the set of integers satisfying the condition

a) A contains 1, since

$$\sum_{k=1}^{1} (2k-1)^2 = 1 = \frac{1(4-1)}{3}.$$

b) If A contains n, then

$$\sum_{k=1}^{n} (2k-1)^2 = \frac{n(4n^2-1)}{3},$$

and the sequel of n is $n+1$, so

$$\sum_{k=1}^{n} (2k-1)^2 + (2n+1)^2 = \sum_{k=1}^{n+1} (2k-1)^2$$

$$= \frac{n(4n^2-1)}{3} + (2n+1)^2$$

$$= \frac{(n+1)[4(n+1)^2 - 1]}{3};$$

therefore A contains the sequel of n.

1.5 Let A be the set of integers satisfying the condition.

a) A contains 1, since
$$\sum_{k=1}^{1} k^3 = 1 = \frac{1(1+1)^2}{4}.$$

b) If A contains n, then
$$\sum_{k=1}^{n} k^3 = \frac{n^2(n+1)^2}{4},$$
and the sequel of n is $n+1$, so
$$\sum_{k=1}^{n} k^3 + (n+1)^3 = \sum_{k=1}^{n+1} k^3$$
$$= \frac{n^2(n+1)^2}{4} + (n+1)^3$$
$$= \frac{n^4 + 6n^3 + 13n^2 + 12n + 4}{4}$$
$$= \frac{(n+1)^2(n+2)^2}{4}.$$

1.6 Let A be the set of integers satisfying the condition.
a) A contains 1 since $(a+b)^1 = a^1 + b^1$.
b) If A contains n and the sequel is $(n+1)$, form
$$(a+b)^{n+1} = (a+b)^n(a+b)$$
$$= (a+b)\left[a^n + \sum_{k=1}^{n-1}\binom{n}{k}a^{n-k}b^k + b^n\right]$$
$$= a^{n+1} + ab^n + (a+b)\sum_{k=1}^{n-1}\binom{n}{k}a^{n-k}b^k + a^n b + b^{n+1}$$
$$= a^{n+1} + \sum_{k=1}^{n}\binom{n}{k}a^{n-k}b^k + b^{n+1},$$
since
$$\binom{n}{k+1} + \binom{n}{k} = \frac{n!}{k!(n-k-1)!}\left[\frac{1}{k+1} + \frac{1}{n-k}\right]$$
$$= \binom{n+1}{k+1}.$$

1.7 $\binom{n}{0}a^n b^0 = a^n$ and $\binom{n}{n}a^{n-n}b^n = b^n$.

1.8*Let A be the set a for which one of the three conditions

   i) $a + c = b$,   ii) $a = b$,   iii) $a = b + c$

holds for all integers b.

a) A contains 1 since if $b = 1$, $a = 1 = b$ and (ii) holds. If $b \neq 1$ it has a unique antecedent c, and $a + c = 1 + c = b$ so (i) holds.

b) If A contains a, then for $a + c = b$, either $c = 1$ and $(a + 1) = b$ [(ii) for the sequel] or c has a unique antecedent and $(a + 1) + (c - 1) = b$ [(i) for the sequel]. For $a = b$,

$$(a + 1) = b + 1 = b + c$$

with $c = 1$ [(iii) for the sequel]. For $a = b + c$,

$$(a + 1) = b + (c + 1)$$

[(iii) for the sequel]. Thus A contains all positive integers.

1.9*Assume that 5a and 5b hold for all elements in A but that there is a non-empty set not contained in A. If so, it must have a first, say N. If it is the first, however, its unique antecedent $N - 1$ must be in A. By assumption then the sequel to $N - 1$, N, is also in A. Therefore, we have a contradiction and the only set not contained in A is empty.

SECTION 1.2.1

1.10 a) $7143 = (-17)(-420) + 3$ (Both remainder conditions)
    b) $-6080 = 42(-145) + 10$
        $-6080 = 42(-144) + (-32)$.

1.11 Shift the decimal point left by a number of places equal to the exponent in the divisor. The quotient is formed by those digits to the left of the decimal point and the remainder by those digits to its right.

Example: $12345/10^3$. Form 12.345 with $Q = 12$, $R = 345$, so that $12345 = (10^3)12 + 345$.

SECTION 1.3

1.12 Base 4: 1, 2, 3, 10, 11, 12, 13, 20, 21, etc.
    Base 7: 1, 2, 3, 4, 5, 6, 10, 11, 12, 13, 14, etc.
    Base 11 (with A = ten): 1, 2, 3, 4, 5, 6, 7, 8, 9, A, 10, 11, etc.
    Base 13 (with A = ten, B = eleven, and C = twelve): 1, 2, 3, 4, 5, 6, 7, 8, 9, A, B, C, 10, 11, 12, etc.

1.13 Sample: In base seven, 14_7 is eleven by counting, and $14_7 = (1 \cdot 7 + 4)_{10} = 11_{10}$.

1.14 A zero digit in the lowest order position.

SECTION 1.3.2

1.15 a) $12345_{10} = 30071_8$
    b) $54321_{10} = D431_{16}$, where D = digit thirteen
    c) $2971_{10} = 11443_7$
    d) $345_{10} = 101011001_2$

1.16 a) $10110101_2 = 181_{10}$

b) $10110101_2 = 265_8$
c) $2346_8 = 1254_{10}$
d) $456_7 = 237_{10}$

SECTION 1.4

1.17 For example: (a)

$$\begin{array}{r} 10010 \\ 1010\overline{)10110101} \\ 1010 \\ \hline 1010 \\ 1010 \\ \hline 1010 \\ \hline 01 = d_0 \end{array} \qquad \begin{array}{r} 1 \\ 1010\overline{)10010} \\ 1010 \\ \hline 1000 = d_1 \end{array} \qquad \begin{array}{r} 0 \\ 1010\overline{)1} \\ 0 \\ \hline 1 = d_2 \end{array}$$

so $10110101_2 = 181_{10}$

SECTION 1.5.1

1.18 If $0 \leq m \leq n$, then the representation is integral (rational). If $m < 0 \leq n$, then the representation is of form

$$\sum_{j=0}^{n} d_j r^j + \sum_{j=-1}^{-|m|} d_j r^j = a + r^{-|m|} b,$$

where a and b are integers. Thus, $a + b/r^{|m|} = (ar^{|m|} + b)/r^{|m|}$ is rational. A similar discussion holds for $m < n < 0$.

1.19 a) $1.00110011\ldots_2 = 1.14631463\ldots_8$
b) $1100.010101110\ldots_2 = 14.256\ldots_8$
c) $11011.001_2 = 33.1_8$

1.20 a) $(\frac{6}{5})_8 = (\frac{6}{5})_{10} = 1.2_{10} (= 1.14631463\ldots)_8$
b) $(\frac{1101}{10})_2 = (\frac{13}{2})_{10} = 6.5_{10}$
c) $3.214_8 = (\frac{419}{128})_{10} = 3.27\ldots_{10}$
d) $1101.11101_2 = (\frac{445}{32})_{10} = 13.90\ldots_{10}$

SECTION 1.6

1.21 Let $a = Ad$ and $m = Md$ so that $(A, M) = 1$ and A and M are relatively prime. Since $a - b = km$,

$$b = a - km = (A - kM)d.$$

Thus, d is a divisor of b as well as m. Since $(A, M) = 1$, however,

$$[(A - kM), M] = 1$$

and, thus, $(b, m) = d$.

1.22 Let $N = d_n 10^n + d_{n-1} 10^{n-1} + \cdots + d_1 \cdot 10 + d_0$. We have $10 \equiv 1 \pmod 9$ so that $10^2 \equiv 1^2 = 1 \pmod 9$ and so on to $10^k \equiv 1^k = 1 \pmod 9$. Hence $d_k 10^k \equiv d_k \pmod 9$ and $N = \sum_{k=0}^{n} d_k 10^k \equiv \sum_{k=0}^{n} d_k \pmod 9$.

1.23

+	0	1	2	3	4
0	0	1	2	3	4
1	1	2	3	4	0
2	2	3	4	0	1
3	3	4	0	1	2
4	4	0	1	2	3

×	0	1	2	3	4
0	0	0	0	0	0
1	0	1	2	3	4
2	0	2	4	1	3
3	0	3	1	4	2
4	0	4	3	2	1

1.24 $3 \equiv 3 \pmod{11}$
$3^2 = 9 \equiv -2 \pmod{11}$
$3^3 = 27 \equiv 5 \pmod{11}$
$3^5 = 3^2 \cdot 3^3 \equiv -10 \equiv 1 \pmod{11}$
$3^{30} = (3^5)^6 \equiv 1^6 = 1 \pmod{11}$
So the remainder is one.

1.25* Let $a = Ad$ and $m = Md$ with $(A, M) = 1$. If $ab \equiv ac \pmod{m}$ then

$$ab - ac = a(b - c) = Ad(b - c)$$
$$= km = kMd.$$

Thus, $A(b - c) = kM$, and since

$$(A, M) = 1,$$

$$b - c = \left(\frac{k}{A}\right)M,$$

where (k/A) is an integer. Thus, $b \equiv c \pmod{M}$.

CHAPTER 2
SECTION 2.1

2.1

a) $(3 + 5) = 0(10) + 8;\quad s = 8, c = 0$
$(9 + 2) = 1(10) + 1;\quad s = 1, c = 1$
$(1 + 4 + 7) = 1(10) + 2;\quad s = 2, c = 1$
$(1 + 0 + 0) = 0(10) + 1;\quad s = 1, c = 0$

b) $(7 - 2) = 0(10) + 5;\quad d = 5, b = 0$
$(2 - 9) = (-1)(10) + 3;\quad d = 3, b = -1$
$(6 - 4 - 1) = 0(10) + 1;\quad d = 1, b = 0$

2.2 Since each sum and carry digit is defined by

$$(c_{i-1} + x_i + y_i) = c_i r + s_i,$$

where $0 \leq s_i < r$, the uniqueness given by the Division Algorithm suffices. A similar argument holds for subtraction.

2.3 $x_0 + y_0 + z_0 = c_0 r + s_0 \leq 3r - 3$ with $0 \leq s_0 \leq r - 1$. Thus, $c_0 r \leq 3r - 3 < 3r$, so that $c_0 \leq 2$. Also

$$x_1 + y_1 + z_1 + c_0 = c_1 r + s_1 \leq 3r - 3 + c_0 \leq 3r - 3 + 2 = 3r - 1.$$

Thus $c_1 r + s_1 \leq 3r - 1$ and $0 \leq s_1 \leq r - 1$ implies $c_1 \leq 2$. A similar argument holds for c_i and $0 \leq s_i < r$.

ANSWERS 269

2.4* We may argue as we did in Exercise 2.3, and the result can be established by induction. If the digits in the ith place are

$$x_{i1}, x_{i2}, \ldots, x_{iN}$$

and the maximum carry from below is $c_{i-1} \le N - 1$, then

$$\sum_{j=1}^{N} s_{ij} + c_{i-1} = c_i r + s \le N(r - 1) + c_{i-1} \le Nr - 1$$

and $0 \le s_i \le (r - 1)$

implies $c_i < N$ or $c_i \le N - 1$. Also try consideration of the N numbers in pairs.

SECTION 2.2

2.5
Decimal	Octal	Binary
$8049 = 8049$	$0151_8 \equiv 10151_8 \pmod{8^4}$, i.e.,	$1000_2 \equiv 11000_2 \pmod{2^4}$,
$2049 \equiv 12049 \pmod{10^4}$	$105_{10} \equiv 4201_{10} \pmod{4096_{10}}$	i.e., $8 \equiv 24_{10} \pmod{16}$

2.6
Decimal	Octal	Binary
$8581 \equiv -1419 \pmod{10^4}$	$6261_8 \equiv -1517_8 \pmod{8^4}$	$1110_2 \equiv -10_2 \pmod{2^4}$,
$6581 = 6581$		i.e., $14_{10} \equiv -2 \pmod{16_{10}}$

SECTION 2.4

2.7
a) $9314 + 4163$ gives $3477 \equiv 753477 \pmod{10^4}$
b) $0111 + 0101$ gives $1100 \equiv 1001100 \pmod{2^4}$ in binary, or $7 + 5$ gives $12 \equiv 76 \pmod{64}$ in decimal
c) $7345 + 7214$ gives $6561 \equiv 36561 \pmod{8^4}$ in octal

2.8
a) $9314 - 4163$ gives $5151 \equiv -694849 \pmod{10^4}$
b) $0111 - 0101$ gives $0010 \equiv 100010 \pmod{2^4}$ in binary, or $7 - 5$ gives $2 \equiv 34 \pmod{16}$
c) $7345 - 7214$ gives $0131 \equiv 20131 \pmod{8^4}$ in octal.

2.9*

	0	1	2	3	4	5	6	7	8	9
0	0	1	2	3	4	5	6	7	8	9
1	1	2	3	4	5	6	7	8	9	0
2	2	3	4	5	6	7	8	9	0	1
3	3	4	5	6	7	8	9	0	1	2
4	4	5	6	7	8	9	0	1	2	3
5	5	6	7	8	9	0	1	2	3	4
6	6	7	8	9	0	1	2	3	4	5
7	7	8	9	0	1	2	3	4	5	6
8	8	9	0	1	2	3	4	5	6	7
9	9	0	1	2	3	4	5	6	7	8

2.10* $|X + Y| < r^n$.

SECTION 2.4

2.11 $k(10^3) - 234 \geq 0$ for $k \geq 1$, thus $766, 1766, 2766, \ldots$

2.12 Same as 2.11.

SECTION 2.5

2.13 (a) $|X| < 10^7$; (b) $|X| < 2^7$; (c) $|X| < 8^7$; (d) $|X| < 12^7$, that is,

(a) $|X| < 10{,}000{,}000$;   (b) $|X| < 128$, etc.

2.14 All are correct except $(+3) + (+6)$ and $(-3) - (-6)$, which cause overflow.

2.15 $|X \pm Y| < r^{n-1}$, so, since $|X \pm Y| \leq |X| + |Y|$, we could impose $|X| < r^{n-1}/2$ and $|Y| < r^{n-1}/2$.

2.16 Octal: 0, 1, 2, 3, indicate positive; 4, 5, 6, 7 indicate negative. $|X| < 4 \cdot 8^{n-1}$ or $|X| \leq 377\ldots7_8$.

2.17 Decimal: 0, 1, 2, 3, 4 indicate positive; 5, 6, 7, 8, 9 indicate negative. $|X| < 5 \cdot 10^{n-1}$ or $|X| \leq 499\ldots9_{10}$.

SECTION 2.6

2.18
```
  (+)  015        (+)  015        (−)  985        (−)  985
  (+)  012        (−)  988        (+)  012        (−1) 988
+      027      +      003             997             973
                                     (−003)          (−027)
```

2.19
```
  (+)  001111     (+)  001111     (−)  110001     (−)  110001
  (+)  001100     (−)  110100     (+)  001100     (−)  110100
  (+)  011011     (+)  000011          111101          100101
                                     (−000011)       (−011011)
```

2.20 No. For example $01111 + 01100 = 11011$ causes overflow.

SECTION 2.7

2.21
```
    a)                        b)                  c)
    002                       428                 712
    397                       692                 426
    399 ≡ 2397 (mod 999)    ①120                  329
                            ↳001                ①467
                              121               ↳001
                                                  468
```

2.22 In Exercise 1.22 we have

$$\sum_{k=0}^{m} d_k 10^k \equiv \sum_{k=0}^{m} d_k \pmod{9},$$

since $10 \equiv 1 \pmod{9}$. Similarly, using modular arithmetic,

$$\rho \equiv 1 \pmod{\rho - 1}$$
$$\rho^k \equiv 1 \pmod{\rho - 1}$$

ANSWERS 271

and
$$\sum_{k=0}^{m} d_k \rho^k \equiv \sum_{k=0}^{m} d_k \pmod{\rho - 1}.$$

In particular, for $\rho = r^n$ we get the stated rule for the modulus $\rho - 1 = r^n - 1$.

2.23

(+)	015	(+)	015	(−)	984	(−)	984
(+)	012	(−)	987	(+)	012	(−)	987
(+)	027		①002		996		①971
			↳ 1		(−003)		↳ 1
		(+)	003				972
							(−027)

2.24

(+)	001111	(+)	001111	(−)	110000	(−)	110000
(+)	001100	(−)	110011	(+)	001100	(−)	110011
(+)	011011		①000010		111100		①100011
			↳ 1		(−000011)		↳ 1
		(+)	000011				100100
							(−011011)

2.25 For decimal:

$12 + 0 = 12$ $\qquad 12 \cdot 0 = 0$

```
   012              012
   999              999
  ①011             ①988
   ↳ 1             ↳ 11
   012              999   (negative zero)
```

Similarly for binary.

2.26 Example: $-15 + 12 = -3$ is formed in a subtractive accumulator as $(-15) - (-12)$ in the form

```
                          984
                          987
        End around    ⊖①997
        borrow          ↳ −1
                          996
                        (−003)
```

Similarly for the other cases.

SECTION 2.8.2

2.27 For example, in a three-digit, open, decimal accumulator we might have Case IIa, with −015 as 985 in A and 012 in X. Applying rules 1, 2, 3, and 4 we have

A	985	
A	015	(Complement)
X	012	
A	027	(Add)
A	973	(No carry, complement)

A similar discussion holds for the other cases.
CHAPTER 3
SECTION 3.2.1
3.1 Let A be the set of all integers for which the extended associative law holds. Then A contains 3, by the basic associative law. Suppose A contains $n \geq 3$. Then

$$a_1 \cdot a_2 \cdot a_3 \cdots a_n$$

is the same for any grouping. Consider every grouping for

$$a_1 \cdots a_n a_{n+1}.$$

It contains, in particular, $(a_1 \cdots a_n) \cdot a_{n+1}$ and $a_1 \cdot (a_2 \cdots a_{n+1})$, which are equal by the basic law. Thus,

$$(a_1 \cdots a_n) \cdot a_{n+1} = a_1(a_2 \cdots a_{n+1}),$$

and each parentheses is independent of grouping, so A contains $n+1$. Why do we not consider $n = 1$ and 2?

3.2 Consider the possible permutations of the elements and extend the commutative law. For example $a_1 a_2 a_3 = a_3 a_2 a_1$, since

$$(a_1 a_2)a_3 = (a_2 a_1)a_3 = a_3(a_2 a_1) = a_3 a_2 a_1.$$

3.3 Replace the Boolean product by the Boolean sum.

3.4 If $a = 1$ then $ab = ac$ is $1 \cdot b = 1 \cdot c$, so $b = c$. If $a = 0$, then $ab = ac = 0$ for both $b = c$ and $b \neq c$. Thus, we can divide out a nonzero factor.

3.5 No general rule.

3.6 For example, with addition on the left, let A be the set of n for which

$$\overline{a_1 + a_2 + \cdots a_n} = \bar{a}_1 \cdot \bar{a}_2 \cdots \bar{a}_n.$$

Then A contains 2, by the basic law. If A contains $n \geq 2$ then we have

$$\overline{a_1 + a_2 + \cdots + a_n + a_{n+1}} = \overline{(a_1 + a_2 + \cdots + a_n) + a_{n+1}}$$

$$= \overline{(a_1 + \cdots + a_n)} \cdot \bar{a}_{n+1} = (\bar{a}_1 \cdot \bar{a}_2 \cdots \bar{a}_n) \cdot \bar{a}_{n+1};$$

but the parentheses are unnecessary, so A contains $n+1$. A similar argument holds for multiplication on the left.

SECTION 3.3.1
3.7 For example, by deMorgan's laws $\overline{xy} = \bar{x} + \bar{y}$, so

$$s = (z + y)(\overline{xy}) = (x + y)(\bar{x} + \bar{y}) = x\bar{x} + x\bar{y} + y\bar{x} + y\bar{y} \quad \text{(Distributive law)},$$

and with $x\bar{x} = y\bar{y} = 0$, we have

$$s = x\bar{y} + \bar{x}y.$$

Similar manipulations verify the other forms.

SECTION 3.3.2
3.8 If $C' = 1$ then $X_i = Y_i = 1$, and $S' = 0$, so $C'' = 0$. If $C'' = 1$ then $S' = 1$ and $C' = 0$.

ANSWERS   273

3.9 We have
$$s = \bar{c}(x\bar{y} + \bar{x}y) + c(\bar{x}\bar{y} + xy)$$
$$= \bar{c}x\bar{y} + \bar{c}\bar{x}y + c\bar{x}\bar{y} + cxy$$

and

$$c' = xy\bar{c} + (x + y)c = xy\bar{c} + xc + yc$$
$$= xy\bar{c} + xc \cdot 1 + yc$$
$$= xy\bar{c} + xc(y + \bar{y}) + yc = xy\bar{c} + xyc + xc\bar{y} + yc$$
$$= xy\bar{c} + xyc + x\bar{y}c + yc(\bar{x} + x)$$
$$= xy\bar{c} + xyc + x\bar{y}c + \bar{x}yc + xyc$$
$$= xy\bar{c} + \bar{x}yc + x\bar{y}c + (xyc + xyc)$$
$$= xy\bar{c} + \bar{x}yc + x\bar{y}c + xyc.$$

3.10 Use a full adder in the lowest-order position with a time delay from the highest order.

SECTION 3.4.2

3.11 Same argument as Exercise 3.8.

3.12 If carry = 1, the sum digit from the half-adder is zero and no borrow is generated. If borrow = 1, the sum digit from the half-adder is one and no carry is generated.

3.13 $d = (x\bar{y} + \bar{x}y)\bar{b} + \overline{(x\bar{y} + \bar{x}y)}b$
$= x\bar{y}\bar{b} + \bar{x}y\bar{b} + (\overline{x\bar{y}})(\overline{\bar{x}y})b = x\bar{y}\bar{b} + \bar{x}y\bar{b} + (\bar{x} + y)(x + \bar{y})b$
$= x\bar{y}\bar{b} + \bar{x}y\bar{b} + \bar{x}xy + \bar{x}\bar{y}b + yxb + y\bar{y}b$
$= x\bar{y}\bar{b} + \bar{x}y\bar{b} + \bar{x}\bar{y}b + xyb.$

Similarly for borrow.

3.14 Similar to Exercise 3.13.

SECTION 3.5

3.15 Since $p_3 c = 1, b = 1$ which sets $p_2 = 1, p_3 = p_4 = 0$ and $p_5 = f = 1$.

SECTION 3.5.1

3.16

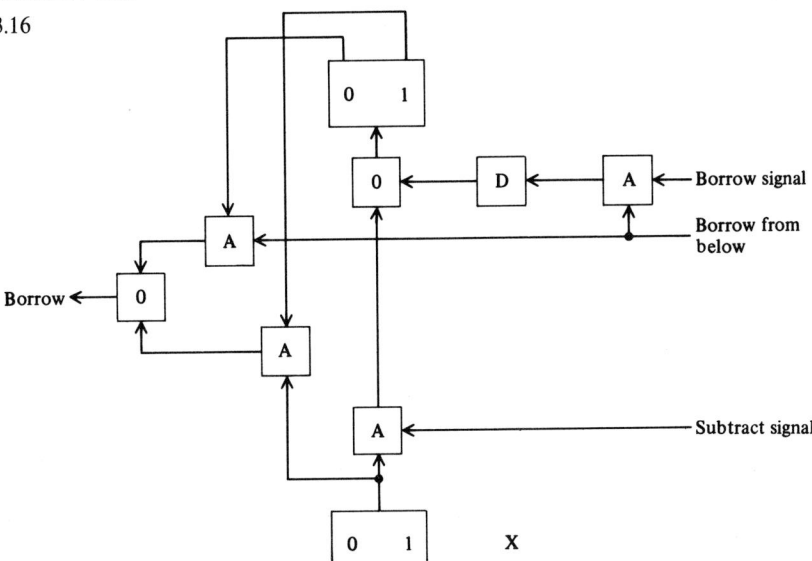

274 ANSWERS

3.17 Compare your result with the additive network given.

SECTION 3.6.2

3.18

(a)	(b)	(c)	Decimal
5 2 1 1	3 3 2 1	7 5 3 −6	Digit
0 0 0 0	0 0 0 0	0 0 0 0	0
0 0 0 1	0 0 0 1	1 0 0 1	1
0 0 1 1	0 0 1 0	0 1 1 1	2
0 1 1 0	0 1 0 0	0 0 1 0	3
0 1 1 1	1 0 0 1	1 0 1 1	4
1 0 0 0	0 1 1 0	0 1 0 0	5
1 0 0 1	1 0 1 1	1 1 0 1	6
1 1 0 0	1 1 0 1	1 0 0 0	7
1 1 1 0	1 1 1 0	0 1 1 0	8
1 1 1 1	1 1 1 1	1 1 1 1	9

The codes are not necessarily unique. For example, 0110 or 0101 would each represent 3 in the 5211 weights.

3.19
a)
```
    0 0 1 0      0 0 1 1      0 0 0 1
    0 1 0 0      0 0 0 0      0 1 1 0
    ───────      ───────      ───────
    0 1 1 0      0 0 1 1      0 1 1 1
       6            3            7
```
No carries, no corrections.

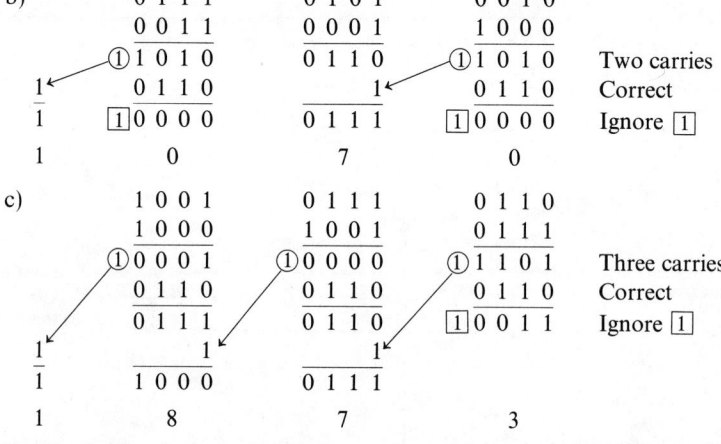

c)
```
                                                        7            3
```

3.20 For example, s_1 is the sum digit from a full adder and c_1 is the carry with inputs x_1, y_1, and c, the carry from below. So:

$$s_1 = \bar{x}_1\bar{y}_1 c + \bar{x}_1 y_1 \bar{c} + x_1 \bar{y}_1 \bar{c} + x_1 y_1 c,$$
$$x_1 = \bar{x}_1 y_1 c + x_1 \bar{y}_1 c + x_1 y_1 \bar{c} + x_1 y_1 c.$$

A similar expression holds for the uncorrected output in the 2, 4, and 8 stages, but these must then be modified for correcting and carry generation.

SECTION 3.6.3

3.23 The codes as listed in the answer to Exercise 3.18 are self-complementing. Note, however, that the choice of 0101 for three in the 5221 code would destroy the self-complementing feature unless we replace 1001 for six by 1010. Similar care with choices must be made for each case.

CHAPTER 4

SECTION 4.2

4.1 It will represent multiplication by r if the highest-order digit is initially zero.

4.2 Same as 4.1.

4.3 It will represent division by r if the lowest-order digit is initially zero.

4.4 Same as 4.3.

Open left shift	*Closed left shift*
a) $345600 = 10(034560)$	Same as open
b) $456000 \equiv 10(345600) \pmod{10^6}$	$456003 \equiv 10(345600) \pmod{10^6 - 1}$
c) $034560 = 10(003456)$	Same as open
d) $456070 \equiv 10(345607) \pmod{10^6}$	$456073 \equiv 10(345607) \pmod{10^6 - 1}$

Open right shift	*Closed right shift*
a) $003456 = 10^{-1}(034560)$	Same as open
b) $034560 = 10^{-1}(345600)$	Same as open
c) $000345 =$ truncated $[10^{-1}(003456)]$	$600345 \neq 10^{-1}[003456]$
d) $034560 =$ truncated $[10^{-1}(345607)]$	$734560 \neq 10^{-1}[345607]$

4.6 a) $560 \equiv 10(456) \pmod{10^3}$
 b) $564 \equiv 10(456) \pmod{10^3 - 1}$

4.7 a) $540 = 10(054)$
 b) $540 = 10(054)$

4.8 An open left shift of 456 gives $560 \equiv 10(456) \pmod{10^3}$.
 A closed left shift of 456 gives $564 \equiv 10(456) \pmod{10^3 - 1}$.
 An open right shift of 456 gives $045 =$ truncated $[10^{-1}(456)]$.
 A closed right shift of 456 gives 645 (no simple relation).
 Similarly, both open and closed left shifts of 054 produce the results of Exercise 4.7.

4.9 We have $Nr^n \equiv N \pmod{r^n - 1}$, but there is no simple relation in the modulus r^n.

4.10 There is no simple relation between octal digits for shifts of one or two positions. For example, if the original number is

$$(d_5 \cdot 2^2 + d_4 \cdot 2 + d_3)8 + (d_2 \cdot 2^2 + d_1 \cdot 2 + d_0) = \phi_1 \cdot 8 + \phi_0$$

gives

$$(d_4 \cdot 2^2 + d_3 \cdot 2 + d_2)8 + (d_1 \cdot 2^2 + d_0 \cdot 2 + d_5) = \phi'_1 \cdot 8 + \phi'_0,$$

and there is no simple relation between ϕ_i and ϕ'_i. A left circular shift of three, however, is a left circular shift of one octally. Thus, octal 73 becomes octal 67 on a binary left shift of one, but 73 becomes 37 on a binary left shift of three.

SECTION 4.4

4.11 Let the shifted result be M; then the circular left shift of $n - s$ guarantees
$$Nr^{n-s} \equiv M \pmod{r^n - 1} \quad \text{or} \quad Nr^n \equiv Mr^s \pmod{r^n - 1}.$$
Also $Nr^n \equiv N \pmod{r^n - 1}$, so $Mr^s \equiv N \pmod{r^n - 1}$.

4.12 If the original number contains s low-order zeros.

SECTION 4.5

4.13 If there are s leading zeros in $|X|$.

4.14 None, since low-order zeros are unmodified in r's complements.

4.15 If there are at least $(s + 1)$ leading sign digits.

4.16 If there are as many high-order (left) or low-order (right) sign digits as the shift count.

4.17 No. Although the final result produces congruence for the modulus r^n, the most significant digits are lost.

4.18 For an open left shift of s positions we have, if N is the original number,
$$N = d_{n-1}d_{n-2} \cdots d_2 d_1 d_0,$$
and the shifted result is
$$S = d_{n-s-1} \cdots d_1 d_0 0 \cdots$$
with s low-order zeros. Since $Nr^s - S = d_{n-1}r^{n+s} + \cdots + d_{n-s}r^n$,
$$S \equiv Nr^s \pmod{r^n}.$$
For an open right shift, however, we have congruence to Nr^{-s} only if there are at least s low-order zeros.

4.19 If $d_{s-1}r^{-1} + d_{s-2}r^{-2} + \cdots + d_0 r^{-s} \leq \frac{1}{2}$.

4.20 If $d_{s-1}r^{-1} + d_{s-2}r^{-2} + \cdots + d_0 r^{-s} > \frac{1}{2}$ when the d_i are the digits of $|X|$ and r's complements are used. Similarly, with the same d_i, if
$$d_{s-1}r^{s-1} + \cdots + (d_0 + 1)r^{-s} > \frac{1}{2}$$
for $(r - 1)$'s complements.

CHAPTER 5

SECTION 5.1

5.1 Let A be the set of Y for which the assertion is true for all X. Then A contains one, since $X \cdot 1 = 0 + X$. If A contains Y then
$$X(Y + 1) = XY + X = XY + X \cdot 1$$
$$= X + X + \cdots + X \, (Y \text{ times}) + X \, (\text{one time})$$
$$= X + X + \cdots + X \, (Y + 1 \text{ times}) \qquad \text{Distributive law.}$$

5.2*We need to show $\overbrace{(X + X + \cdots + X)}^{Y \text{ times}} = \overbrace{(Y + Y + \cdots + Y)}^{X \text{ times}}$. Let A be the set of

Y for which this is valid for all X. Then A contains 1 since

$$X = \overbrace{1 + 1 + \cdots + 1}^{X \text{ times}}.$$

(Establish this by induction.) If A contains Y then

$$X(Y + 1) = XY + X = \overbrace{(X + X + \cdots + X)}^{Y \text{ times}} + X = \overbrace{(X + X + \cdots X)}^{Y + 1 \text{ times}}.$$

But

$$(Y + 1)X = (1 + Y)X = X + YX$$
$$= X + \overbrace{(Y + Y + \cdots + Y)}^{X \text{ times}}$$
$$= \overbrace{(1 + 1 + \cdots + 1)}^{X \text{ times}} + \overbrace{(Y + Y + \cdots + Y)}^{X \text{ times}}$$
$$= (Y + 1) + (Y + 1) \cdots + (Y + 1).$$

5.3 $(XY)Z = \overbrace{(X + X + \cdots + X)}^{Y \text{ times}} Z$

$$= \overbrace{(X + X + \cdots + X)_Y + (X + X + \cdots + X)_Y + \cdots + (X + X + \cdots + X)_Y}^{Z \text{ times}}$$
$$= (X + X + \cdots + X)_Z + \cdots + (X + \cdots + X)_Z$$
$$= (XZ)Y.$$

A similar argument permits us to interchange Y and Z to form $(YZ)X = X(YZ)$, so $(XY)Z = X(YZ) = XYZ$.

5.4 $X(Y + Z) = \overbrace{(X + X + \cdots + X)}^{(Y + Z) \text{ times}}$

$$= \overbrace{(X + X + \cdots + X)}^{Y \text{ times}} + \overbrace{(X + X + \cdots + X)}^{Z \text{ times}} = XY + XZ$$

5.5 Note that $X + X + \cdots$ zero times, added to zero, is zero.

5.6 Since $(x - X^*) = k_1 M$ and $(y - Y^*) = k_2 M$ and $(X^*Y^* - R^*) = k_3 M$,
$$xy - R^* = xy - xY^* + xY^* - X^*Y^* + X^*Y^* - R^*$$
$$= x(y - Y^*) + (x - X^*)Y^* + (X^*Y^* - R^*)$$
$$= (xk_2 + k_1 Y^* + k_3)M.$$

5.7 In all cases if the addition is carried out consistently for the appropriate modulus (as a 2n-digit open or closed accumulation).

5.8 $X^{**} + Y^{**} = X^* + Y^* + (k + l)r^n(r^n - 1)$

5.9 Since $|X| + |Y| < 2r^{n-1} \le r^n$ (because $r \ge 2$), an $(n + 1)$-digit accumulator would suffice.

5.10
$$rX = r(X_{n-1}r^{n-1} + \cdots + X_1 r + X_0)$$
$$= X_{n-1}r^n + X_{n-2}r^{n-1} + \cdots + X_1 r^2 + X_0 r + 0.$$

5.11 $\frac{15}{95} = \frac{3}{19}$

5.12 Use the smaller number as the multiplier.

5.13* Let $X = X_{n-1}X_{n-2}\ldots X_0$ and $Y = Y_{n-1}Y_{n-2}\ldots Y_0$. The number of shifts in each case is the same. The number of additions, however, is

$$\sum_{k=0}^{n-1} X_k \quad \text{or} \quad \sum_{k=0}^{n-1} Y_k.$$

Unless these sums are equal, fewer additions will be required one way than will be required in the other.

SECTION 5.3

5.14 For addition, the sum would have to be shifted right k places (circular shifting). For multiplication, if the low-order digits of the multiplier are used first, $2k$ circular right shifts (for each digit) would be required, followed by a circular left shift of k digits. If the final right shift is omitted, these would reduce to $2k - 1$ right and $k - 1$ left shifts. Open shifting would not be appropriate. Extension facilities would fill in with $2k$ sign digits.

5.15 No. The first two steps would form

$$(XY_{n-1}r^{n-1} + XY_{n-2})r^{n-2} = XY_{n-1}r^{2n-3} + XY_{n-2}r^{n-2}.$$

SECTION 5.3.2

5.16 If X and Y are the integral operands, the fractional interpretation would be Xr^{-n} and Yr^{-n}. Thus, the fractional interpretation of XY as XYr^{-2n} is correct. If we provide for a sign digit to the left of the base point in X and Y and XY, this is no longer true.

5.17 Digits to the left of the base point.

SECTION 5.4

5.18 The machine representation of X is $X^* = kM_1 + X$, where, if $|X| < r^{n-1}$,

$$k = 0 \text{ if } X^*_{n-1} = 0 \quad \text{and} \quad k = 1 \text{ if } X^*_{n-1} = r - 1.$$

To form X^{**} we need an extension E such that

$$X^{**} = X^* + E \equiv X \pmod{M_2}.$$

This will be valid if $E = 0$ (for $k = 0$) or $E = r^{2n} - r^n$ (for $k = 1$).

SECTION 5.4.1

5.19 For example, with $M_1 = 2^4 - 1$, we have Case IV and a correction $2^4|X| - |X|$.

ANSWERS   279

The binary product is 1100, 1011, 1111. For the eight-digit closed accumulator this reduces to 11001011. If we add $2^4|X|$ in the closed accumulator, we have 00001100, and subtraction of $|X| = 4$ yields correctly $+8 : 00001000$.

5.20 Use two's complementation, open shifting, and omit the last step. The result, 00001111, is correctly 15 with X and Y interpreted as -5 and -3.

SECTION 5.4.2

5.21*We wish to enter $k(r^n - 1) + X$, where $k = 0$ if $X \geq 0$ and $k = 1$ if $X < 0$. In the fractional accumulator this would be entered as $[k(r^n - 1) + X]r^n + E$ where E is the extension. To obtain the proper version of X we would right-circular-shift by n positions. The circular shift would produce

$$k(r^n - 1) + X + Er^n = k(r^{2n} - 1) + X.$$

Hence $Er^n = k(r^{2n} - r^n)$. This is all zeros if $k = 0$, and n digits $(r - 1)$ followed by n zeros if $k = 1$. To form E we shift this right n places. Thus, we must fill in the low-order positions with sign digits.

5.22 Digits to the left of the base point.

SECTION 5.5

5.24 We can interpret the product $XY = \overbrace{(X + \cdots + X)}^{Y \text{ times}}$ as a multiply–add X by one a total of Y times.

CHAPTER 6

SECTION 6.1

6.1 Since $x = 2x - Nx^2$, we have $Nx^2 - x = 0$ which has the two solutions $x = 0$ and $x = 1/N$.

6.2 With all numbers in binary, $x_{n+1} = x_n(10 - Nx_n) = x_n(10 - 11x_n)$. Thus,

$$x_1 = (0.1)(10 - 11(0.1)) = (0.1)(0.1) = 0.01,$$
$$x_2 = 0.0101,$$
$$x_3 = 0.01010101,$$

and
$$x_4 = 0.0101010101010101.$$

6.3 The digits generated are the same.

6.4 For example, using x_2 we form $(101)(0.0101) = 1.1001$. The true expansion is $1.101010\ldots$.

6.5*Let $x_{n+1} = \phi(x_n)$. Then

$$\left|x_{n+1} - \frac{1}{N}\right| = \left|\phi(x_n) - \phi\left(\frac{1}{N}\right)\right|$$
$$= |\phi'(\xi)|\left|x_n - \frac{1}{N}\right|.$$

Since $\phi'(\xi) = 2 - 2N\xi$,

$$|\phi'(\xi)| < 1 \quad \text{if} \quad \frac{1}{2N} < x_0 < \frac{3}{2N}.$$

6.6 Arithmetic in binary (closed accumulator)

$$\begin{array}{r} 0\ 1\ 0\ 1 \\ 1\ 1\ 0\ 0 \\ \hline ①0\ 0\ 0\ 1 \\ 1 \\ \hline 0\ 0\ 1\ 0 \end{array} \quad \text{count} = 1 = \text{quotient}$$

Remainder

Multiply remainders by 2 and repeat division by subtraction.

SECTION 6.2

6.7 In ordinary long division we guess a value of q_{n-1} and correct if necessary, rather than checking all values. The partial remainders are merely the results of the successive subtractions in long division.

6.8 Use ordinary long division.

6.9 There are only two possible quotient digits.

6.10 If $D = 0$, successive subtractions would never produce a negative result. If a maximum value of $q_{n-1} = r - 1$ has not produced a negative result, there is overflow.

SECTION 6.2.1

6.11 Replace addition of complements of D by direct subtraction of D.

6.12 Yes, if we interpret $Q = 0.1110$, $D = 0.1111$, $N = 0.11010011$, and $R = 0.00000001$. That is, in $N = QD + R$, we can rewrite the equation as

$$Nr^{-2n} = (Qr^{-n})(Dr^{-n}) + Rr^{-2n}.$$

SECTION 6.3

6.16 For example, if $N < 0$, $D < 0$ we can rewrite the relation

$$|N| = |Q||D| + R' \quad (R' \geq 0)$$

as

$$-|N| = |Q|(-|D|) - R' = (|Q| + 1)(-|D|) + R,$$

where $R = |D| - R' \geq 0$.

6.17 We rewrite $|N| = |Q| \cdot D + R'$, $R' \geq 0$, as

$$N = -|N| = (-|Q|)D - R'$$
$$= -(|Q| + 1)D + D - R'$$
$$= -(|Q| + 1)D + R,$$

with $R = D - R' \geq 0$. If $R' = 0$, then $R = D$ violates $R < D$.

6.18 If a is the sign digit for N (+ = 0) and b is the sign digit for D, then the sign digit for Q is c if $a + b \equiv c \pmod{2}$.

SECTION 6.3.1

6.19 For example, convert the counterexample given, to binary.

6.20*While $N \leq (r^{n-1} - 1)|D|$ guarantees $N < (r^{n-1})|D|$, the latter condition does not guarantee the former.

ANSWERS 281

6.21* For example, if $N = (2^n - 1)|D| + 1$, then
$$N = (2^n - 1)|D| + 1 < (2^n - 1)|D| + |D| = 2^n|D|$$
if $|D| \geq 2$, but $N > (2^n - 1)|D|$.

CHAPTER 7
SECTION 7.1

7.1 $(12.3457)10^4) = 123457$ and $(3.2653)(10^4) = 32653$.
$$(X + Y)10^4 = 156110$$
so
$$X + Y = 15.6110.$$
$$(XY)10^8 = 4031241421$$
so
$$XY = 40.31241421.$$
$$(X)10^7 = (3780)Y \cdot 10^4 + 28660,$$
so
$$\frac{X}{Y} = 3.780.$$

7.2 For example, we still align digits for addition:

$$\begin{array}{r} 12.3457 \\ 3.2653 \\ \hline 15.6332 \end{array} \quad \text{interpreted} \quad \begin{array}{rl} 123457 & = (X)8^4 \\ 32653 & = (Y)8^4 \\ \hline 156332 & = (X + Y)8^4 \end{array}$$

A similar consideration of scale factors is used for multiplication and division.

SECTION 7.2

7.3 Use 10^8, 0271828182 or 0271828183 if rounding is used.

7.4 Use 10^{-1}, 0271828182 or 0271828183 if rounding is used.

SECTION 7.2.1

7.5 $p = 1, s = 8, n = 10$ so $p + s = n - 1$. Loss of significant digits.

7.6 Since $2^4 \leq 17.25 < 2^5$, $p = 5$, $p + s = 5 + s \leq 8 - 1 = 7$, so that $s \leq 2$. If $s = 2, 17.25_{10} = 10001.01_2$ and $X^* = 01000101$. For the negative number we use 10111010.

SECTION 7.2.2

7.7 $p + s = 5 + s \leq 0, s \leq -5$. For $s = -5, 01000101$.

7.8 From $p + s \leq n - 1$, $p + s$ is the number of digits retained (exclusive of sign). Thus, the equality gives the maximum number $n - 1$. The value of $p + s$ is not the number of digits retained in nonintegral scaling. See, for example, Exercise 7.7 where $p + s$ may have the value 0.

7.9 $p + s = p + 10 \leq 7, p \leq -3, |X| < 2^{-3}$.

SECTION 7.3.1

7.10 $(1234.56)10^2 = 0123456$ and $(65.4321)10^4 = 0654321$. For addition we must use 10^2 as the common scale factor.

7.11 $2^{10} \leq 1234.56 < 2^{11}$ so we can scale at 2^{12}.

$2^6 \leq 65.4321 < 2^7$ so we can scale at 2^{16}.

For addition we use common scale factor 2^{12}.

7.12 Since scaling is correct we note that $p + s = p + 45 \leq 47$. Thus $|X| < 2^2$ and $|Y| < 2^2$. This guarantees $|X + Y| < 2 \cdot 2^2 = 2^3$. To be certain of correctness we would have to rescale both at 2^{44}.

7.13*For X we know that $s \leq n - 1 - p$. Since Y must be scaled at r^s for addition, it would appear as zero if $p' + s \leq 0$ (no digits retained) where $|Y| < r^{p'}$. In particular, if $s = n - 1 - p$, then for $|Y| < r^{p+1-n}$, Y would appear as zero. This is true for Y such that $|Y/X| < r^{2-n}$.

SECTION 7.3.2

7.14 We know that $2^3 \leq |X| < 2^4$ and $2^{-5} \leq |Y| < 2^{-4}$, so that $2^{-2} \leq |XY| < 1$. To guarantee no overflow in a 48-bit register we would have to have XY scaled at 2^s with $s \leq 47$. Thus, we would have to rescale Xr^x and Yr^y so that $x + y \leq 47$ before multiplication in a 48-bit register. In 96 bits we can correctly form $XY \, 2^{94}$, but would then need a right shift of at least 47 to scale down to 48-bit size.

7.15*If we are required to shift right by enough to yield only sign digits ($+0$ or -0) we have an apparent zero product. On the other hand if X and Y are correctly scaled we know that $|XY| < r^p$ where $p + s + t \leq 2(n - 1)$. Since the right shift required is $\rho \geq p + s + t - (n - 1)$, the equality would yield a right shift $\rho \geq n - 1$. Thus, we would get an apparent zero only by overshifting.

SECTION 7.3.3

7.16 Since X and Y are optimally scaled, we have, from Exercise 7.14,

$$2^7 < \frac{X}{Y} < 2^9.$$

Thus, we can correctly scale X/Y at 2^{38}. If we form

$$\frac{(X)2^{43+\lambda}}{(Y)2^{51}} = \left(\frac{X}{Y}\right)2^{\lambda-8},$$

a left shift of X prior to division of $\lambda - 8 \leq 38, \lambda \leq 46$.

7.17*For $Y = 0$, there is no integer p for which $|X/Y| < r^p$, so no scale factor is correct. Any case where an inadequate upper bound is used on a quotient can produce division overflow—a computer analogy to division by zero. This division overflow is sensed in most modern computers.

SECTION 7.3.4

7.18 Establish that for $n > 0$ each x_n satisfies $2^{-1} < x_n < 1$.

SECTION 7.5

7.22 In binary $2.5 = 10.1$ and $0.25 = 0.01$. The normalized integers are $01010\ldots0$ (28 bits) and $010\ldots0$ (28 bits). The exponents are thus -25_{10} and -28_{10} or -11001_2 and -11100_2. The complement form for these is 11100110_2 and 11100011_2, respectively.

7.23 The digits of the number are unchanged. The exponents become 2 and -1 or 00000010_2 and 11111110_2, respectively.

7.24*If $r^{p-1} \leq |X| < r^p$, the equality $p + s = n - 1$ guarantees optimal scaling. It also guarantees that the number of digits retained in X is $(n - 1)$. The definition of the normalized version X^* likewise guarantees this.

SECTION 7.5.1

7.25 We record
$$X = (0123456, -2) \quad \text{and} \quad Y = (0654321, -4).$$
For addition we must replace Y by $(0006543, -2)$.

7.26 The inequality $r^{-1} \leq |X^*| < 1$ guarantees $(n - 1)$ digits of $|X|$.

CHAPTER 8

SECTION 8.2

8.1 a) $|E| = 7 \cdot 10^{-6} + (\frac{1}{7})10^{-6}$
   b) $|E| = 6 \cdot 10^{-6} + (\frac{1}{7})10^{-6}$
   c) $|E| = (\frac{11}{30})10^{-4}$
   d) $|E| < 0.00022$

8.2 a) Less than 5×10^{-5} (0.005%)
   b) Less than 4.3×10^{-5} (0.0043%)
   c) $(\frac{11}{260})10^{-4}$ (less than (0.0004%))
   d) Less than 1.6×10^{-4} (0.016%)

8.3 Suppose we approximate $|X|$ by α digits. Then, the omitted portion is
$$\sum_{j=\alpha+1}^{\infty} d_{p-j} r^{p-j} = d_{p-\alpha-1} r^{p-\alpha-1} + \sum_{j=\alpha+2}^{\infty} d_{p-j} r^{p-j}$$
and
$$d_{p-\alpha-1} \geq \frac{r}{2}.$$
Hence, the omitted portion is greater than or equal to
$$d_{p-\alpha-1} r^{p-\alpha-1} \geq \frac{r}{2} r^{p-\alpha-1} = \frac{r^{p-\alpha}}{2}$$
and according to the rule, rounding should be up.

8.4 No. For example, if we round $1.1211\ldots$ (base 3) to its integral portion, the fact that $1 < \frac{3}{2}$ would suggest rounding down. On the other hand
$$0.1211\ldots = \frac{1}{9} + \frac{1}{3}\left[1 + \frac{1}{3} + \frac{1}{3^2} + \cdots\right] = \frac{1}{9} + \frac{1}{2} > \frac{1}{2}.$$

8.5 Base 10: $2.142857142\ldots$, cut to 2.1428, rounded to 2.1429
   Base 8: $2.11111\ldots$, cut to 2.1111, rounded to 2.1111
   Base 2: $10.001001001\ldots$, cut to 10.001, rounded to 10.001.

8.6 Base 10: Cut error $< 10^{-4}$; rounded error $\leq (\frac{1}{2})10^{-4}$; true errors $(\frac{4}{7})10^{-4}$; $(\frac{3}{7})10^{-4}$.

Base 8: Cut error $< 8^{-4}$; rounded error $\leq (\frac{1}{2})8^{-4}$; true errors both $(\frac{1}{7})8^{-4}$.
Base 2: Cut error $< 2^{-3}$; rounded error $\leq 2^{-4}$; true errors both $(\frac{2}{7})2^{-4}$.

8.7 Base 10:   97.16666...   rounded to 97.1666
     Base 8:   75.23535...   rounded to 75.2353
     Base 2:   101.010011101011...   rounded to 101.001

8.8 Base 10:   2.83333...   rounds to 02.8333   complemented as 97.1666
     Base 8:   2.5424242...   rounds to 02.5424   complemented as 75.2353
     Base 2:   10.101100010...   rounds to 010.110   complemented as 101.001

SECTION 8.2.1

8.9 Base 10: 10^{-4}; $(\frac{1}{2})10^{-4}$; $(\frac{4}{15})10^{-4}$; $(\frac{1}{5})10^{-4}$
     Base 8: 8^{-4}; $(\frac{1}{2}) \cdot 8^{-4}$; $(\frac{1}{15})8^{-4}$
     Base 2: 2^{-4}; 2^{-5}; $(\frac{2}{15})2^{-4}$
     Base 10/Base 8: $3(\frac{4}{5})^4$, about 1.2
     Base 8/Base 2: $\frac{1}{8}^3$

8.10 For fixed α we have bounds $r^{-\alpha+1}$ and $(r')^{-\alpha+1}$, and the ratio is $(r/r')^{-\alpha+1}$. For $\alpha = 1, r > r'$ implies $r/r' > 1$, but for $\alpha > 1$, $(r/r')^{-\alpha+1} < 1$. That is, ordinarily, the larger the base the smaller the relative error for a fixed number of digits.

8.11 If we equate approximate errors (for either cutting or rounding) $r^{-\alpha+1} = (r')^{-\alpha'+1}$ or $-\alpha + 1 = (-\alpha' + 1)\log_r r'$ and $\alpha = 1 + (\alpha' - 1)\log_r r'$ for which we would use the next larger integer. In particular, if $r = 10$ and $r' = 2$, $\log_{10} 2 = 0.3$ or about one third. That is, it takes approximately one third as many decimal digits to produce the same accuracy as binary digits.

SECTION 8.2.2

8.12 $\frac{22}{7} = 3.142857...$; $\pi = 3.141592...$

If we cut to 3.142 we have four significant digits. If we round to 3.143 we have only three.

8.13 Seven, by either cutting or rounding.

8.14 Three.

SECTION 8.2.3

8.15 For $2^{p-1} \leq |X| < 2^p$ with optimal scaling at 2^s, the maximum number of significant bits is $p + s = 59$, so the maximum guaranteed accuracy is 2^{-58} with cutting. In order to have $2^{-58} \leq 10^{-\alpha+1}$, we must have $\alpha \leq -18.4$, so that integral values of $\alpha \leq 18$ give maximum guarantee of 18 decimal digits.

8.16 Fourteen decimal digits.

8.17 About $(\frac{1}{2})10^{-14}$. About 10^{-322} to 10^{294}.

SECTION 8.3

8.18 a) $(\frac{1}{3})10^{-4}$; $(\frac{5}{26})10^{-4}$
     b) $(\frac{1}{3})10^{-4}$; $(\frac{5}{14})10^{-4}$
     c) $(\frac{4}{3})10^{-4}$; $(\frac{5}{2})10^{-4}$
     d) $(\frac{1}{3})10^{-4}$; 10^{-5}

SECTION 8.4.1

8.19 $\dfrac{|y - \bar{y}|}{|y|} = 1 = 100\%$; $\dfrac{|x - \bar{x}|}{x} = 0.75 = 75\%$.

8.20 True solution $x = 9$, $y = -36$, $z = 30$. One method of solution with two digits yields $z = 2.9$, an error of about 90%. Another method yields a zero determinant and no solution.

SECTION 8.4.2

8.21 The leading octal digit for positive numbers is in binary (000, 001, 010, 011) and for negative numbers is (100, 101, 110, 111).

8.22 Restrict operands to $|X| \le 5(10^{n-1}) - 1$. All nonnegative numbers will begin with 0, 1, 2, 3, or 4. Negative numbers, represented by complements, will begin 9, 8, 7, 6, or 5.

8.23 As in Exercise 8.22, restrict operands to $(r/2)(r^{n-1}) - 1$.

8.24*Use the algorithms of Chapter 2, with carry detection from each block of k digits. Compare with the use of binary addition in decimal codes in Chapter 3.

INDEX

INDEX

absent, in Boolean algebra, 68, 69
absolute value, 3, 4, 44, 59, 61
   and sign operands, 59, 61
   of true result, 44
accumulate, *see* accumulator
accumulator, additive, 42, 43
   binary, 82
   closed (or circular), 49, 50 (defined), 52, 54, 82, 134, 135, 140, 163
   double-length, 203–209, 213, 214
      in scaling division, 206–209, 213, 214
      in scaling multiplication, 203–205
   flip-flop, 93, 94
      additive, 93, 94
      subtractive, 94
   fractional, 131 (defined), 133–136, 140, 142, 157 (in division)
   integral, 130 (defined), 131, 132, 139, 140, 150 (in division)
   open, 38 (defined), 39, 82, 135–140, 155
   subtractive, 43
adder, 76–80, 99, *see also* half-adder, full adder
   decimal, 99
   full, 78–80
   half, 76–79
addition, 3, 27, 32–34, 36, 37, 42, 49, 57, 60–63, 95
   base r, 95
   with complement operands, 61–63
   fixed point, *see* fixed point
   floating-point, *see* floating-point instructions, 57
   of integers, 3
   of residue classes, 27, 42, 49
additive accumulator, *see* accumulator

additive law of exponents, 190
   in locating base point, 190
add signal, 92–94
and operation, 69
antecedent, 2, 3, 263
arithmetic, 12, 29, 39–43, 46–49, 67–104
   base r, 12
   modulo r^n, 39–43, 46–49
   nonnumeric aspects of, 67–104
   of residue classes, 29
arithmetic registers, *see* registers
associative law, 3, 71, 73
   in Boolean algebra, 71, 73
   for integers, 3

base, 7 (defined), *see also* radix
base point, 16 (defined), 131, 188, 192, 193, *see also* radix point
   in fractional operations, 193
   in integral operations, 193
   location of in machine number, 131
   as reference point, 192, 193
   rules for location, 188, 190
basic algorithm for multiplication, 137–145
   amended for complements, 142
   in multiply-add, 145
binary addition, 75–87
   logic of, 75
binary arithmetic, 12, 75–87
binary code, 95
binary coded decimal, 96, *see also* code
binary complement division algorithm, *see* division algorithm
binary operation (in Boolean algebra), 68, 71

binary system, 7 (defined), 11
binomial coefficient, 29
binomial theorem, 29
bistable element, 87, 88
Boolean algebra, 68, 69, 70, 71–77
   negation, 69, 70, 74
   operational laws, 71
   product, 68, 70
   sum, 68, 70
borrow, 12, 32, 34 (defined), 35, 36, 38
   end-around, 49 (defined), 52, 53, 54
   as quotient, 35

cancellation law, 31
   for congruences, 31
   restricted, 31
carry, 12, 32, 33 (defined), 34, 36, 38, 49
   as residue, 33, 34
   end-around, 49
   signal, 92–94
carry signal, 92–94
casting out nines, 31
circular accumulator, *see* accumulator, closed
closed, 68
closed accumulator, *see* accumulator
code, 95–103
   binary, 95
   binary coded decimal, 96
   complement in, 100–102
   decimal carry in, 98–103
   8-4-2-1, 96–99
   excess three, 100, 101
   ring, 95
   self-complementing, 100
commutative law, 3, 71, 73
   in Boolean algebra, 71, 73
   for integers, 3
comparison, 61
   facility, 61
complement, 42 (defined modulo r^n), 49 (defined modulo $r^n - 1$)
   addition and subtraction of, 60–62
   cutting in, 239, 240
   extended, 238 (defined)
   formation of, 45–47
   in multiplication, 138
   as negative number, 42–45
   nines, 49
   one's, 49, 90, 138
   operands, 59, 62
   radix, 47, 169–185
   radix-less-one, 49, 169–185
   rounding in, 239, 240
   seven's, 49
   ten's, 47
   the, 42 (defined modulo r^n), 49 (defined modulo $r^n - 1$)
   two's, 47, 138
complements, 42 (defined modulo r^n), 43, 45, 49 (defined modulo $r^n - 1$), 50–54
   formation of, 45, 47, 54
complement division, *see* division
complement operands, *see* complement
complete residue system, *see* residue system
congruence, 23 (defined), 24, 40, 159
   in arithmetic, 40
   in division, 159
circular accumulator, *see* accumulator
circular shift, *see* shift
closed accumulator, *see* accumulator
closed shift, *see* shift
continuum of real numbers, 28
control, 57, 60–62
   sequence, 60–62
conversion, 10, 11, 13, 21, 22
   algorithm, 13
   of base of radix, 10, 11, 13, 21, 22
   of rational numbers, 21
correction term, 141, 142, 143, 144
   for fractional multiplication, 143
   for integral multiplication, 141, 142
   partial, 144
counting, 154, 155, 156, 157, 165, 166, 176–180
   to obtain quotient digits, 154, 155, 156, 157, 165, 166, 176–180
cut (a number), 235–239
   error in, *see* error
   extended complement, 239
cutting, 235–239
   error in, *see* error
   extended complement, 239

decimal full adder, *see* adder
decimal point, 188–190
   alignment in addition and subtraction, 188
   rules for location, 188–190
decimal system, 7
deMorgan's laws, 75
   in Boolean algebra, 75
determinative property, 24
difference digit, 32, 34 (defined), 35
   as residue, 34, 35
digit, 6–9, 244, 245
   base r, 6–9
   maximal, 6–9
   minimal, 6–9
   significant, 244, 245
digital representation, 6, 15
   of integers, 6
   of rational numbers, 15
distributive law, 3, 72, 73
   in Boolean algebra, 72, 73
divident, 4, 30, 149, 151–166, 205, 206, 213
   as product, 149, 151
   in scaling, 205, 206, 213
division, 4, 149–187, 205, 206
   algorithm, 4, *see also* division algorithm
   with complement operands, 167–185
   fixed point, *see* fixed point
   floating-point, *see* floating-point
   as inverse, 149–185
   with negative operands, 165–185
   in scaling, 205, 206
   by shifting and subtracting, 150–187
division algorithm, 4, 5, 15, 30, 33, 35, 36, 149, 152–156, 160, 161, 181–185, 206–209, 228
   in addition, 33, 36
   binary complement, 181–185
   in floating-point, 228
   fractional mechanization of, 157, 160, 161
   integral mechanization of, 161
   in machine division, 152–156
   in scaling, 206–209
   in subtraction, 35, 36
   uniqueness of, 5
divisor, 4, 30, 149, 151–166

digits, 151
   as multiplicand, 149, 151, 157
double-length accumulator, *see* accumulator
double precision, *see* precision
duodecimal system, 11 (defined)

eight's complement, *see* complement
end-around borrow, *see* borrow
end-around carry, *see* carry
end-around shift, *see* shift
equivalent numbers, 27
error, 234–262, 235 (defined)
   absolute, 235
   in arithmetic operations, 248, 250
   bias, 242
   in cut number, 235, 240, 243
   relative, 262, *see also* relative error
   in rounded number, 235, 240, 243
   round-off, 234–262
   truncation, 234
exact computations, 250–252
   to increase precision, 251, 252
excess-three code, *see* code
exclusive *or*, *see or* operation
exponent overflow, *see* overflow
exponent underflow, 220 (defined)
extended complement, *see* complement
extension (of machine numbers), 124, 125, 127, 139, 140, 141
   facility, 139, 140, 141, 150, 163
extension facility, 139–141, 150, 163
   in division, 150, 163

false, 68, 74
   in Boolean algebra, 68, 74
fixed point, 188–217, 246
   arithmetic, 188–217
   operands, 193
   operations relative to error, 246
flip-flop, 87, 88 (defined), 89–92
   accumulator, 92
   dual input, 90
   single input, 91
floating-point, 193, 217–233
   addition and subtraction, 221–224
   division, 228–233

multiplication, 224–228
operand, 193, 217–233
operation, 217–223
zero in, 220
fractional accumulator, *see* accumulator
fractional operation, 193, 212–217, *see also* accumulator
  in addition and subtraction, 212, 213
  in division, 213, 214
  in multiplication, 213, 214
  in scaling, 193, 212–217
fractional part, 17
  of rational number, 17
full adder, 76–80, 99
  decimal, 99
  function table, 79
  as two half-adders, 78–80

half-adder, 76–80, 93, 256
  binary, 76
  in flip-flop accumulator, 93
  to form full adder, 78–80
  function table, 77
  in multiple precision, 256
half-subtractor, 84, 85
  borrow in, 84, 85
  difference digit in, 84, 85
  function table, 84, 85

inclusive *or*, see *or* operations
inequality, 3
integer, 1–3
  arithmetic of, 1, 2
  counting property of, 1
  natural, 1
  negative, 1
  nonnegative, 1
  positive, 1
  properties of, 2, 3
integral accumulator, *see* accumulator
integral operation, 193, 194, 198, 202–211, *see also* accumulator
  combined, 209–211
  in division, 205–209
  in multiplication, 202–205
  in scaling, 193, 194, 198, 202–211
inverter, *see* negation

irrational numbers, 1, 17, 18
  digital representation, 17, 18

k-fold precision, *see* precision

left shift, *see* shift
logical circuitry, 68

machine operand, *see* machine representation
machine representation, 27, 39, 40, 41, 50, 51
  as an arithmetic result, 41, 51
  as an operand, 39, 40, 41, 50, 51
mathematical induction, 2, 6
maximal digit, 6
minimal digit, 6
modular arithmetic, *see* residue class arithmetic
modulus, 23 (defined), 24, 25, 125, 126
  in double-length register, 125, 126
multiple precision, *see* precision
multiplicand, 121, 128, 130, 132, 133, 137, 141, 143, 149, 150, 157
  as divisor, 149, 150, 157
multiplication, 3, 27, 50, 121–148, 203–205
  basic algorithm for, 137
  in closed accumulator, 134, 135
  fractional, 136
  of integers, 3
  machine representation of, 123, 124
  in open accumulator, 122, 123
  parallel, 146
  by repeated addition, 121, 122
  of residue classes, 27, 50
  scaling in, 203–205
  serial, 146
  by shifting, 127–129, 132
multiplier, 121, 128, 132, 136, 137, 141, 143, 149, 152
  as quotient, 149, 152
  register, 121
multiply-add operation, 144–146

natural number, *see* integer
negation (in Boolean algebra), 69 (defined), 70, 74

negative factors, 137, 138, 139, 140, 141, 142
   with complements, 139, 140, 142
negative numbers, 42
   complement representation of, 42
negative zero, *see* zero
nonintegral operation, 196, 211–217
   in scaling, 196, 211–217
nonmaximal digit, 7, 8
nonnumeric aspects of arithmetic, *see* arithmetic
normalization, 218 (defined), 219–226, 228–230
   in addition and subtraction, 221
   condition for fractional case, 218–220
   condition for integral case, 218–222
   in division, 228–230
   internal, 223
   in multiplication, 224–226
   of operand, 218
not operation, *see* negation

octal system, 7 (defined), 11
open, 68
open accumulator, *see* accumulator
open shift, *see* shift
optimal scaling, *see* scaling
or operation, 69
   exclusive, 69
   inclusive, 69
ordered pair, 14, 219
   as floating-point number, 219
   as rational number, 14
ordering, 3
   of integers, 3
overflow, 45 (defined), 136, 158, 159, 164–185, 200, 201, 220, 221, 229
   in addition, 221
   adjustment, 177–185
   check, 167, 177
   division, 164, 165
   exponent, 220
   by improper scaling, 200
   position (for division), 136, 158, 159, 167
   quotient, 229
overflow position, *see* overflow

parallel addition, 81, 82, 87, *see also* addition
parallel multiplication, 146, *see also* multiplication
parallel subtraction, 81, 82, 87, *see also* subtraction
partial product, 129 (defined), 130, 132, 133, 134, 136
partial remainder, 154, 157–161, 164–185
   compared with true remainder, 171
   in complement division, 168–185
   in multiple precision, 259
   sign of, 158–161, 164–185
partial sum, 16
periodic (digital) representation, 17
places, number of, 243
positional representation, 6, 7
   of integers, 6, 7
positive whole number, *see* integer
precision, 250–257
   double, 252
   end-around carry in multiple, 257
   increased, 250, 251
   k-fold, 253 (defined), 254
   multiple, 252–255
   single, 252
present, in Boolean algebra, 68, 69
product, 121–149
   as dividend, 149
   true, 122
pulse, 89, *see also* add signal *and* carry signal

quinary system, 7 (defined)
quotient, 4, 149, 151–153, 159–164, 169–185, 205–209, 229
   digits, 151, 153, 159–164, 169–185
   as multiplier, 149, 152
   as n-digit integer, 156, 168, 169–185
   overflow, 229
   in scaling, 205–209

radix, 7 (defined)
radix point, 16 (defined), 131, 188, 190, 192, 193, *see also* base point
   in fractional operation, 193
   in integral operation, 193

294 INDEX

location of in machine number, 131
   as reference point, 192, 193
   rules for location, 188, 190
radix complement, *see* complement
rational numbers, 14–18, 28, 29
   approximation of, 28
   arithmetic of, 29
   digital representation of, 15–18
   as ratio, 14
real numbers, 1, 14, 18, 20, 28, 29
   approximation of, 28
   arithmetic of, 29
   digital representation of, 20
register, 37, 58–60
   arithmetic, 58–60
   storage, 58, 59
reflexive property, 24
relative error, 235, 243, 246, *see also* error
   bounds for, 243
   in machine representation, 246
   round-off, 235
relatively prime, 25
remainder, 4, 5, 26, 30, 149–185, 205, 206
   condition, 5, 30, 168, 173–175
   in residue class, 26, 159
   restricted, 27
   in scaling, 205, 206
remainder condition, *see* remainder
repeating representation, *see* periodic representation
residue, 34, 35
   as borrow, 35
   as carry, 34
residues, *see* residue system
residue class, 23, 25 (defined), 26, 27, 40–43, 159
   addition, 42, 43
   arithmetic of, 27, 28, 42, 43
   in division, 159
   in machine representation, 40, 42, 43
residue system, 26
   complete, 26
   least nonnegative, 26
right shift, *see* shift
ring code, 95, *see also* code
round (a number), 235–240
   down, 236, 240

error in, *see* error
   in extended complement, 239
   up, 236, 239
rounded digital representation, 19
   in approximation, 19
rounding, 194, 236–242, *see also* round
   compared to cutting, 242
   down, 236, 240
   up, 236, 237, 239
round-off error, *see* error

scale factor, 192–217, *see also* scale factor exponent
   for addition and subtraction, 199–202
   for division, 205–209
   for integral operation, 194
   for multiplication, 202–205
   for nonintegral operation, 196
   optimal, 196–217
   permissible, 195 (defined), 196–198, 205–217
scale factor exponent, 192–217, *see also* scale factor
   common, for addition and subtraction, 199, 200
   for division, 205–209
   for multiplication, 202–205
   optimal, 196–217
   permissible, 195–200, 217
   useful, 195
scaling, 192–217, *see also* scale factor
self-complementing code, 101, *see also* code
sequel, 2, 3, 263
serial addition, 81, 83, 87
serial computation, 255
   in multiple precision, 255
serial multiplication, *see* multiplication
serial subtraction, 87
shift, 105–120, *see also* shifting
shifting, 105–120, 127–132, 157–162, 165
   circular, 106–108, 113–115
   closed, 106–108
   in division, 157–162, 165
   end-around, 107
   end-off, 106
   left, 105–112, 127–132, 157–162, 165

in multiplication, 127–132
open, 106
partial products, 130
right, 107, 113–120, 157–162, 165
sign digit, 46 (defined), 159, 160, 168, 169, *see also* sign indicator
  in division, 159, 160, 168, 169
sign discriminator, *see* sign indicator
sign indicator, 59, 60, 62, 63, 137, 159, 160, 168, 169, 254, *see also* sign digit
  in division, 159, 160, 168, 169
  in multiple precision, 254
  in multiplication, 137
significant digits, 244, 245 (defined)
subtraction, 3, 27, 32–37, 42, 49, 57, 61–63, 95
  base r, 95
  with complement operands, 62, 63
  of integers, 3
  instruction, 57
  of residue classes, 27, 42, 49
subtractive accumulator, *see* accumulator
subtractor (full), 85–86, 94
  in flip-flop accumulator, 94
  function table, 86
  as half-adder and half-subtractor, 86
  as two half-subtractors, 84–85
subtractor (half), 84, 85
sum digit, 32, 33 (defined), 34–36, 49, 52–54
  end-around, 49, 52, 53, 54
  as residue, 34
symmetric property, 24

ten's complement, *see* complement
terminating (digital) representation, 17, 18
transitive property, 24
transmission (of digits), 58–61, 134

additive, 58–61, 134
subtractive, 58
trinary system, 7 (defined)
triple precision, *see* precision
true, 68, 74
true result, 44, 51–53, 122–127
  as difference, 64
  modulo r^n, 44
  modulo $r^n - 1$, 51–53
  as product, 122–127
  as sum, 64
true zero, *see* zero
truncated digital representation, 19, *see also* truncation
  in approximation, 19
truncation (of digital representation), 19, 192, 194, 197, 204
two's complement, *see* complement

unary operation, 68
  in Boolean algebra, 68
underflow, 220 (defined)
  exponent, 220
uniqueness, 8, 28, 64
  of borrow, 64
  of carry, 64
  of difference digit, 64
  of digital representation, 8
  of residue class, 26
  in residue class arithmetic, 28
  of sum digit, 64

whole number, *see* integer

zero, 1, 2, 54, 220–224
  floating-point form, 220–224
  negative, 54
  symbol for, 1, 2
  true, 54

QA
76.5
S 752

AUG 25 1971